THE GEOHELMINTHS:
ASCARIS, TRICHURIS AND HOOKWORM

World Class Parasites

VOLUME 2

Volumes in the World Class Parasites book series are written for researchers, students and scholars who enjoy reading about excellent research on problems of global significance. Each volume focuses on a parasite, or group of parasites, that has a major impact on human health, or agricultural productivity, and against which we have no satisfactory defense. The volumes are intended to supplement more formal texts that cover taxonomy, life cycles, morphology, vector distribution, symptoms and treatment. They integrate vector, pathogen and host biology and celebrate the diversity of approach that comprises modern parasitological research.

THE GEOHELMINTHS:
ASCARIS, TRICHURIS AND HOOKWORM

edited by

Celia V. Holland
Department of Zoology, University of Dublin

and

Malcolm W. Kennedy
Division of Environmental and Evolutionary Biology
Institute of Biomedical and Life Sciences
University of Glasgow

SPRINGER SCIENCE+BUSINESS MEDIA, LLC

Library of Congress Cataloging-in-Publication Data

The geohelminths : ascaris, trichuris, and hookworm / Celia V. Holland and Malcolm W. Kennedy.
p. cm. – (World class parasites ; v. 2)
ISBN 978-1-4419-4922-6 ISBN 978-0-306-47383-8 (eBook)
DOI 10.1007/978-0-306-47383-8
1. Ascariasis. 2. Trichuriasis. 3. Hookworm disease. I. Holland, Celia. II. Kennedy, M. W. (Malcolm W.) III. Series

RA644.A82 G46 2001
616.9'654—dc21

2001050285

Printed on acid-free paper.

This book is dedicated to the memory of

Anne Keymer

(1957 - 1993)

Biologist and friend

TABLE OF CONTENTS

Section 4 - Genetics - mice, worms and people

Section 5 - Interaction between geohelminth infections and other diseases

LIST OF CONTRIBUTORS

Marco Albonico Ivo de Carneri Foundation, Milan, Italy

Zvi Bentwich R. Ben-Ari Institute of Clinical Immunology and AIDS Center, Kaplan Medical Center, Hebrew University Hadassah Medical School, Rehovot 87100, Israel

John Blangero Department of Genetics, South West Foundation for Biomedical Research, San Antonio, Texas 78245-0549, USA

Peter R. Boag Victorian Institute of Animal Science, Attwood, Victoria 3049, Australia and Department of Veterinary Science, The University of Melbourne, Werribee, Victoria 3030, Australia

Jaap Boes Danish Bacon and Meat Council, Axelborg, Axeltorv 3, DK-1609 Copenhagen V, Denmark

Gadi Borkow R. Ben-Ari Institute of Clinical Immunology and AIDS Center, Kaplan Medical Center, Hebrew University Hadassah Medical School, Rehovot 87100, Israel

Janette E. Bradley School of Life and Environmental Sciences, University of Nottingham, Nottingham, NG7 2RD, UK

Alan Brown Boots Science Building, School of Pharmacy, University of Nottingham, NG7 2RD, UK

Neil B. Chilton Department of Veterinary Science, The University of Melbourne, Werribee, Victoria 3030, Australia

Philip J. Cooper Department of Infectious Diseases, St George's Hospital Medical School, Cranmer Terrace, Tooting, London SW17 ORE, UK and Laboratorio de Investigacion, Hospital Pedro Vicente Maldonado, Pedro Vicente Maldonado, Pichincha Province, Ecuador

Helen Faulkner School of Life and Environmental Sciences, University of Nottingham, Nottingham, NG7 2RD, UK

Robin E. Gasser Department of Veterinary Science, The University of Melbourne, Werribee, Victoria 3030, Australia

Helen Guyatt Wellcome Trust Research Laboratories-Kenya Medical Research Institute, PO Box 43640, Nairobi, Kenya and Centre for Tropical Medicine, University of Oxford, John Radcliffe Hospital, Oxford OX1 3QU, UK.

J. M. Hawdon Department of Microbiology and Tropical Medicine, George Washington University, Washington, D.C., USA

Celia Holland Department of Zoology, Trinity College, Dublin 2, Ireland

Peter J. Hotez Department of Microbiology and Tropical Medicine, George Washington University, Washington, D.C., USA

Gregers Jungersen Danish Veterinary Laboratory, Bülowsvej 27, DK-1790 Copenhagen V, Denmark

Malcolm W. Kennedy Division of Environmental and Evolutionary Biology, Institute of Biomedical and Life Sciences, University of Glasgow, Graham Kerr Building, Glasgow G12 8QQ, Scotland

Jane Kvalsvig School of Anthropology, Psychology and the Centre for Social Work, University of Natal, Durban, South Africa.

Antonio Montresor Parasitic Diseases and Vector Control, World Health Organization, 1211 Geneva 27, Switzerland

Susan E. Newton Victorian Institute of Animal Science, Attwood, Victoria 3049, Australia

David I. Pritchard Boots Science Building, School of Pharmacy, University of Nottingham, Nottingham, NG7 2RD, UK

Rupert J. Quinnell School of Biology, University of Leeds, UK

Helen Roberts Laboratory of Evolutionary Genetics, Department of Biology, UCL, London NW1 2HE, UK

Lorenzo Savioli Parasitic Diseases and Vector Control, World Health Organization, 1211 Geneva 27, Switzerland

Lani S. Stephenson Division of Nutritional Sciences, Savage Hall, Cornell University, Ithaca, NY 14853 USA

Anita H. J. van den Biggelaar Department of Parasitology, Leiden University Medical Center, Albinusdreed 2, 2333 ZA Leiden, The Netherlands

Derek Wakelin School of Life and Environmental Sciences, University of Nottingham, Nottingham, NG7 2RD, UK

Sarah Williams-Blangero Department of Genetics, South West Foundation for Biomedical Research, San Antonio, Texas 78245-0549, USA

Maria Yazdanbakhsh Department of Parasitology, Leiden University Medical Center, Albinusdreed 2, 2333 ZA Leiden, The Netherlands

Xingquan Zhu Department of Veterinary Science, The University of Melbourne, Werribee, Victoria 3030, Australia

PREFACE

The soil-transmitted nematode parasites, or geohelminths, are so-called because they have a direct life cycle, which involves no intermediate hosts or vectors, and are transmitted by faecal contamination of soil, foodstuffs and water supplies. They all inhabit the intestine in their adult stages but most species also have tissue-migratory juvenile stages, so the disease manifestations they cause can therefore be both local and systemic. The geohelminths together present an enormous infection burden on humanity. Those which cause the most disease in humans are divided into three main groupings, *Ascaris lumbricoides* (the large roundworm), *Trichuris trichiura* (whipworm), and the blood-feeding hookworms (*Ancylostoma duodenale* and *Necator americanus*), and this book concentrates on these. These intestinal parasites are highly prevalent worldwide, *A. lumbricoides* is estimated to infect 1471 million (over a quarter of the world's population), hookworms 1277 million, and *T. trichiura* 1049 million. The highly pathogenic *Strongyloides* species might also be classified as geohelminths, but they are not dealt with here because the understanding of their epidemiology, immunology and genetics has not advanced as rapidly as for the others. This is primarily because of the often covert nature of the infections, with consequent difficulties for analysis. If there is ever a second edition of this book, then there will hopefully be much to say about this infection.

Despite the considerable numbers of geohelminth infections, the public health perception has traditionally been that these intestinal parasites contribute comparatively little to overt disease. That perception has changed through new understanding of the parasites' epidemiology and their contribution to covert chronic disease conditions. For instance, the numbers of worms recovered from populations of hosts exhibit an overdispersed or aggregated distribution – most hosts harbour few or no worms whereas a small proportion of hosts carry very heavy burdens. These heavily infected individuals are therefore important from a public health perspective because they represent the main source of infection, but probably also represent the clinically most affected subpopulation. The manifestations of severe disease include fatal intestinal obstruction or pulmonary allergic reactions in ascariasis, severe anaemia in hookworm infections, and chronic dysentery and rectal prolapse in trichuriasis. Evidence has also accumulated that moderate infections of intestinal nematodes contribute significantly to chronic conditions such as growth retardation, which regresses after treatment. An even more insidious and alarming effect upon cognitive development has been suggested by a growing number of intervention studies, and a recent focus upon the interrelationships between intestinal nematodes and other microparasitic diseases has revealed further potential benefits of large scale de-worming programmes. So, there is still a great need to understand more fully the true cumulative impact of these infections.

Chemotherapy still remains the most effective way of reducing the intensity of geohelminth infections so as to decrease the associated morbidity and mortality. The anthelmintic drugs currently available for the treatment of human intestinal parasites are relatively safe and effective, and evidence for the development of resistance to these drugs is still scarce, although there is clearly no room for complacency given the development of resistance to similar drugs against nematodes of veterinary importance. For the moment, however, there is little to say about drug development and the development of resistance as far as human geohelminths are concerned, so the subject is not dealt with here with any emphasis. The development of new drugs is essential, but, sadly, advances are more likely to come through the veterinary imperative than from human medicine for brutal market force reasons.

The understanding of epidemiological patterns has contributed to new approaches to control. For example, age-targeted chemotherapy of children focuses upon those individuals with the highest worm burdens in the community and those most at risk from developmental morbidity. An added advantage has been that children can be treated at school, thereby increasing the cost-effectiveness of treatment programmes. Moreover, a broad-spectrum approach is being considered whereby simultaneous treatment for lymphatic filariasis and intestinal nematodes is employed. Economic analysis is now an important tool used to assess the cost of infection (in terms of morbidity, lost productivity and lost human potential) and the cost of intervention, and the development of cost-effective programmes is essential for any progress to be made in developing countries in which the budgets available for healthcare are small.

Another important concept in intestinal nematode biology is that of predisposition. For an individual host, worm burden does not show a random pattern upon reinfection, but exhibits consistency in the re-acquisition of low and high worm burdens. The mechanisms behind this phenomenon are likely to be multiple and it has proved difficult to unravel their relative contributions under field conditions in humans. The use of a number of different animal models and, in particular, the recently described *Ascaris* predisposition pig model, are likely therefore to be particularly illuminating, particularly now that the gap between laboratory animal experimentation and work on humans is closing. For instance, a recent human pedigree study has revealed evidence for a strong genetic component in the observed variation in *Ascaris* worm burden from person to person, and host genetics are also being examined in a variety of mouse models. Further analysis of both mice and men will hopefully soon reveal precisely which genes are involved in endowing susceptibility or resistance to intestinal nematode infections. Furthermore, the genetics of the parasites also requires understanding, both in terms of strains, geographical variation, and even at the household level.

The immune response of the host to intestinal parasites has received considerable attention and, although an understanding of the individual responses mounted by the host has improved, the protective role of the different effector mechanisms is still less well understood. In particular, the relationship between infection, the production of IgE and the manifestations of atopy requires further exploration, as does the balance between immune-mediated resistance to infection and immunopathology. Parasitic nematodes, *Ascaris* in particular, are renowned for their elicitation of powerful IgE and T helper type 2 (Th2) responses, and how these (or their absence) relate to allergic reactions is currently a focus of research, particularly in view of the dramatic increase in allergies over recent decades. The most illuminating recent studies in this regard come from immunoepidemiological studies on filariasis and schistosomiasis in humans. We have, therefore, taken the (perhaps rash) step of including a chapter on these aspects from the perspective of schistosomiasis (neither a nematode, nor intestinal!), which we argue will greatly contribute to the debate and provide direction for similar future studies on geohelminths and atopy/allergy.

In summary, intestinal nematode infections are an important, prevalent and preventable public health problem, which contribute to considerable human debilitation worldwide. The challenge of their control lies in the need to raise awareness of their morbid effects and to find cost-effective and operationally realistic ways of treating the populations that are infected by them. Furthermore, aspects of their biology provide the opportunity to investigate important fundamental processes including the genetic basis of susceptibility to chronic infectious diseases and their relationships with other diseases like HIV/AIDS. Simultaneous studies on human hosts living in endemic areas and the use of appropriate animal models will help to unravel these complex host-parasite relationships.

The literature on all aspects of geohelminth infections is extensive, and the purpose of this book is not to review the field comprehensively, but to present chapters by selected experts, who were asked to review a particular area and to take a prospective view in order to identify new and emerging approaches and ideas. The understanding of the genetics, epidemiology and immunology of intestinal helminths has taken dramatic leaps forward over the past decade, and we hope that this book will contribute to a wider understanding and stimulate further development of the field for both practical and theoretical purposes.

Celia Holland
Malcolm Kennedy

Dublin and Glasgow July 2001

ACKNOWLEDGEMENTS

We are compelled to extend our particular thanks to Marina Pearson, Zoology Department, Trinity College, Dublin, for her superb help and skills in formatting and collating the manuscripts, Alison Boyce, also of the Zoology Department, for excellent technical assistance with the figures, and Joanne Tracy and Dianne Wuori of Kluwer Academic Publishers for editorial support and guidance.

Chapter 1
DISTRIBUTIONS AND PREDISPOSITION: PEOPLE AND PIGS

Celia Holland[1] and Jaap Boes[2]
[1]*Department of Zoology, Trinity College, Dublin 2, Ireland*
[2]*Zoonoses Research Group, Veterinary and Food Advisory Service, Danish Bacon and Meat Council, Copenhagen, Denmark*
e-mail: cholland@tcd.ie

> "As has been the subject of recent emphasis many times elsewhere, collaboration between immunology and epidemiology is the necessary basis for future progress in this area (the mechanisms of predisposition). Specifically, long-term studies of nutritional status and exposure-related variables are required together with measurement of parasite-specific humoral and cellular immune responses during periods of reinfection following drug treatment in patients of all ages and initial infection levels (Keymer & Pagal, 1990)"

1. INTRODUCTION

The number of parasites a host carries is fundamental to our understanding of helminth parasite epidemiology. Worm burden is now known to influence the pathogenicity of the infection including effects upon nutritional status and cognitive function (see Chapters 3 and 4), to contribute to the regulation of infection and to impact upon the development of the most effective strategies for control (see Chapter 2). Three key epidemiological patterns which relate to worm burden have been described and studied intensively during the last two decades - these are (i) the frequency distribution of worms per host in a population, (ii) the relationship between host age and worm burden and (iii) the correlation between worm burdens during periods of reinfection. Presently we have good empirical information to describe these patterns for *Ascaris lumbricoides*, *Trichuris trichiura* and hookworm spp from a variety of geographical locations. The mechanisms which contribute to these observed patterns remain much more elusive and are likely to involve the interplay of exposure, acquired immunity and innate resistance. In this chapter we provide a historical perspective on the studies which have been undertaken to describe these patterns. We then assess the information available on the causative mechanisms behind reinfection and predisposition in humans and outline the difficulties inherent in the design of such studies. Finally we parallel the developments in humans with those in animal models and highlight the possibilities of using some new models which will be amenable to experimental manipulation of epidemiology, nutrition, immunology and genetics.

2. A HISTORICAL PERSPECTIVE ON AGGREGATION AND PREDISPOSITION IN HUMAN HELMINTH INFECTIONS

In his seminal work, 'A quantitative approach to parasitism', Crofton described the frequency distribution of parasites in a host population as clumped or overdispersed and best described mathematically by the negative binomial (Crofton, 1971). The pattern of overdispersion among helminth parasites within their hosts is now known to be widespread in both human and other animal hosts (see Anderson & May, 1979; Crompton, Keymer & Arnold, 1984; Shaw & Dobson, 1995). The first paper to detail this phenomenon and its significance in humans was that of Croll & Ghadirian (1981). They described endemic communities wherein most hosts harbour few or no parasites and the so-called 'wormy persons' carrying very heavy burdens. The worm burdens of *Ascaris lumbricoides*, *Trichuris trichiura* and the two species of hookworm, *Ancylostoma duodenale* and *Necator americanus* were counted after anthelmintic treatment of subjects from three Iranian villages and all distributions were overdispersed. Ironically, given later developments, in this study no significant correlation was found between pre-treatment and post-treatment worm burdens 12 months later.

Seeking an explanation for this observed frequency distribution (Figure 1.1) was to become one of the major concerns for parasite epidemiologists in the decades that followed. The practical implications were significant and related to the possibility of selectively treating the so-called 'wormy persons' in order to reduce morbidity and mortality in that group and to modify the transmission dynamics of the community as a whole (Anderson & Medley, 1985; Asaolu, Holland & Crompton, 1991).

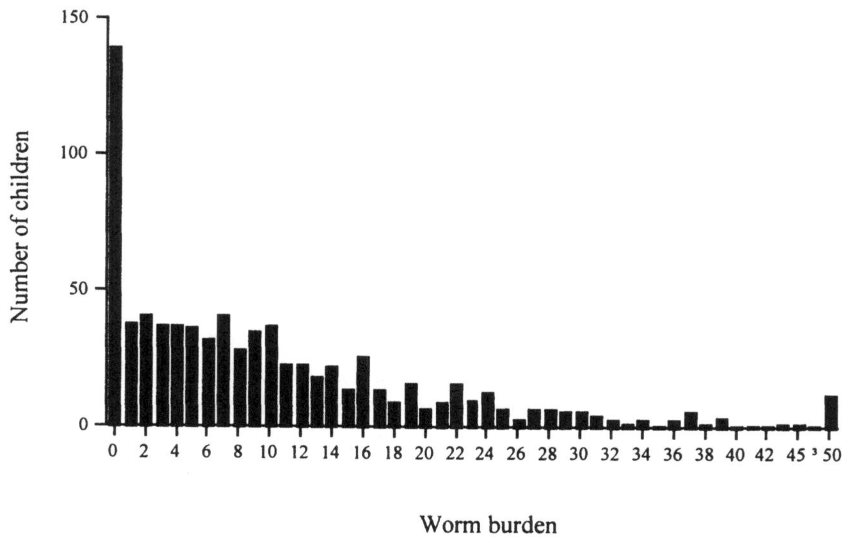

Fig. 1.1. Frequency distribution of numbers of *Ascaris lumbricoides* per child in Ile-Ife, Nigeria (n = 808).

After the early paper of Croll and Ghadirian, further studies on the epidemiology of the three important species of human helminths followed and, most importantly, a secondary phenomenon was described. Longitudinal studies of the patterns of reinfection in individual patients after chemotherapeutic treatment were performed and an assessment was made of the degree to which individuals who were lightly or heavily infected, required similar burdens (Anderson, 1986). This led to the description of this consistency in reinfection pattern as 'predisposition'. Predisposition was described for *A. lumbricoides* (Elkins, Haswell-Elkins & Anderson, 1986), *T. trichiura* (Bundy *et al.* 1987a) and hookworm (Schad & Anderson, 1985). Evidence for multiple species predisposition (*Ascaris*, *Trichuris*, hookworm and *Enterobius*) was then provided by Haswell-Elkins, Elkins & Anderson (1987). A third important epidemiological pattern concerns the relationship between helminth parasite intensity and age. Changes in the average intensity with age are convex in form with intensity peaking in the 5 to 15 year old age classes for *Ascaris* and *Trichuris* and in the older age classes for hookworm (Bundy *et al.* 1987b; Haswell-Elkins *et al.* 1988; Holland *et al.* 1989). This age-relationship can influence the observed overdispersion and predisposition and requires careful interpretation and sample selection in order to take account of its contribution.

In an important review of the phenomenon of predisposition, Keymer & Pagal (1990) collated the evidence for predisposition, the data available which might throw light on its cause, and its significance with respect to the epidemiology and control of human helminths. The authors reiterated the interrelationship between overdispersion and predisposition i.e. that 'wormy persons' are in fact predisposed to their condition, and raised the question of the causative factors behind the two phenomena.

In 1986 Anderson emphasized the need for large sample sizes, careful statistical analysis and standardization by age and sex, in studies of predisposition. Keymer & Pagal reviewed 12 studies all published during the mid to late 1980s (some earlier studies were found to be biased with respect to several important variables which probably explains their inconclusive results (see Croll & Ghadirian, 1981)). All these studies except one, yielded evidence of predisposition, but the relationship between initial and final infection levels is seldom strong; with a few exceptions, the value of Kendall's tau correlation coefficient is rarely over 0.30. Additional factors - such as the influence of age on intensity, the way intensity was measured (direct worm counts versus egg counts) and the duration of the reinfection period - were also identified as important contributors to the detection of predisposition. For statistical and biological reasons, predisposition may be easier to detect in certain age classes for example children for *Ascaris* and *Trichuris* and adults for hookworm. More recently, Peng *et al.* (1998) used a novel method to explore predisposition, namely natural reinfection over a one-year period in the absence of chemotherapeutic intervention.

3. THE NATURE VERSUS NURTURE DEBATE AS IT APPLIES TO HUMAN HELMINTHIASES

The relative contributions of 'exposure' versus 'susceptibility', or 'ecology' versus 'immunology', to the epidemiological patterns of human helminths have been discussed and reviewed by a number of authors (Bundy, 1988; Keymer & Pagal, 1990; Bundy & Medley, 1992). Exposure includes the contribution of individual behaviour patterns and sociocultural and socioeconomic factors including malnutrition. Susceptibility includes host genetics and the immune status of the host which may be influenced by the host genotype or by phenotypic factors such as nutrition, reproductive status or the presence of other infections (Keymer & Pagal, 1990). It is highly likely that multiple factors which vary in space and time will influence the observed epidemiological patterns.

Evidence that predisposition could be maintained over multiple rounds of treatment (Holland *et al.* 1989), suggested it was unlikely that predisposition was generated by treatment effects. Whether predisposition is a feature of long term causal factors, such as host genetics and host socioeconomic status, or short term factors, such as the host acquired immune response, is obviously important for the design of appropriate control strategies. McCallum (1990) used probability theory to demonstrate that predisposition is weak, subject to the influence of transient factors and that both short and long term factors make an equal contribution to the observed heterogeneity. The author advocated the collection of empirical evidence over time in order to validate the theoretical predictions. Recently, Quinnell *et al.* (2001) assessed reinfection and predisposition to *N. americanus* in a rural village in Papua New Guinea over an eight year period. Interestingly predisposition could be detected six to eight years after a single round of chemotherapy but was not detectable after repeated chemotherapy. The authors concluded that differences in host susceptibility are likely to influence predisposition but that longer-term variation in either exposure or susceptibility limits the period over which significant predisposition can be detected.

Clearly, understanding the relative influence of exposure, immunity and genetics on the observed patterns of overdispersion (or intensity) and predisposition in individual patients is a difficult task. One of the major difficulties associated with the epidemiology of soil-transmitted nematodes is the measurement of exposure to infection. This is in contrast to the schistosomes where quantitative indices of exposure have been developed (for example see Chandiwana, Woolhouse & Bradley, 1991). Wong, Bundy & Golden (1988) developed a method for measuring soil-derived silica in faeces as a measure of geophagia and hence a proxy for exposure to geohelminths. Despite demonstrating heterogeneities in geophagia which correlated with the observed intensity relationship of *Trichuris*, the relationship at the individual level was not explored. Furthermore, this is a relatively complex and time-consuming method for routine use. Few studies have examined the impact of human behaviour on geohelminth infection, but in one of a series of elegant papers on the epidemiology of *Ascaris* in children from S.E. Madagascar, Kightlinger, Seed & Kightlinger (1998) demonstrated that intensity of infection was influenced by gender-related behavioural factors and environmental factors that contribute to exposure.

In contrast, considerably more attention has been paid to the relationship between the human humoral immune response and geohelminth intensity (for example for *A. lumbricoides* see Haswell-Elkins *et al.* 1989, 1992; *T. trichiura* Bundy *et al.* 1991, Lillywhite *et al.* 1991; Needham *et al.* 1992 ; hookworm Pritchard *et al.* 1990)(see Chapters 6, 8 and 9). Many of these studies found evidence for strong antibody responses to infection, but did not always yield convincing evidence for any protective function in contrast to the observations made for schistosomiasis (Hagan *et al.* 1991). In contrast, the data on human cytokine responses to geohelminths is very sparse ; MacDonald *et al.* (1994) compared the production of TNFα in lamina propria and peripheral blood in TDS (*Trichuris* dysentry syndrome - associated with heavy infection (see Stephenson, Holland & Cooper, 2000)) and control patients and demonstrated elevated levels in the infected subjects compared to controls. Furthermore, a recent paper by Cooper *et al.* (2000) provides the first information on cellular immunity in ascariasis (see Chapter 6).

3.1 Factors which influence re-infection and predisposition to soil-transmitted helminths in humans

Over a decade ago, Keymer & Pagal (1990) stated that sufficient information on predisposition was available for the erection and testing of specific hypotheses concerning causal mechanisms. These authors did acknowledge the ethical constraints concerning interventions in humans and advocated the concurrent use of laboratory models, where variables such as nutritional status, genetic background, immunocompetence and behaviour can be subject to experimental control. Perhaps not suprisingly, the number of studies that have attempted to unravel the factors which may contribute to predisposition in humans is relatively small. Furthermore, most of these studies have concentrated upon differences in susceptibility rather than exposure given the difficulties in assessing the latter aspect quantitatively (see Table 1.1).

The data provided in Table 1.1 highlight a number of important features - these include sample size and study design, relative helminth intensity and its measures and types of host factors investigated. Probably one of the key factors which underpins the successful assessment of predisposition and its relationship with host factors is sample size. If a predisposition study is designed to identify individuals who show consistency in worm burden, and then assign them to particular worm burden groups, the initial sample size needs to be very large indeed to accommodate the relatively weak correlations between worm burdens. For example, the Nigerian study by Holland *et al.* (1989) began with 808 children who provided post-treatment samples but this number fell to 580 by the third treatment round. After selection of three groups of children - who were predisposed to remain uninfected, lightly infected and heavily infected - the total sample size narrrowed to 120 and a subsequent study, using these subjects was criticized for its low sample size (Holland *et al.* 1992).

In the study by Palmer *et al.* (1995), the persons providing initial worm counts numbered 1,765 (see Hall, Anwar & Tomkins, 1992), but fell to 880 after three rounds of treatment, and the final sample size for the immunological investigations was 84. In contrast, in the pedigree study by Williams-Blangero *et al.* (1999), the design of the study did not necessitate the selection of worm burden groups, but performed a pedigree analysis in a population which manifested predisposition ; as a result the sample size remained high and showed little reduction over the two year study (see Table 1.1). What these figures emphasize is how difficult it is to carry out investigation of the causation of predisposition. Furthermore for both biological and sampling reasons, predisposition to ascariasis is easier to detect and investigate in children rather than adults. Despite this, excluding adults from the investigation ignores valuable information, particularly if the proposition being tested is that exposure to infection is more important in children and differential susceptibility more marked in adults.

The relationship between reinfection with *A. lumbricoides* and a variety of risk factors for exposure was explored in preschool children (Henry, 1988)(see Table 1.1). Reinfection was significantly reduced among children who had access to a household water supply and a latrine, compared

Table 1.1. A summary of the studies on the mechanisms underlying predisposition in intestinal helminths

Reference	Location	Sample size (age range in years)	Helminth species	Intensity measure	Duration of reinfection	Pre-disposition[7]	Drug treatment	Host parameters
Henry 1988[1]	St Lucia	205-188* (< 3)	*Ascaris lumbricoides*	EPG	3 rounds of six and a half months	No	Piperazine phosphate	Water/sanitation Family size/crowding Income Education mother
Holland *et al*[2] 1992 McSharry *et al*[2] 1992	Ile-Ife Nigeria	808 - 580* $82^{>}$ $92^{>}$ (5-16)	*Ascaris lumbricoides*	Worms	3 rounds of six months	Yes	Levamisole	HLA class I A,B + C Total IgE, specific IgE, other antibody isotypes Inflammation indicators
Pritchard *et al* 92[3] Quinnell *et al* 95[3] Pritchard *et al* 95[3]	Madang province Papua New Guinea	202 109 72 (all except < 3 + > 48)	*Necator americanus*	EPG Worms Biomass Fecundity	2 rounds of twenty four months	No	Pyrantel pamoate	IgM, IgG, IgA, IgE IgD Eosinophilia
Hagel *et al* 1993[4]	Caracas, Venezuela	98 (5-10)	*Ascaris lumbricoides*	EPG	1 round of 8 months	No	Oxantel/ Pyrantel	Total IgE Specific IgE Eosinophilia

Reference	Location	Sample size (age range in years)	Helminth species	Intensity measure	Duration of reinfection	[7]Pre-disposition	Drug treatment	Host parameters
Palmer *et al* 1995[5]	Dhaka, Bangladesh	1765-880* 84> (mean age 9)	*Ascaris lumbricoides*	Worms biomass	3 rounds of six months	Yes	Pyrantel pamoate	Total IgE, specific IgE, IgA and other antibody isotypes
Williams Blangero *et al* 1999[6]	Dolakha E. Nepal	1007-965^ 910> (3-85)	*Ascaris lumbricoides*	Worms biomass EPG	1 round of 12 months	Yes	Albendazole	Pedigree analysis

Footnotes

1. EPG determined but only prevalence data presented (% infected, reinfected).
2. Children classified as follows: predisposed to remain uninfected (no worms on either of the two occasions), those with consistently light infections (1 to 24 worms in phase 1 and 1 to 8 worms in phase 2) and those with consistently heavy infections (> 25 at phase 1 and >9 at Phase 2). Cut-offs were based upon the mean ± SD of the worm burdens in Phase 1 and 2 (Holland *et al*, 1989).
3. The mean worm burden at round 1 was 25.3 in 1988 and 13 in 1990 : Predisposition was established in another study (see Quinnell *et al*, 1993)
4. At the beginning of the study mean EPG was 620±40 and 580±40 in those children who became reinfected and those that did not.
5. Based upon worm burdens over 3 rounds of treatment, cases were the third most heavily infected (110 ± 37 worms total) controls were the third most lightly infected children (18 ± 8)
6. Year 1 mean worm burden 2.67 ± 4.33, year 2 2.67 ± 4.48
7. Predisposition established

* Sample size at round 1 and round 3

^ Sample size at round 1 and round 2

> Sample size for immunological or genetic studies

to those with access to tap water alone or public water standpipes. Furthermore, crowding (persons per room) and sanitation were revealed to be the most significant factors in whether children became reinfected or not. Unfortunately, no quantitative data on the intensity of *Ascaris* is provided in this paper although a quantitative method for the calculation of eggs per gram faeces (epg) is described in the methods.

Three studies focused upon the relationship between the humoral immune response (with particular emphasis upon the IgE isotype) and reinfection or predisposition to *A. lumbricoides*. The earliest study by Hagel *et al.* (1993), did not establish predisposition but did compare groups of children who did and did not exhibit reinfection with the parasite. The intensity of infection based upon epg was low, not overdispersed and did not differ between those children who subsequently became reinfected or remained uninfected (Table 1.1). A comparison of the total IgE and parasite-specific IgE responses of reinfected versus non-reinfected children, revealed an inverse correlation between the two responses. A significant association was found between reinfection and high pretreatment total IgE levels but low levels of specific IgE. The authors concluded that specific IgE may have a protective role against *Ascaris* and other helminth infections.

Palmer *et al.* (1995), in a carefully designed case-control study, compared consistently lightly-infected subjects with those consistently-heavily infected (Table 1.1). A range of antibody isotypes were measured, including total IgG, IgG1, IgG2, IgG3, IgG4, IgA, total IgE and parasite-specific IgE. Children who were predisposed to heavy infection showed higher concentrations of antibody isotypes compared to children predisposed to light infections. In contrast to the findings of Hagel and colleagues, the concentrations of total IgE and parasite-specific IgE in this study mirrored the infection intensity of the subjects. The authors do not rule out an effector role for these antibodies, but suggest that the utilization of more specific antigens may rule out polyspecific responses to numerous antigens which may mask any epitope-specific protective responses.

This point was borne out in a study by McSharry *et al.* (1999) who compared a range of serum factors in children predisposed to remain uninfected, lightly infected and heavily infected. These groups of children showed few differences in measures of socioeconomic status and lived in environments where samples of soil contained eggs of *Ascaris*, assumed to be those of *A. lumbricoides*. Three different sources of *Ascaris* antigen were used in immunological assays but only the defined allergen (*Ascaris* ABA-1 as a bacterial recombinant protein), provided evidence for a significant relationship between predisposition status and parasite-specific IgE. A subgroup of children, who responded to the ABA-1 allergen, was selected and a relationship between reduced rABA-1 specific IgE titre and increasing parasite load was detected. Subjects were further divided according to high or low levels of IgE antibody using a threshold median value, and a distinct pattern emerged. The putatively immune group tended to have higher levels of rABA-1 specific IgE and the susceptible groups had low levels (Figure 1.2). Significantly higher levels of inflammatory indicators - such as serum ferritin, eosinophil cationic protein and c-reactive protein - were detected in the putatively immune group. The authors concluded that IgE responses, in conjunction with innate inflammatory responses, associate statistically with natural immunity to ascariasis.

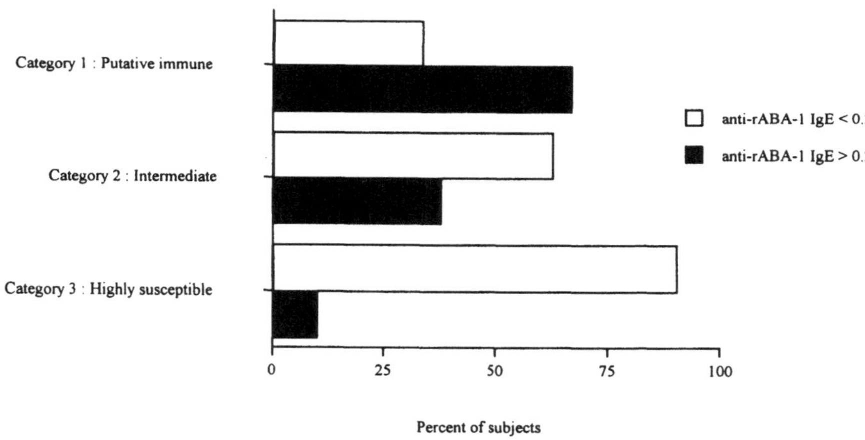

Figure 1.2. The relationship between predisposition and IgE antibody response against r-ABA-1 allergen of *Ascaris* (Adapted from McSharry *et al.* 1999)

Pritchard and co-workers (Pritchard *et al.* 1992; Quinnell *et al.* 1995 and Pritchard *et al.* 1995) reported the relationship between *N. americanus* infection and humoral antibody responses in subjects who experienced reinfection after chemotherapy. Subjects were not assigned to groups based upon reinfection or predisposition, but analyzed longitudinally over a three-year-period. After controlling for the effects of age, the authors demonstrated that correlation coefficients between levels of IgG antibody against adult worm excretory-secretory (ES) materials and worm burden declined significantly with age and did not persist after reinfection. The trend was the same for anti-larval IgG response, but the pattern persisted after reinfection. The switch from positive to negative correlation in adults appears consistent with a protective role (Quinnell *et al.* 1995). Furthermore, the effect of the humoral immune response on the weight and fecundity of the parasite was investigated in the same subjects. After controlling for the effects of age and parasite burden, a significant negative correlation between total and specific IgE and the weight and fecundity of *Necator* worms was detected at initial treatment and after reinfection (Pritchard *et al.* 1995), which the authors suggest may reflect a T helper type 2 (Th2) response.

Remarkably little work has been performed on the relationship between susceptibility to human helminths and host genetics in contrast to genetic studies in laboratory and other animals. Evidence for familial predisposition had already been provided (Forrester *et al.* 1990, Chan *et al.* 1994a) but when Chan *et al.* (1994b) dissected out the correlation of parasite intensities within families they failed to detect a trend consistent with a significant role for host genetic factors. An association study examined the distribution of major histocompatibility complex (MHC) alleles in HLA-A, HLA-B and HLA-C among groups of Nigerian children who were selected for predisposition to remain uninfected, lightly infected and heavily infected with *A. lumbricoides* (Holland *et al.* 1992) (Table 1.1). None of the children who were predisposed to remain uninfected possessed the A30/31 antigens, and the frequency of the occurrence of this antigen combination was significantly higher in the children observed to be consistently infected. Williams-Blangero *et al.* (1999) subsequently stated that such association studies have potentially major statistical problems which can lead to false-positive results and advocate the use of large extended pedigrees crossing multiple households. In their large scale study (see Chapter 10), which involved 1261 subjects, all of whom belonged to the same pedigree, these authors demonstrated a strong genetic component accounting for between 30% and 50% variation in worm burden. Sharing a household accounted for only 3 to 13% of the total phenotypic variance. The average worm burden in this population is low (Table 1.1) and it would be of interest to undertake a similar pedigree analysis in a population experiencing a much higher infection pressure.

To conclude, what emerges primarily from these studies is that only small pieces of the jigsaw are being put in place to explain the factors which may contribute to the observed epidemiological patterns. Despite the exhortations of many people working in the field to collect long-term data on both exposure and susceptibility-related factors in individual patients, this has proved to be exceedingly difficult in practice. In our own experience in

Nigeria - despite collating information on socioeconomic status,*Ascaris* eggs in soil adjacent to the households, immune factors and host genetics in individual subjects - it proved impossible to quantify the relative contribution of the various factors to the observed predisposition due to methodological and sample size limitations. Echoing this experience, Woolhouse (1993) highlighted some of the difficulties in interpreting complex immunoepidemiological patterns under field conditions. In attempting to design simple mathematical models to predict the relationships between parasite burden, rates of re-infection, exposure and immunity, he emphasizes the contribution of stochastic variation, measurement error, low sample size and host-age as potential confounding factors.

4. MODELLING PREDISPOSITION

To study the multiple factors likely to be involved in predisposition experimental manipulation is desirable, but for obvious reasons humans cannot be subjected to experimentation. As Keymer & Pagel (1990) pointed out, studies in laboratory animals could be carried out to complement studies in human communities. The advantage of an animal model for predisposition is a greater degree of control over the different parameters under study, such as genetic background, nutritional status, immunocompetence and behaviour.

4.1 Rodent models

To date, rodent models are the most frequently used and best described host-parasite systems. These models have obvious advantages as mice, rats, guinea pigs and rabbits are relatively easy to keep and handle, they are not expensive, and reproduce rapidly and in large numbers. However, the choice of host animal depends on the parasite under study - roundworm, pinworm or hookworm - and other host-parasite models have been suggested (see Boes & Helwigh, 2000).

For the study of the four important nematodes of humans (*Ascaris*, *Trichuris*, both hookworm species), only one natural equivalent rodent model has been used: the mouse-*Trichuris muris* model. *Ascaris* rarely completes its lifecyle in rodents (with the possible exception of guinea pigs and rabbits) and therefore represents an abnormal host-parasite relationship. Instead, migration of larvae may be used as a model in which to study intestinal immunity in the early phase of *Ascaris* infection (Slotved *et al.* 1998). Hookworms of rodents are not considered true hookworms in the sense that they do not suck blood and thus cannot give rise to similar morbidity as in humans (Behnke, 1990), nor are they exposed to the same immune components. However, both *Heligmosomoides polygyrus* in mice and *Nippostrongylus brasiliensis* in rats have been used as laboratory models of hookworm infection. These and other rodent-parasite models have been used to study parasite aggregation, but the two essential models that have dealt with predisposition are the mouse-*Trichuris muris* model (Wakelin & Blackwell, 1988) and the mouse-*H. polygyrus* model (Scott, 1988a; Scott & Tanguay, 1994).

In the mouse-*Trichuris* model, certain mouse strains are predisposed to trichuriasis, being unable to express protective immunity. In a series of experiments with controlled nutritional and behavioural factors, this genetically determined variation in immune responsiveness could easily be demonstrated (Else, Wakelin & Roach, 1989). Mice exposed to a complete primary infection were fully susceptible when challenged after the removal of the primary infection by anthelmintic. In addition to host factors, two parasite-induced effects were investigated: worm size did not influence the immune responsiveness of mice, but the ability of the host to expel the parasite by day 21 after infection appeared crucial. Strains of mice that express protective immunity before day 21 do not exhibit differential responsiveness (Else & Wakelin, 1988). The factors responsible for this immunomodulatory effect were not identified, but it was suggested that any delay in the initiation of a protective immune response - whether determined by genetic variation in immune response or by behavioural factors - may leave the host exposed to immunosuppressive parasite stages, resulting in the build-up of heavy chronic infections (Else *et al.* 1989).

Scott (1988a) was able to demonstrate predisposition in mice infected with the nematodes *H. polygyrus* and *Aspicularis tetraptera*. The author made four interesting observations: (1) correlations between worm burdens at treatment and after reinfection were improved when data were analysed by age class; (2) correlations tended to be higher for mice that were mature at the beginning of the study compared to juvenile mice; (3) predisposition was not detected when egg count data were used; and (4) predisposition to *H. polygyrus* and *A. tetraptera* were independent. A follow-up study demonstrated that, in contrast to data from human studies, reinfection levels in mice during a second and third reinfection period were not correlated with initial worm load (Scott, 1988b).

Tanguay & Scott (1992) further developed the mouse-*H. polygyrus* model to study the importance of host heterogeneity in generating parasite aggregation. Heterogeneity in acquired resistance and, less consistently, host behaviour were found to contribute significantly to variability in parasite burden. Rather surprisingly, the authors did not find that worm burdens were more variable in outbred mice compared to inbred mice, but in resistant strains variability in worm burdens after challenge infection was higher than after primary infections. The authors concluded that the relative contributions of innate resistance, acquired resistance and behaviour in generating variable worm burdens are likely to vary spatially and temporally (Tanguay & Scott, 1992).

4.2 Pig models

Economic and practical considerations as well as immunological, physiological, anatomical and metabolic similarities have led several authors to propose the pig as a model for human parasite infections (e.g. Stephenson, 1987; Willingham & Hurst, 1996). Recently, a pig-*Ascaris* model and a pig-*Trichuris* model have been developed (Boes & Helwigh, 2000).

Boes *et al.* (1998) demonstrated that the degree of aggregation of *A. suum* in continuously exposed pigs on pasture is very similar to that of *A. lumbricoides* in humans. In addition, initial worm burdens and those resulting from reinfection were significantly correlated (Fig 1.3.) indicating that individual pigs are predisposed to heavy or light infection (Boes *et al.* 1998). Following up on these results, Coates (2000) conducted a study in which groups of pigs were trickle inoculated with low or high doses of *A. suum* eggs, then treated with anthelmintic followed by a reinfection period. At both dose levels pigs were predisposed to heavy or light infection, and worm burdens were heavily overdispersed. The degree of aggregation was not significantly different between groups of pigs exposed to high or low doses of infection (Coates, 2000). It was concluded that the pig-*Ascaris* model using continuous exposure is a suitable model for *A. lumbricoides* population dynamics in humans in endemic areas. Similar degrees of predisposition are recorded in both the mouse and pig models, with correlation coefficients of similar magnitude to those found in human studies, which typically do not exceed 0.50 (Keymer & Pagel, 1990).

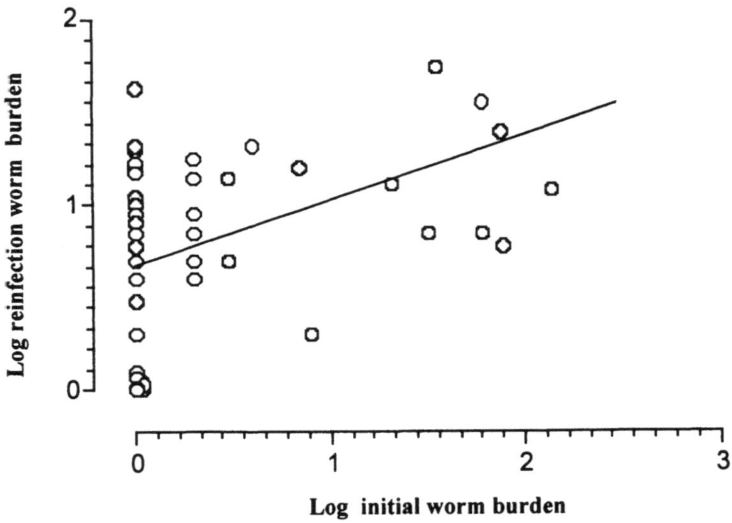

Figure 1.3: Evidence for predisposition to *Ascaris suum* infection in continuously exposed pigs The data show that initial worm burdens for individual pigs are significantly correlated with worm burdens acquired following anthelmintic treatment and a period of reinfection.

Models that certainly deserve more attention are those of *Trichuris suis* in the pig as a model for *T. trichiura* in humans, and possibly a pig-hookworm model. The pig-*T. suis* model has been used successfully to study the effect of nutritional deficiencies on helminth infection (Johansen *et al.*, 1997; Pedersen *et al.* 2001), but population dynamic studies including predisposition in continuously exposed pigs have yet to be performed. The only significant large animal model of hookworm infection that has been developed is the canine model (Behnke, 1990) but it can be argued that an omnivore model (pigs) is to be preferred to a carnivore model, because of the many physiological similarities shared by pigs and humans. The possibility of developing such a model deserves attention, not least because hookworm causes more morbidity in humans than *Ascaris* and *Trichuris* (Crompton, 2000).

4.3 Sample size and heterogeneity

In the model of *A. suum* in continuously exposed pigs, the experimental animals used were 50 triple crossbred pigs (Boes *et al.* 1998). In contrast, measurement of overdispersion and predisposition to infection in human populations is usually based on large sample sizes (see Section 3.1). However, both the degree of parasite aggregation and predisposition in continuously exposed pigs were comparable to that reported for *A. lumbricoides* in human field studies, showing that this relatively small-scale application of the pig-*Ascaris* model was useful and appropriate (Boes, 1999). Based on these results, Coates (2000) calculated that to reliably measure aggregation and predisposition in pigs continuously exposed to *A. suum*, a minimum group size of 30 animals would be required. Compared to human populations where there is considerable genotypic and phenotypic heterogeneity, the experimental groups of pigs used by Boes *et al.* (1998) and Coates (2000) were not very heterogeneous (all males, same age, all healthy, well fed), although in pig terms they could have been more homogenous (e.g. if inbred animals) However, demonstrating the phenomenon using small group sizes of pig is only one of the important issues and it seems unlikely that such groups will suffice to investigate the mechanisms underlying predisposition because of host heterogeneity. Even inbred pig strains may show a heterogeneous response to infection with *A. suum* compared to outbred pigs (L. Eriksen, personal communication) but perhaps a well-defined inbred pig strain could be used to identify the genes involved in the anti-*Ascaris* immune response (see Behnke *et al.* 2000). The results of a newly launched collaborative pig genome project in China and Denmark may be able to contribute to this and are awaited with great interest. In addition, pedigree studies involving large numbers of genetically well-defined pigs (e.g. the pig population in Denmark) and carried out as was done by Williams-Blangero *et al.* (1999) in human populations (see Chapter 10) is worthy of consideration.

4.4 Genetics

Variation in immunocompetence most likely has a genetic basis, but is also influenced by phenotypic factors such as nutrition, reproductive state and concurrent infections (Keymer & Pagel, 1990). Interestingly, the very similar degree of aggregation and predisposition in the pig-*Ascaris* model -

which employed healthy, well fed castrated male pigs infected only with *A. suum* – and in various human studies (Boes, 1999), seems to suggest that the heterogeneous response to infection seen in this study, was basically genetically determined. However, the possibility of behavioural differences, which have been shown to be of some influence in the mouse-*H. polygyrus* model (Tanguay & Scott, 1992), could not be ruled out.

Despite the fact that it is now well known that resistance to infection is variable within host species, little progress has been made in defining the genes responsible. The known loci of genes that are linked to gastrointestinal nematode infections are all MHC associated – although background genes also exert considerable influence on infection patterns (Wakelin, 1992) – and resistance to gastrointestinal nematodes is heritable (Behnke *et al.* 2000). On the other hand, parasites themselves are probably genetically and antigenically heterogeneous (Grant, 1994; Kennedy, 1995; Fraser & Kennedy, 1991) and hosts may vary in their susceptibility to parasite evasive strategies (immunomodulation) (Behnke *et al.* 2000).

The obvious approach to the study of predisposition based on evidence generated in laboratory mice (Else *et al.* 1989; Tanguay & Scott, 1992) is to undertake genetic studies in well-defined strains of animals. Behnke *et al.* (2000) carried out a series of genetic studies on resistance of mice to *H. polygyrus* as a model for identification of homologous genes in domestic animals. They were able to show that the F1 progeny of a susceptible and a resistant strain behaved much like the latter, but expelled the infection at an even earlier stage than the resistant parent strain, indicating gene complementation. The authors intend to phenotype F2 and eventually F6 progeny from crosses between resistant and susceptible strains for parasitological and immunological traits. It is expected that data from this project will facilitate breeding for resistance to parasites and increase understanding of genetic resistance.

4.5 Immunology

As is clear from the studies in humans cited above, an immunological explanation for predisposition to *Ascaris* infection using serum antibody responses seems unlikely to be straightforward. In this regard, it is interesting that IgE levels have been found to differ between individual humans that were susceptible or resistant to *A. lumbricoides* infection (Palmer *et al.* 1995; McSharry *et al.* 1999), and that immune recognition of certain *Ascaris* allergens is under MHC control in rodents (Kennedy, Fraser & Christie, 1991), indicating genetic differences in immune reactivity. It would be interesting to investigate the possible role of IgE in *A. suum* infection, but although porcine IgE has been isolated (Roe *et al.* 1993) no studies measuring total or specific IgE in pigs have been published to date.

Little is known about the cellular response of pigs to *Ascaris* larvae and adult worms, but in mouse models it has been shown that it is Th2 mediated and that cytokines play an important role (Behnke *et al.* 2000). Studies in the mouse-*Trichuris* model have demonstrated a crucial role for the activation of distinct T-helper cells in determining expulsion of intestinal worm burdens. Mice that do not expel their worms mount an inappropriate dominant Th1 response and will be susceptible to challenge infection, while mice mounting a dominant Th2 response will expel their worms and be resistant to challenge (see reviews by Grencis, 1996, and Artis & Grencis, 2001). This balance between Th1 and Th2 responses is influenced critically by the kinetics of infection and by cytokine excretion (see Chapter 8), and deserves further investigation in the pig model. In addition, recent reports indicate that the role of B cells and antibodies may be more important in resistance to nematode infections than was hitherto assumed (Blackwell & Else, 2001).

Using the pig model, Roepstorff *et al.* (1997) showed that although each pig became infected with *A. suum* upon experimental inoculation, the majority of pigs expelled the worms between days 14 and 21, resulting in the well-known overdispersed distribution of adult worms. The mechanism behind worm expulsion still has not been revealed, but the key processes resulting in predisposition to either high or low intensity of infection are likely to occur in this expulsion interval, and the pig model is certainly a promising candidate for further study.

5. CONCLUDING REMARKS

In conclusion, there is considerable scope for study of predisposition in animal models. Immunological studies should be followed by genetic studies with the aim to explain why certain immunological events occur in some individuals while they fail to happen in other individuals in the same population. The observation that a protective immune response is the result of a balance of immune factors rather than an all-or-nothing event, combined with the observation that under field conditions hosts change predisposition status suggests that it will not be easy to disentangle the influence of host susceptibility and exposure to infection. And even if, in laboratory models, the genetic background for differences in resistance and susceptibility within the same host population is defined, the next problem will be to identify the contribution of perturbations such as variability in exposure, behaviour, nutrition and others under field conditions. And finally, the question remains which hosts are of most interest: those that eventually end up harbouring worms, or those that remain worm free - even after one or more rounds of deworming and reinfection.

REFERENCES

ARTIS, D. & GRENCIS, R.K. (2001). T helper cytokine responses during intestinal nematode infection: Induction, regulation and effector function. In: *Parasitic Nematodes – Molecular biology, biochemistry and immunology*. (eds. Kennedy, M.W. & Harnett, W.) pp. 311-371. CABI Publishing.

ASAOLU, S.O., HOLLAND, C.V. & CROMPTON, D.W.T. (1991). Community control of *Ascaris lumbricoides* in rural Oyo State, Nigeria: mass, targeted and selective treatment with levamisole. *Parasitology* **103**, 291-298.

ANDERSON, R.M. (1986). The population dynamics and epidemiology of intestinal nematode infections. *Transactions of the Royal Society of Tropical Medicine and Hygiene* **80**, 686-696.

ANDERSON, R.M. & MAY, R.M. (1979). Population biology of infectious diseases: Part I. *Nature*, London **280**, 361-367.

ANDERSON, R.M. & MEDLEY, G. F. (1985). Community control of helminth infections of man by mass and selective chemotherapy. *Parasitology* **90**, 629-660.

BEHNKE, J.M. (1990). Laboratory animal models. In: *Hookworm infection: Current status and new directions* (ed. Schad, G.A. & Warren, K.S.), pp. 105-128. Taylor and Francis, London.

BEHNKE, J.M., LOWE, A., MENGE, D., IRAQI, F. & WAKELIN, D. (2000). Mapping genes for resistance to gastrointestinal nematodes. *Acta Parasitologica* **45**, 1-13.

BLACKWELL, N.M. & ELSE, K.J. (2001). B cells and antibodies are required for resistance to the gastrointestinal nematode parasite *Trichuris muris*. *Infection and Immunity* **69**, 3860-3868.

BOES, J. (1999). Population biology of *Ascaris suum* with special emphasis on worm burden distributions in pigs. Ph.D. Thesis, Royal Veterinary and Agricultural University, Copenhagen.

BOES, J. & HELWIGH, A.B. (2000). Animal models of intestinal nematode infections of humans. *Parasitology* **121** (Suppl.), S97-S111.

BOES, J., MEDLEY, G.F., ERIKSEN, L., ROEPSTORFF, A. & NANSEN, P. (1998). Distribution of *Ascaris suum* in experimentally and naturally infected pigs and comparison with *Ascaris lumbricoides* infections in humans. *Parasitology* **117**, 589-596.

BUNDY, D.A.P., COOPER, E.S., THOMPSON, D.E., DIDIER, J.M., ANDERSON, R.M. & SIMMONS, L. (1987a). Predisposition to *Trichuris trichiura* in humans. *Epidemiology and Infection* **98**, 65-71.

BUNDY, D.A.P., COOPER, E.S., THOMPSON, D.E., ANDERSON, R.M. & DIDIER, J.M. (1987b). Age-related prevalence and intensity of *Trichuris trichiura* infection in a St. Lucian community. *Transactions of the Royal Society of Tropical Medicine and Hygiene* **81**, 85-94.

BUNDY, D.A.P. (1988). Population ecology of intestinal helminth infections in human communities. *Philosophical Transactions of the Royal Society of London* **B321**, 405-420.

BUNDY, D.A.P. & MEDLEY, G.F. (1992). Immuno-epidemiology of human geohelminthiasis: ecological and immunological determinants of worm burden. *Parasitology* **104** (Suppl.), S105-S119.

BUNDY, D.A.P., LILLYWHITE, J.E., DIDIER, J.M., SIMMONS, I. & BIANCO, A.E. (1991). Age-dependency of infection status and serum antibody levels in human whipworm (*Trichuris trichiura*) infection. *Parasite Immunology* **13**, 629-638.

CHAN, L., BUNDY, D.A.P. & KAN, S.P. (1994a). Aggregation and predisposition at the familial level. *Transactions of the Royal Society of Tropical Medicine and Hygiene* **88**, 46-48.

CHAN, L., BUNDY, D.A.P. & KAN, S.P. (1994b). Genetic relatedness as a determinant of predisposition to *Ascaris lumbricoides* and *Trichuris trichiura* infection. *Parasitology* **108**, 77-80.

CHANDIWANA, S.K., WOOLHOUSE, M.E.J. & BRADLEY, M. (1991). Factors affecting the intensity of reinfection with *Schistosoma haematobium* following treatment with praziquantel. *Parasitology* **102**, 73-83.

COATES, S. (2000). Modelling the population dynamics of *Ascaris suum* in pigs. Ph.D. Thesis, University of Warwick, Coventry, UK.

COOPER, P.J., CHICO, M.E., SANDOVAL, C., ESPINEL, I., GUEVARA, A., KENNEDY, M.W., URBAN, J.F., GRIFFIN, G.E. & NUTMAN, T.B. (2000). Human infection with *Ascaris lumbricoides* is associated with a polarized cytokine response. *Journal of Infectious Diseases* **182**, 1207-1213.

CROFTON, H.D. (1971). A quantitative approach to parasitism. *Parasitology* **63**, 343-364.

CROLL, N.A. & GHADIRIAN, E. (1981). Wormy persons : Contributions to the nature and patterns of overdispersion with *Ascaris lumbricoides, Ancylostoma duodenale, Necator americanus* and *Trichuris trichiura*. *Tropical and Geographical Medicine* **33**, 241-248.

CROMPTON, D.W.T. (2000). The public health importance of hookworm disease. *Parasitology* **121** (Suppl.), S39-S50.

CROMPTON, D.W.T., KEYMER, A.E. & ARNOLD, S.E. (1984). Investigating over-dispersion ; *Moniliformis* (Acanthocephala) and rats. *Parasitology* **88**, 317- 331.

ELKINS, D.B., HASWELL-ELKINS, M. & ANDERSON, R.M. (1986). The epidemiology and control of intestinal nematodes in the Pulicat Lake region of Southern India. I Study design and pre- and post- treatment observations on *Ascaris lumbricoides* infection. *Transactions of the Royal Society of Tropical Medicine and Hygiene* **80**, 774-792.

ELSE, K.J. & WAKELIN, D. (1988). The effects of H-2 and non-H-2 genes on the expulsion of the nematode *Trichuris muris* from inbred congenic mice. *Parasitology* **96**, 543-550.

ELSE, K.J., WAKELIN, D. & ROACH, T. (1989). Host predisposition to trichuriasis: The mouse-*Trichuris muris* model. *Parasitology* **98**, 275-282.

FORRESTER, J.E., SCOTT, M.E., BUNDY, D.A.P. & GOLDEN, M.H.N. (1990). Predisposition of individuals and families in Mexico to heavy infection with *Ascaris lumbricoides* and *Trichuris trichiura*. *Transactions of the Royal Society of Tropical Medicine and Hygiene* **84**, 272-276.

FRASER, F.M. & KENNEDY, M.W. (1991). Heterogeneity in the expression of surface-exposed epitopes among larvae of *Ascaris lumbricoides*. *Parasite Immunology* **13**, 219-325.

GRANT, W.N. (1994). Genetic variation in parasitic nematodes and its implications. *International Journal for Parasitology* **24**, 821-830.

GRENCIS, R.K. (1996). T Cell and cytokine basis of host variability in response to intestinal nematode infections. *Parasitology* **112** (Suppl.), S31-S37.

HAGEL, I., LYNCH, N.R., DI PRISCO, M.C., ROJAS, E., PEREZ, M. & ALVAREZ, N. (1993). *Ascaris* reinfection of slum children : relation with the IgE response. *Clinical and Experimental Immunology* **94**, 80-83.

HALL, A., ANWAR, K. & TOMKINS, A.M. (1992). Intensity of reinfection with *Ascaris lumbricoides* and its implications for parasite control. *Lancet* **339**, 1253-1257.

HAGAN, P., BLUMENTHAL, U.J., DUNN, D., SIMPSON, A.J.G. & WILKINS, H.A. (1991). Human IgE, IgG4 and resistance to reinfection with *Schistosoma haematobium*. *Nature*, London **349**, 243-245.

HASWELL-ELKINS, M.R., ELKINS, D.B. & ANDERSON, R.M. (1987). Evidence for predisposition in humans to infection with *Ascaris*, hookworm, *Enterobius* and *Trichuris* in a South Indian fishing community. *Parasitology* **95**, 323-337.

HASWELL-ELKINS, M.R., ELKINS, D.B., MANJULA, K., MICHAEL, E. & ANDERSON, R.M. (1988). An investigation of hookworm infection and reinfection following mass anthelmintic treatment in the South Indian fishing community of Vairavankuppam. *Parasitology* **96**, 565- 577.

HASWELL-ELKINS, M.R., KENNEDY, M.W., MAIZELS, R.M., ELKINS, D.B. & ANDERSON, R.M. (1989). The antibody recognition profiles of humans naturally infected with *Ascaris lumbricoides*. *Parasite Immunology* **11**, 615-621.

HASWELL-ELKINS, M.R., LEONARD, H., KENNEDY, M.W,. ELKINS, D.B. & MAIZELS, R.M. (1992). Immunoepidemiology of *Ascaris lumbricoides* : relationships between antibody specificities, exposure and infection in a human community. *Parasitology* **104**, 153-159.

HENRY, F.J. (1988). Reinfection with *Ascaris lumbricoides* after chemotherapy : a comparative study in three villages with varying sanitation. *Transactions of the Royal Society of Tropical Medicine and Hygiene* **82**, 460-464.

HOLLAND, C.V., ASAOLU, S.O., CROMPTON, D.W.T., STODDART, R.C., MACDONALD, R. & TORIMIRO, S.E.A. (1989). The epidemiology of *Ascaris lumbricoides* and other soil-transmitted helminths in primary school children from Ile-Ife, Nigeria. *Parasitology* **99**, 275-285.

HOLLAND, C.V., CROMPTON, D.W.T., ASAOLU, S.O., CRICHTON, W.B., TORIMIRO, S.E.A. & WALTERS, D.E. (1992). A possible genetic factor influencing protection from infection with *Ascaris lumbricoides* in Nigerian children. *Journal of Parasitology* **78**, 915-916.

JOHANSEN, M.V., BØGH, H.O., GIVER, H., ERIKSEN, L., NANSEN, P., STEPHENSON, L.S. & KNUDSEN, K.E.B. (1997). *Schistosoma japonicum* and *Trichuris suis* infections in pigs fed diets with high and low protein. *Parasitology* **115**, 257-264.

KENNEDY, M.W. (1995). Genetic control of the immune repertoire in nematode infections. *Parasitology Today* **5**, 316-324.

KEYMER, A.E. & PAGAL, M. (1990). Predisposition to helminth infection. In *Hookworm disease : Current status and New directions* (ed. Schad, G.A. & Warren, K.S.), pp177-209. Taylor & Francis, London.

KIGHTLINGER, L.K., SEED, J.R. & KIGHTLINGER, M.B. (1998). *Ascaris lumbricoides* intensity in relation to environmental, socioeconomic, and behavioural determinants of exposure to infection in children from Southeast Madagascar. *Journal of Parasitology* **84**, 480-484.

LILLYWHITE, J.E., BUNDY, D.A.P., DIDIER, J.M., COOPER, E.S. & BIANCO, A.E. (1991). Humoral immune responses in human infection with the whipworm *Trichuris trichiura*. *Parasite Immunology* **13**, 491-507.

MACDONALD, T.T., SPENCER, J., MURCH, S.H., CHOY, M.-Y., VENUGOPAL, S., BUNDY, D.A.P. & COOPER, E.S. (1994). Immunoepidemiology of intestinal nematode infections. 3. Mucosal macrophages and cytokine production in the colon of children with *Trichuris trichiura* dysentry. *Transactions of the Royal Society of Tropical Medicine and Hygiene* **88**, 265-268.

MCCALLUM, H.I. (1990). Covariance in parasite burdens : the effect of predisposition to infection. *Parasitology* **100**, 153-159.

MCSHARRY, C., XIA, Y., HOLLAND, C. & KENNEDY, M.W. (1999). Natural immunity to *Ascaris lumbricoides* associated with Immunoglobulin E antibody to ABA-1 allergen and inflammation indicators in children. *Infection and Immunity* **67**, 484-489.

NEEDHAM, C.S., BUNDY, D.A.P., LILLYWHITE, J.E., DIDIER, J.M., SIMMONS, I. & BIANCO, A.E. (1992). The relationship between *Trichuris trichiura* transmission intensity and the age-profiles of parasite-specific antibody isotypes in two endemic communities. *Parasitology* **105**, 273-283.

PALMER, D.R., HALL, A., HAQUE, R. & ANWAR, K.S. (1995). Antibody isotype responses to antigens of *Ascaris lumbricoides* in a case-control stuidy of persistently heavily infected Bangladeshi children. *Parasitology* **111**, 385-393.

PEDERSEN, S., SAEED, I., FRIIS, H. & MICHAELSEN, K.F. (2001). Effect of iron deficiency on *Trichuris suis* and *Ascaris suum* infections in pigs. *Parasitology* **122**, 589-598.

PENG, W., ZHOU, X., CUI, X., CROMPTON, D.W.T., WHITEHEAD, R.R., XIONG, J., WU, H., YANG, Y., WU, W., XU, K. & YAN, Y. (1998). Transmission and natural regulation of infection with *Ascaris lumbricoides* in a rural community in China. *Journal of Parasitology* **84**, 252-258.

PRITCHARD, D.I., QUINNELL, R.J., SLATER, A.F.G., MCKEAN, P.G., DALE, D.D.S., RAIKO, A. & KEYMER, A.E. (1990). Epidemiology and immunology of *Necator americanus* infection in a community in Papua New Guinea : humoral responses to excretory-secretory and cuticular antigens. *Parasitology* **100**, 317-326.

PRITCHARD, D.I., WALSH, E.A., QUINNELL, R.J., RAIKO, A., EDMONDS, P. & KEYMER, A.E. (1992). Isotypic variation in antibody responses in a community in Papua New Guinea to larval and adult antigens during infection, and following reinfection, with the hookworm *Necator americanus*. *Parasite Immunology* **14**, 617-631.

PRITCHARD, D.I., QUINNELL, R.J. & WALSH, E.A. (1995). Immunity in humans to *Necator americanus* : IgE, parasite weight and fecundity. *Parasite Immunology* **17**, 71-75.

QUINNELL, R.J., SLATER, A.F.G., TIGHE, P., WALSH, E.A., KEYMER, A.E. & PRITCHARD, D.I. (1993). Reinfection with hookworm after chemotherapy in Papua New Guinea. *Parasitology* **106**, 379-385.

QUINNELL, R.J., WOOLHOUSE, M.E.J., WALSH, E.A. & PRITCHARD, D.I. (1995). Immunoepidemiology of human necatoriasis : correlations between antibody responses and parasite burdens. *Parasite Immunology* **17**, 313-318.

QUINNELL, R.J., GRIFFIN, J., NOWELL, M.A., RAIKO, A. & PRITCHARD, D.I. (2001). Predisposition to hookworm infection in Papua New Guinea. *Transactions of the Royal Society of Tropical Medicine and Hygiene* **95**, 139-142.

ROE, J.M., PATEL, D. & MORGAN, K.L. (1993). Isolation of porcine IgE and preparation of polyclonal antisera. *Veterinary Immunology and Immunopathology* **37**, 83-97.

ROEPSTORFF, A., ERIKSEN, L., SLOTVED, H-C. & NANSEN, P. (1997). Experimental *Ascaris suum* infections in the pig: Worm population kinetics following single inoculations with three doses of infective eggs. *Parasitology* **115**, 443-452.

SCHAD, G.A. & ANDERSON, R.M. (1985). Predisposition to hookworm infection in humans. *Science* **228**, 1537-1540.

SCOTT, M.E. (1988a). Predisposition of mice to *Heligmosomoides polygyrus* and *Aspiculuris tetraptera* (Nematoda). *Parasitology* **97**, 101-114.

SCOTT, M.E. (1998b). Effect of repeated anthelminitc treatment on ability to detect predisposition of mice to *Heligmosomoides polygyrus* and *Aspiculuris tetraptera* (Nematoda) infections. *Parasitology* **97**, 453-458.

SCOTT, M.E. & TANGUAY, G.V. (1994). *Heligmosomoides polygyrus*: a laboratory model for direct life-cycle nematodes of humans and livestock. In: *Parasitic and infectious diseases. Epidemiology and ecology* (ed. Scott, M.E. & Smith, G.), pp. 279-300. Academic Press Ltd., London.

SHAW, D.J. & DOBSON, A.P. 91995). Patterns of macroparasite abundance abd aggregation in wildlife populations : a quantitative review. *Parasitology* **111** (Suppl.), S111-S133.

SLOTVED, H-C., ERIKSEN, L., MURRELL, K.D. & NANSEN, P. (1998). Early *Ascaris suum* migration in mice as a model for pigs. *Journal of Parasitology* **84**, 16-18.

STEPHENSON, L.S. (1987). The design of nutrition-parasite studies. In: *The impact of helminth infections on human nutrition: schistosomes and soil-transmitted helminths* (ed. Stephenson, L.S. & Holland, C.V.), pp. 21-46. Taylor & Francis, Philadelphia.

STEPHENSON, L.S., HOLLAND, C. & COOPER, E.S. (2000). The public health significance of *Trichuris trichiura*. *Parasitology* **121**, S73-S95.

TANGUAY, G.V. & SCOTT, M.E. (1992). Factors generating aggregation of *Heligmosomoides polygyrus* (Nematoda) in laboratory mice. *Parasitology* **104**, 519-529.

WAKELIN, D. (1992). Genetic variation in resistance to parasitic infection: experimental approaches and practical applications. *Research in Veterinary Science* **53**, 139-147.

WAKELIN, D. & BLACKWELL, J. (1988). *Genetics of resistance to infection.* Taylor & Francis, London.

WILLIAMS-BLANGERO, S., SUBEDI, J., UPADHAYAY, R.P., MANRAL, D.B., RAI, D.R., JHA, B., ROBINSON, E.S. & BLANGERO, J. (1999). Genetic analysis of susceptibility to infection with *Ascaris lumbricoides*. *American Journal of Tropical Medicine and Hygiene* **60**, 921-926.

WILLINGHAM, A.L. & HURST, M. (1996). The pig as a unique host model for *Schistosoma japonicum* infection. *Parasitology Today* **12**, 132-134.

WONG, M.S., BUNDY, D.A.P. & GOLDEN, M.H.N. (1988). Quantitative assessment of geophagous behaviour ass a potential source of exposure to geohelminth infection. *Transactions of the Royal Society of Tropical Medicine and Hygiene* **82**, 621-625.

WOOLHOUSE, M.E.J. (1993). A theoretical framework for immune responses and predisposition to helminth infection. *Parasite Immunology* **15**, 583-594.

Chapter 2
CONTROL STRATEGIES

Lorenzo Savioli[1], Antonio Montresor[1] and Marco Albonico[2]
[1] *World Health Organization, Geneva, Switzerland*
[2] *Ivo de Carneri Foundation, Milan, Italy*
e-mail: saviolil@who.ch

1. INTRODUCTION

Geohelminth infections represent a serious public health problem in countries where sanitation and hygienic conditions are insufficient to respond to the needs of the population, and where effective drugs for their control are neither widely available nor accessible to the population in need.

In countries where an improvement of the sanitation condition as a natural component of the country's economic progress had taken place, a parallel progressive decline of the prevalence of geohelminth infections was invariably observed. Where universal or targeted deworming programmes accompanied such economic growth, the results were obtained in a much shorter time span and they were long-term. However, where periodic chemotherapy was available, even in the absence of sanitation improvement and economic growth, important control of morbidity was obtained.

Control strategies should aim to control morbidity due to geohelminth infections in the first place, and to control their transmission where conditions are such to allow a comprehensive effort in preventive measures. Different approaches have been implemented in endemic countries according to the local health relevance of the problem, and to their resources. Results obtained from control programmes in endemic areas are continuously monitored to design appropriate strategies for the control of geohelminth infections.

2. THE EXPERIENCE FROM JAPAN AND KOREA

Japan has achieved successful and sustained control of geohelminth infections and has led the way in this effort.

In 1949, a nation-wide survey of faecal samples reported an overall prevalence of 73.0% for intestinal nematodes: *A. lumbricoides* (62.9%), *T. trichiura* (50%), and hookworms (3.5%). Non-governmental Organizations (NGOs) took the initiative, private laboratories were established, stool examinations were carried out and treatment with anthelminthic drugs began. School children regularly underwent mass stool examination and positive cases received treatment twice a year. In 1955, the Japanese Association of Parasite Control was founded and, the government passed the School Health Law in 1958 and issued guidance on control technologies. The cellophane thick smear method (to become the Kato Katz technique) was invented and was widely adopted for stool examinations. By 1990, the prevalence of *A.lumbricoides* dropped to 0.9%, *T. trichiura* to 0.25% and hookworms to 0%.

A similar experience occurred in Korea between 1969 and 1995. In this case the programme focused on selective treatment of infected schoolchildren and the significant results obtained are presented in Figure 2.1.

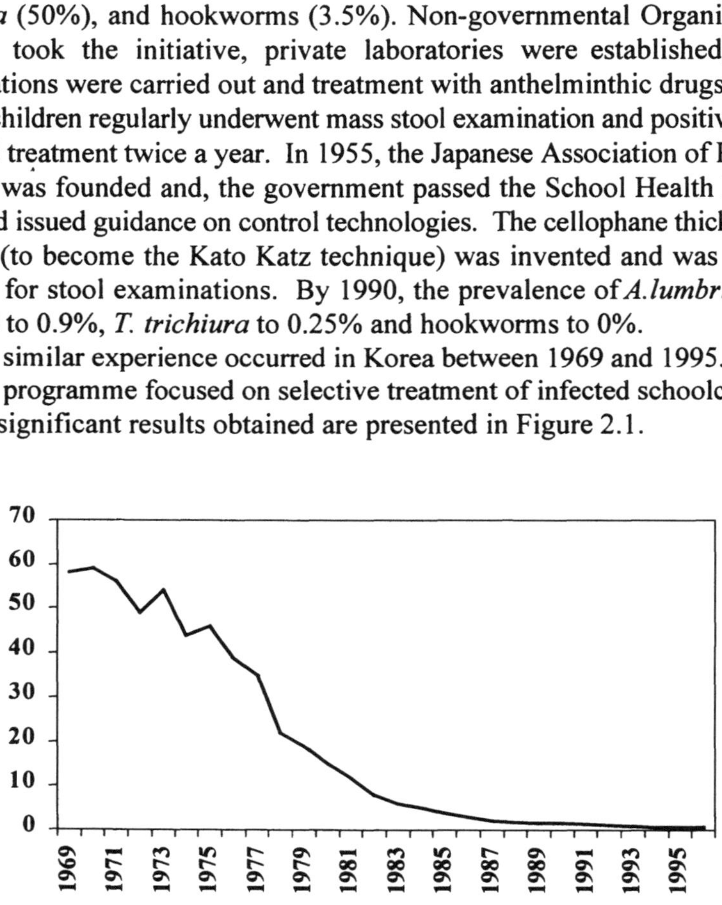

Figure 2.1. Decrease in the prevalence of *A.lumbricoides* in schoolchildren between 1969 and 1965 in the Republic of Korea (Ministry of Health and Social Affairs, 1996)

The relevance of economic development (linked to an improvement in sanitation standards) to permanently solve the public health problem caused by geohelminths is confirmed by the fact that in other countries significant reductions in prevalence have been obtained virtually without control activity: for example in Italy between 1965 and 1980 the prevalence of trichiuriasis dropped from 65% to less than 5% and the prevalence of ascariasis from 10% to 0% (de Carneri, 1989).

When geohelminths control measures are applied in a situation where economic development is ongoing, the results in terms of decline in prevalence and health improvement, are rapid and definitive. In addition, the control of morbidity due to geohelminths infection can in itself contribute to the economic development of the country by boosting the capacity of schoolchildren to grow and learn better, by increasing the physical fitness of adults and the health of adolescent girls and women of child-bearing age (see Chapters 3 and 4).

3. THE 'REALITY' IN DEVELOPING COUNTRIES

Unfortunately, this situation of economic development does not apply in most of the 'developing countries' where during the last decades they have faced a progressive deterioration in their economic situation and a concomitant decline in sanitation and hygiene standards (The World Health Report, 1999). In this context of limited resources, the population is more vulnerable to the damage caused by the geohelminths and the need for control activities is greater (see Chapter 5). However, control is more logistically difficult and the results are, therefore, less dramatic.

4. EPIDEMIOLOGICAL BASIS OF THE WHO STRATEGY

To select appropriate control measures and to evaluate the outcomes correctly an understanding of the important epidemiological patterns of the geohelminths is required.

4.1 Children and women harbour peak worm burdens (Bundy *et al.* 1992):

Worm burdens peak in children and women and in addition these groups experience intense metabolism and physical growth, resulting in increased nutritional needs. This explains why pre-school children, schoolchildren and women of child bearing age are particularly vulnerable to the nutritional deficits related to the infections and are considered the population groups at greater risk of morbidity due to geohelminths (see Chapter 3).

4.2 Heavy intensity infections are the major source of morbidity:

Morbidity is directly related to worm burden (Bundy *et al.* 1992). For example, in the case of hookworms, the amount of blood lost in the faeces (as an indicator of morbidity) is directly positively associated with hookworm egg count (as a measure of worm burden) (Stoltzfus *et al.* 1996).

4.3 Until environmental and/or behavioural conditions have changed, the prevalence of infection will tend to return to original pre-treatment levels

Re-infection occurs because infective stages will continue to contaminate the environment. Therefore the population will get re-infected, but repeated treatment can ensure that they have fewer worms, for shorter periods. This will significantly reduce the potential damage caused by these infections. (Guyatt *et al.* 1993).

The challenge is to develop an appropriate and cost-effective control strategy, which would ensure as a priority the reduction of morbidity in the high-risk groups. This is done by reducing to minimal levels the proportion of heavily infected individuals and can be achieved by periodic distribution of deworming drugs accompanied by health education campaigns. At the same time, according to the available resources, other complementary control measures such as social mobility, information, education and communication, and improvement of sanitation should be promoted in order to sustain the benefits of periodic treatment and to achieve long lasting control of transmission of infection.

5. THE WHO STRATEGY FOR HELMINTH CONTROL

The WHO strategy in 'developing countries' (World Health Assembly (WHA) 54.19) is therefore based on the delivery to the three high risk groups, pre-school children, schoolchildren and women of child-bearing age, of:

- periodic treatment (in order to keep the worm burden low)
- health education (in an attempt to reduce at risk behaviour and to prevent re-infection)

If possible, these interventions should be accompanied by an improved access to safe water and sanitation.

Practical approaches are suggested to adapt the strategy to the different epidemiological situations and to deliver this intervention to the high-risk groups at low cost:

5.1 Community diagnosis instead of individual diagnosis

This approach entails periodic checking of the parasitological and nutritional status in samples of the population to evaluate the necessity for an intervention and frequency of the application required. The same approach can be applied to monitor the results obtained.

5.2 Community treatment instead of the individual treatment

This approach applies in areas with high transmission of geohelminth infections and is recommended due to the safety and low cost of the drug used. Where appropriately applied, it reduces the laboratory work and programme cost significantly. This approach also provides treatment for individuals that, due to the limited sensitivity of the laboratory diagnosis used in some endemic countries, would have been recorded negative despite being infected.

5.3 Use of the existing infrastructure to deliver the intervention

This approach eliminates the need for new infrastructures to deliver the intervention, and suggests that schools, Maternal and Child Health (MCH) clinics and vaccination campaigns could easily be used as a means to reach the groups at risk.

Among the global targets for 2010 endorsed by the World Health Assembly in resolution WHA 54.19 in 2001 the goal of attaining regular deworming, of at least 75% up to 100% of all schoolchildren at risk of morbidity has particular relevance.

WHO is presently advocating a Partnership for Parasite Control with UN organizations, bilateral agencies, non-governmental organizations and the private sector to co-ordinate the global effort to combat morbidity due to geohelminths infection and schistosomiasis. The strategy endorsed envisages country planning, training and capacity building, support for national drug supply, resource mobilisation and donor relationships, and surveillance monitoring and evaluation and takes advantage of the existing structure to deliver the control measures.

6. EXPERIENCE IN SEYCHELLES

The Seychelles archipelago comprises 115 islands with 73,000 inhabitants, but most live on the main islands of Mahe, Praslin and La Digue. GDP per head is US$ 7000. Education covers over 95% of the school-eligible age group and only 5% of the population lack latrines. The Ministry of Health devised a plan of action with the objective of reducing the intensity of intestinal nematode infections to a level which no longer constituted a public health problem. The specific control objectives within a three-year span were: (i) reduction of intensity (epg) of infections with *A. lumbricoides* by 60%, and of *T. trichiura* and hookworm infections by 30% in school-age children, (ii) reduction in the target population of prevalence of *S. stercoralis* infection by 30% and (iii) reduction in the target population of prevalence of amoebiasis of 40%.

School children and pregnant women represented the target groups. Sixty percent of children were infected with one or more intestinal parasites, with significant variation by region. *T. trichiura* was the most common parasite with a prevalence of 53.3%, followed by *A. lumbricoides* with a prevalence of 17.7%. Hookworm infections were present in 6.3% of school children and in 8.6% of pregnant women.

School children were dewormed every four months in the first year, with a coverage rate of 99.4%. Mebendazole (500 mg tablet), given as a single dose, was the anthelminthic chosen by the Ministry of Health due to the high prevalence of *T. trichiura*. Treatment was delivered by teachers under the supervision of staff from the nearest health centre. Due to the low prevalence of infection in pregnant women, selected treatment was given to positive cases as diagnosed by a routine stool examination. Treatment was administered after the first trimester of pregnancy.

Print media (newspaper, posters, leaflets) and electronic media (radio, television, audio-visual aids) were extensively used to increase public information and awareness on intestinal parasites control. Since the start of the programme, education about preventive measures on intestinal parasites was included in the school curriculum. Mobile health teams (environmental health officers, school health nurses), in collaboration with Social Education teachers, organized sessions and disseminated health messages in all schools. The radio advertised the programme's activities and general preventive measures. TV and the national newspaper were also involved in advertising chemotherapy campaigns. A video on prevention and control of intestinal parasites produced in the Seychelles was widely distributed to schools, health centres and broadcast by local TV. Leaflets and posters on the prevention and control of intestinal parasitic infection were designed in Creole and printed locally.

After three chemotherapy campaigns, a parasitological evaluation showed that the cumulative prevalence of intestinal parasites dropped from 60.5% to 33.8% in the children. The mean egg counts was reduced by 85%, 53% and 32% from the baseline value, for *A. lumbricoides*, *T. trichiura* and hookworm, respectively (Albonico *et al.* 1996).

A recent report, showed that after seven years of control activities, intestinal parasitic infections in the Seychelles have reached such a low level indicating that transmission as well has morbidity control have been successfully achieved (Shamlaye, 2001).

7. EXPERIENCE IN NEPAL

Nepal has 23 million inhabitants, of which 700,000 live in Kathmandu. GDP per head is US$ 165 and 75% of the population lacks latrines.

Since 1990 the World Food Programme (WFP) has been providing daily mid-day snacks to 250,000 schoolchildren in 16 districts with the aim of developing the country's human resources.

In 1996 a school survey showed a very high prevalence of geohelminths in all the districts investigated: 74.2% of the children tested were infected with at least one of the geohelminths and 9.3% presented with heavy infections. Since 1998, WFP-Nepal, in collaboration with WHO, has been including deworming (with locally produced albendazole) within the School Feeding Programme. In November 2000, an epidemiological survey was conducted by a MoH-WFP-WHO team to monitor and evaluate the impact of the programme.

The results of the survey when compared with the baseline data from other available nutritional information in the country showed a remarkable impact on the health of the children periodically treated. The prevalence was reduced by 20% but more importantly, the heavy infections had virtually disappeared, being confined to children who had recently arrived in the areas and were, therefore, not yet covered by the intervention.

Comparison of haemoglobin levels in schools covered by WFP activities and the national data showed a significant difference in the number of children with anaemia and severe anaemia. Only 10% of children were anaemic (Hb<11g/dl) (compared to the expected 58%) and, most importantly, no severe anaemia (Hb<7g/dl) was detected. This is probably due to the combined action of food fortification and deworming. Convinced by these results, other organizations started to include de-worming in their activities: United Nations High Commission for Refugees (UNHCR), in collaboration with Centers for Diseases Control and Prevention (CDC), Atlanta, Caritas and the Japanese NGO, Association of Medical Doctors of Asia (AMDA), started, in July 2001, to deworm more than 50,000 children including those under five years of age.

In addition, the Ministry of Health of Nepal in collaboration with UNICEF included de-worming among the routine interventions for pregnant women after the first trimester of pregnancy.

8. INTEGRATED APPROACH

In all endemic countries, and particularly in countries with limited resources available, strategies for the control of parasitic infections are being re-considered in order to optimise human and financial resources and make the best use of personnel, expertise, surveillance and data collection, health infrastructure and communication system. This approach of integrated control has enabled a broader range of health problems to be tackled more effectively and at affordable and sustainable costs. Integrated disease control is the merging of resources, services and intervention sat different levels and between sectors to improve health outcomes.

Since 1997, with the support of WHO, a few countries have developed programmes based on an integrated approach to disease control. Their communicable disease control activities have been integrated within their national public health system, based on a single plan of action drawn up, endorsed by WHO and approved by governments (WHO, 1998).

Geohelminth infections are particularly suitable for this kind of intervention as their control approach can be adopted to combat other diseases such as schistosomiasis and lymphatic filariasis. The Programme for Elimination for Lymphatic Filariasis, based on regular treatment of communities with single dose drugs such as ivermectin and albendazole which are also effective against geohelminths, creates an excellent opportunity for integration. Indeed, control of geohelminth infections can be the port of entry to control other endemic communicable and non-communicable diseases (WHO, 1996). This is the approach that was successfully adopted by JOICFP (Japanese Organization for International Cooperation in Family Planning) which utilised mass screening and treatment of intestinal nematodes to stimulate people's interest in family planning and in environmental and family hygiene (Yokogawa, 1985).

8.1 Experience in Zanzibar

Zanzibar, with about 800,000 inhabitants, comprises the islands of Unguja and Pemba and is one of the countries assisted by WHO to prepare and implement a plan of action for integrated disease control. A number of favourable conditions were present to make the implementation of the control strategy possible. First of all, the epidemiological situation was well known with very intense transmission of malaria, schistosomiasis, filariasis and intestinal parasitic infections (including *S. stercoralis*). *A. lumbricoides*, *T. trichiura* and hookworm infections were widespread with a total prevalence of 94.4%. (Renganathan *et al.* 1995)

There were vertical control programmes for each disease, and successful programmes such as control of helminth infections, led to the building up of the integrated approach. At the same time, there was a Health Sector Reform focussed on the decentralization of the health system and there was a well-established School Health Programme. The implementation of the integrated approach was based on the combined administration of drugs and health education through the schools and the community. In view of the launching of the national Lymphatic Filariasis Elimination Programme, mass treatment was planned with the following proposed annual schedule:

Time 0	praziquantel + albendazole to all school children
4th month	ivermectin + albendazole to all population
8th month	mebendazole to all school children

What facilitated the integrated control in Zanzibar was the close and effective collaboration between the Ministry of Health and Ministry of Education which enabled the successful implementation of the control activities in schools, as well as the social mobilisation and community awareness. Another important facility was the availability and involvement of the Public Health Laboratory which is closely collaborating with the District Health Management Team to promote monitoring and evaluation of control programmes, including geohelminths, as well as supervision at the peripheral level, and implementation of operational research according to the Ministry of Health priorities, on-the-job and local training of health staff.

In addition, the Helminth Control Programme in Zanzibar tested a successful and inexpensive outreach approach to treat the school-age children non-enrolled in schools, with a coverage of 89% (98.9 % of school children enrolled, plus 60% of those non-enrolled) (Montresor *et al.* 2001).

10. CONCLUSIONS

Control strategies for geohelminth infections follow different approaches according to the epidemiological characteristics of each endemic area, such as pattern of transmission and rate of re-infection, prevalence and intensity of infection, and prevalent parasite species. Although general guidelines have been recommended for targeting communities in endemic areas (Montresor *et al.* 1998), there is no pre-packed package, and each country should adapt the recommended approach to its peculiar eco-epidemiological and socio-economical conditions. Available resources and health priorities are important determinants to choose the most cost-effective approach to control geohelminthiasis.

In a limited number of countries that are really 'developing', like Seychelles, Iran and South Africa, it may be possible to replicate the experience from Japan and Korea (long-term elimination of the problem- no need of further intervention). For the rest of the 'developing' countries, such as Nepal and in Sub-Saharan Africa, the objective is less ambitious (morbidity control in at risk groups) bus still necessary and relevant for the health of the groups at risk.

Endemic countries should evaluate the need for integrated control of geohelminthiasis with the objective to improve effectiveness and reduce cost of control programmes. Priority areas for integration at the national level and partners and opportunities for integrated geohelminth control should be identified. A recent workshop on integrated control of parasitic infections in East Mediterranean Countries (WHO, 2001) made the following recommendation: "*Where the health system allows, integration of parasitic and communicable diseases should be implemented at all levels: inter- sectoral (Health, Interior, Agriculture, Education), regional, district and primary health care level. Special efforts should be made to strengthen the inter-sectorial collaboration and coordination between Ministries at central level, and the intra-sectorial co-ordination within departments of the MoH.*"

The WHO strategy for control of geohelminth infections is designed to meet the need of endemic countries and to promote tools for diagnosis and disease control which are appropriate and sustainable. An essential component is the monitoring and evaluation which enables managers of helminth control programmes and health planners to quantify the benefits of the intervention and to adapt the control strategy according to its outcome. Targets are reachable and measurable with recommended standardised techniques which allow the comparison between different countries (Montresor *et al.* 1999).

REFERENCES

ALBONICO, M., SHAMLAYE, N., SHAMLAYE, C., SAVIOLI, L. (1996). Control of intestinal parasitic infections in the Seychelles: a comprehensive and sustainable approach. *Bulletin of the World Health Organization* **74**, 577-586.

BUNDY, D.A.P., HALL A., MEDLEY, G.F. & SAVIOLI, L. (1992). Evaluating measures to control intestinal parasitic infections. *World Health Statistics Quarterly* **45**, 168-179.

DE CARNERI, I. (1989). Parasitologia generale ed umana, [in Italian]. Casa Editrice Ambrosiana Milano 44-45.

GUYATT, H.L., BUNDY, D.A.P. & EVANS D. (1993).. A population dynamic approach to the cost-effectiveness analysis of mass anthelminthic treatment: effects of treatment frequency on *Ascaris* infection. *Transactions of the Royal Society of Tropical Medicine and Hygiene* **87,** 570-5.

MINISTRY OF HEALTH AND SOCIAL AFFAIRS, KOREAN ASSOCIATION FOR PARASITE ERADICATION. (1996). Prevalence of intestinal parasitic infection in Korea, sixth report, Monographic series [in Korean] KAPE, Seoul.

MONTRESOR, A., CROMPTON, D.W.T., BUNDY, D.A.P,, HALL, A. & SAVIOLI, L. (1998). *Guidelines for the evaluation of soil-transmitted helminthiasis and schistosomiasis at community level.* Division of Control of Tropical Diseases. WHO/CDS/SIP98.2. Geneva.

MONTRESOR, A., CROMPTON, D.W.T., BUNDY, D.A.P,, HALL, A, & SAVIOLI, L. (1999). *Monitoring helminth control programmes.* Communicable Diseases Prevention and Control. WHO/CDS/CPC/SIP/99.. Geneva.

MONTRESOR, A., RAMSAN, M., CHWAYA, H.M., AMEIR, H., FOUM, A., ALBONICO, M., GYORKOS, T. & SAVIOLI, L. (2001). Extending anthelminthic coverage to non-enrolled school-age children using a simple and low-cost school-based method. *Tropical Medicine & International Health,* In press.

RENGANATHAN, E., ERCOLE, E., ALBONICO, M., DE GREGORIO, G., ALAWI, K.S., KISUMKU, U.M. & SAVIOLI L. (1995). Evolution of operational research studies and development of a national control strategy against intestinal helminths in Pemba Island, 1988-92. *Bulletin of the World Health Organization* **73,**183-190.

SHAMLAYE, N. (2001). Experince and progress in controlling disease due to helminth infections in Seychelles In: *Controlling Disease due to Soil-Transmitted Helminths* (eds. Crompton, D.W.T. & Nesheim, M.C.). World Health Organization, In press.

STOLTZFUS, R.J., ALBONICO, M., CHWAYA, H.M., SAVIOLI, L., TIELSCH, J., SCHULZE, K. & YIP, R. (1996). Hemoquant determination of hookworm-related blood loss and its role in iron deficiency in African children. *American Journal of Tropical Medicine and Hygiene* **55,** 399-404.

THE WORLD HEALTH REPORT. (1999). *Making a difference.* World Health Organization, Geneva.

WORLD HEALTH ORGANIZATION. (1996). *Report of the WHO informal consultation on the use of chemotherapy for the control of morbidity due to soil-transmitted nematodes in humans. Geneva 29 April to 1 May 1996.* Division of Control of Tropical Diseases. WHO/CTD/SIP.96.2. Geneva.

WORLD HEALTH ORGANIZATION. (1998). *Integrating Disease Control: the challenge*. Division of Control of Tropical Diseases. WHO/CTD/98.7. Geneva.

WORLD HEALTH ORGANIZATION. (2001). *Report of the WHO Regional Workshop on the integrated control of parasitic infections*. Tunis *22-24 April 2001*. Division of Control of Tropical Diseases. WHO-EM/CTD/2001. Alexandria, In press.

YOKOGAWA, M. (1985). JOICFP'S experience in the control of ascariasis within an integrated programme. In: *Ascariasis and its Public Health Significance* (eds. Crompton, D.W.T., Nesheim, M.C. & Pawloski, Z.S.). pp 265-277. Taylor and Francis, London and Philadelphia.

Chapter 3
PATHOPHYSIOLOGY OF INTESTINAL NEMATODES

Lani S. Stephenson
Division of Nutritional Sciences, Savage Hall, Cornell University, Ithaca, NY 14853 USA
e-mail: lss5@cornell.edu

1. INTRODUCTION

An estimated 1,472 million persons harbour *Ascaris lumbricoides*, 1,298 million are infected with hookworm, and about 1,049 million have *Trichuris trichiura* (Crompton, 1999). Intestinal helminth infections exert an enormous toll on human health, development, and prosperity. Hookworm, *Ascaris* and *Trichuris* infections can interfere with appetite, growth, physical fitness, physical activity, work capacity, cognitive development (see Chapter 4) and school performance in malnourished populations. The estimated number of disability-adjusted life-years (DALYs) lost globally because of hookworm infection is 22.1 million, while the estimates for *Ascaris* and *Trichuris* are 10.5 million and 6.4 million, respectively (World Bank, 1993; Chan *et al.* 1994; Chan, 1997; de Silva, Chen & Bundy, 1997a). The DALY for these three nematodes combined is a whopping 39.0 million life-years, while that for malaria, which is inherently more overtly disabling, is similar, at 35.7 million life-years lost (Stephenson, Latham & Ottesen, 2000a).

Furthermore, infected persons, particularly children and girls and women of childbearing age, can benefit substantially from treatment (Stephenson, Latham & Ottesen, 2000a; Crompton, 2000; O'Lorcain & Holland, 2000; Stephenson, Holland & Cooper, 2000, Crompton, 2001). Hookworm anaemia, if untreated, is especially pernicious during pregnancy and in very young children, and it can lead to a vicious cycle of low birth weight and stunting in subsequent generations that perpetuates malnutrition and its sequelae (Roche & Layrisse, 1966; Crompton & Stephenson, 1990; WHO, 1996; Seshadri, 1997; Stephenson *et al.* 2000b). In addition, there might be an entirely new justification for aggressive treatment and control of these infections if the recently described effects they have on potentiating HIV infections in affected populations can also be further substantiated and extended (see Chapter 16).

Much of the pathophysiology of these parasites is nutritional in nature, and their geographic distributions overlap with those of the four most common forms of malnutrition. The four most important forms of malnutrition worldwide (protein-energy malnutrition, iron deficiency and anaemias, vitamin A deficiency, and iodine deficiency disorders) affect hundreds of millions of people, especially children and women and girls of childbearing age (Table 3.1) (ACC/SCN, 2000; Stephenson, Latham & Ottesen, 2000b). Deficiencies of zinc, folate, vitamin B12 and other nutrients are also important in a number of areas.

2. PARASITES AND MALNUTRITION: MECHANISMS

Figure 3.1 shows a conceptual framework for how intestinal nematode and some other parasitic infections may influence nutritional status, and with it, a person's physical, cognitive, educational and overall societal development (Stephenson, Latham & Ottesen, 2000a). Most nutrients essential for humans may be negatively affected, including sodium, potassium, and chloride, especially in cases of vomiting and diarrhoea. However, energy *intake* is the most important and most commonly compromised nutritional variable in children and other vulnerable groups, including pregnant women. This decrease in food intake is a consequence of both appetite inhibition (anorexia) due to infections, and food withdrawal as misguided therapy for children and adults (Scrimshaw & SanGiovanni, 1997). The reduction in energy intake can vary from 10-85% in young children (Molla *et al.* 1983, Bentley *et al.* 1991). When people consume less food energy, they also usually reduce their intake of essential micronutrients.

All forms of chronic intestinal inflammation lead to growth failure, either by secondary effects on nutrient balance or by more direct effects on metabolism (Cooper, 1991). Children with intense infections of *T. trichiura*, who often suffer severe depressions of growth in height, have the symptoms and signs associated with chronic colitis of any cause. Intestinal inflammation is also thought to be an important mechanism contributing to

Table 3.1. Major forms of malnutrition in developing countries, 1995-2000

Form of malnutrition	Population group & % affected	Number (millions)
Protein-energy malnutrition		
Children:	27% of Under-5's (2000)	149.63
Underweight	32% of Under-5's (2000)	181.92
Stunting (chronic)	9% of Under-5's (1995)	50.59
Wasting (acute)		
Adults:		
Moderate Underweight		245
Severe Underweight		93
Anaemias	Pregnancy ~56% Under-5's ~42% Children 5-14y ~53%	2 000
Iron Deficiencies	43% all women; 34% all men	3 580
Iodine Deficiency Disorders	15% of world's	834
Goitre	population	49.6
Cretinoids (some mental	All ages	
impairment)	All ages	16.5
Cretinism		
Vitamin A Deficiency		
Subclinical	42% of Under-5's globally	251
Xerophthalmia (clinical)	0.5% of Under-5's globally	2.85

Mortality rates in children <5yr are two and a half times higher in those moderately underweight, and five times higher in the severely underweight. About 50% of deaths among these children were associated with malnutrition, and malnutrition was the direct cause for about 370,000 <5yr deaths in developing countries. (Adapted from Stephenson, Latham & Ottesen (2000b; data sources: *World Health Report* 1998, *Fourth Report on the World Nutrition Situation*, and ACC/SCN, 2000.)

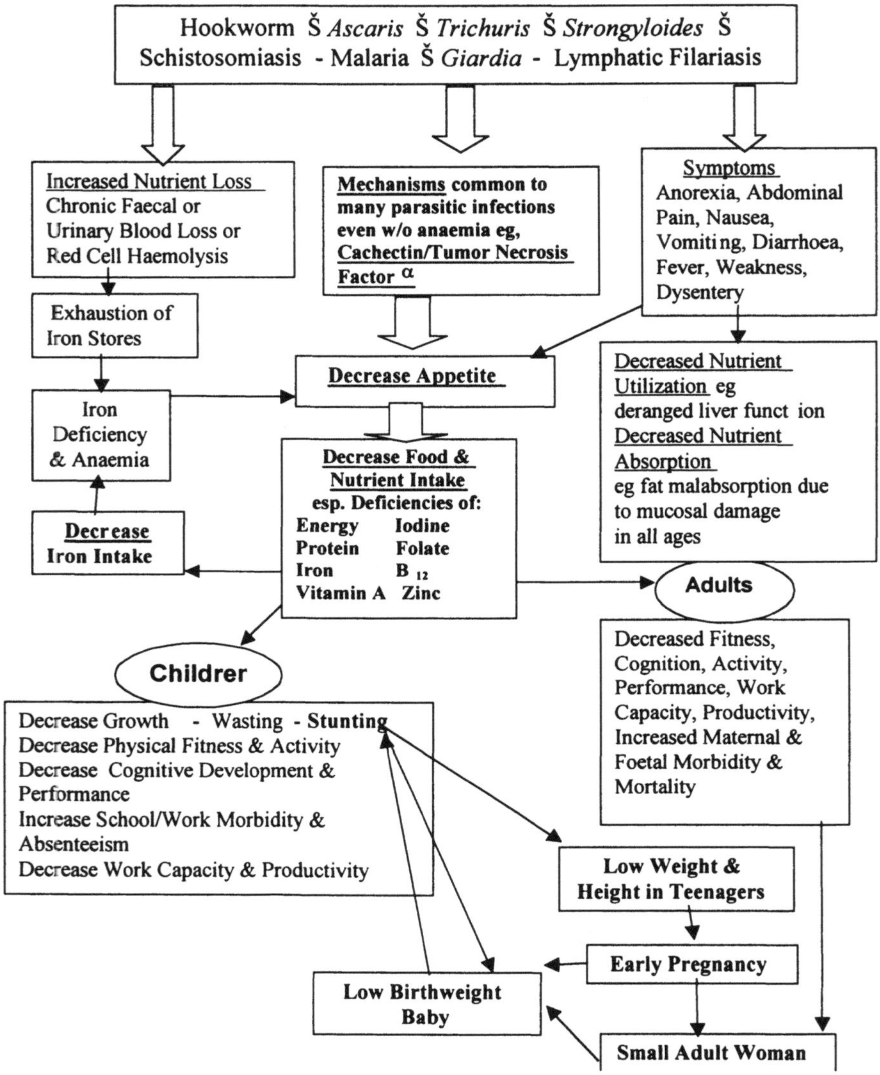

Figure 3.1. How parasites cause/aggravate malnutrition and retard development. Adapted from Stephenson & Holland, 1987; ACC/SCN, 1992; and Stephenson, Latham & Ottesen, 2000.

poor growth in hookworm-infected children (Cooper, 1991) and occurs in *Ascaris* infection as well (see Section 4). Co-infections of intestinal nematodes and bacteria or viruses can also act synergistically to worsen nutritional status. For example, necrotic proliferative colitis, due to *Campylobacter jejuni*, occurred only in weanling pigs previously inoculated with *Trichuris suis* at 8 wk of age, but not in animals without *T. suis* infection. The mechanism was thought to be whipworm-induced suppression of mucosal immunity to the resident bacteria (Mansfield & Urban, 1996) and may very well occur in children as well.

3. HOOKWORM

As of 1990, an estimated 7% of the world's preschool age children (41 million), 26% of school age children (239 million), and 44.3 million of the developing world's 124.3 million pregnant women harboured hookworm infection (WHO, 1996; Michael *et al.* 1997). At least 50% of pregnant women and over 40% of preschool-age children in developing countries are likely to be clinically anaemic (de Benoist, 1999). Data from child growth studies (Stephenson, 1993; Stephenson *et al.* 1993a,b) and one study on weight gain in treated hookworm-infected pregnant women in Sierra Leone (Torless, 1999) suggest that even relatively light hookworm infections may decrease growth and therefore weight gain in pregnancy. Some clinical signs and potential nutritional outcomes of hookworm infection are listed in Table 3.2.

3.1 Loss of blood, including iron and other nutrients

The blood-sucking activity of hookworms in the gut is considered to cause a daily blood loss of from 0.03 to 0.15 ml per worm (Table 3.3). *Ancylostoma duodenale* causes about five times as much blood loss per worm as does *Necator americanus*, but the key issue for the host is total worm load and total blood loss. Some of the iron lost in to the lumen of the small intestine may be re-absorbed farther down the GI tract, but bleeding continues even after feeding stops because the worms produce anticoagulants (Hotez & Cerami, 1983). Erythrocytes labeled with ^{59}Fe or ^{51}Cr have been used to estimate faecal blood loss in hookworm infected persons (Martinez-Torres *et al.* 1967). It is clear that blood loss and hence the probability of

Table 3.2. Clinical features of hookworm infection and potential nutritional outcomes

Stage of infection	Clinical features	Potential nutritional outcome
Cutaneous invasion and subcutaneous migration of filariform larvae	"Ground itch": dermatitis blisters macules papules	? ↓ food intake
Migration of larvae through lungs, bronchi and trachea to oesophagus	Coughing Wheezing Bronchitis Pneumonitis Eosinophilia Wakana disease associated with: nausea vomiting	? ↓ food intake ↓ food intake ↑ nutrient excretion
Attachment of adult worms to mucosa of upper small intestine	Epigastric pain and tenderness Peptic symptoms: anorexia vomiting Acute infection: nausea vomiting diarrhoea occasional massive intestinal haemorrhage Chronic infection: chronic intestinal blood loss and chronic iron deficiency anaemia leading to high output cardiac failure exertional dyspnoea apathy Protein-losing enteropathy: hypoproteinemia oedema	↓ food intake ↓ food intake ↓ nutrient absorption (if clays with calcium are eaten) ↓ food intake ↑ nutrient excretion ↑ nutrient excretion ↑↑ nutrient loss (iron, vitamin B_{12}, folic acid, protein) ↑ nutrient loss (iron, vitamin B_{12}, folic acid, protein) ↓ food intake Not known if due to ↑ nutrient loss or ↓ intake of dietary protein

Adapted from Holland (1987); clinical features adapted from Banwell & Schad (1978) and Beaver, Jung & Cupp (1984).

developing iron deficiency anemia (IDA) increase as intensity of infection increases (Figure 3.2) (see Crompton, 2000; Crompton & Stephenson, 1990; Stoltzfus *et al.* 1996). Measurements of faecal blood loss from hookworm infected school children in an area with very high prevalences of both anemia and hookworm showed that on average faecal hemoglobin loss increased by 0.825mg/g of feces for each additional 1000 eggs per gram of feces (epg) (Stoltzfus *et al.* 1996). The feeding activity of hookworms also causes a loss of blood plasma and its constituents in to the gut, and in heavy infections hypoalbuminemia and other nutrient deficiencies may develop (Pawlowski, Schad & Stott, 1991).

Table 3.3. Blood loss due to hookworm infection

	N. americanus	*A. duodenale*
Intestinal blood loss in ml per worm per day, mean (range)	0.03 (0.01 - 0.04) 0.04 (0.02 – 0.07)	0.15 (0.05 - 0.30) 0.20 (0.14 - 0.26)
Number (range) of worms causing a blood loss of 1ml/day	25 (14 - 50)	5 (4-7)
Blood loss (ml/day) per 1000 epg stool, mean ± SD	1.3 (0.82 - 2.24) 2.2 ± 1.01	2.2 (1.54 - 2.86) 4.4 ± 2.16
Iron loss (mg/day) per 1000 epg stool	0.45 0.65	0.76 1.35
Worm burden responsible for 1000 epg stool	32	11

Adapted from Crompton (2000); data from Holland (1987; 1989), and Pawlowski, Schad & Stott (1991) who give details of sources of information and techniques used. Female worms responsible for egg production probably require more blood for food than males.

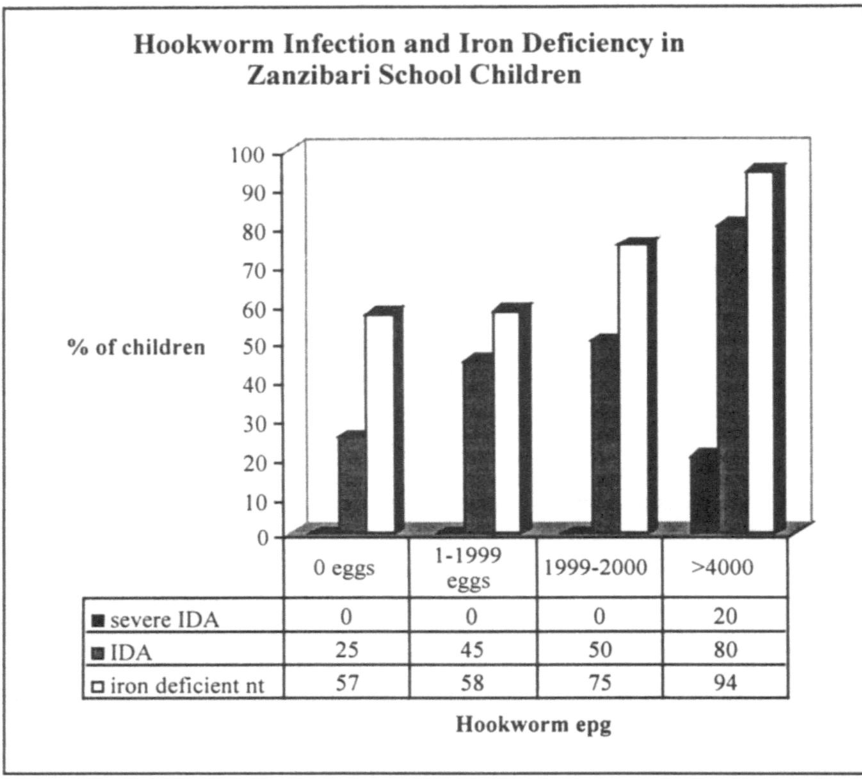

	0 eggs	1-1999 eggs	1999-2000	>4000
severe IDA	0	0	0	20
IDA	25	45	50	80
iron deficient nt	57	58	75	94

Figure 3.2. Relationship between intensity of hookworm infection (mainly *Necator americanus*) and degree of iron deficiency in 203 Zanzibari school children. Severe IDA (iron deficiency anemia) = Hb (hemoglobin) <70 g/l and serum ferritin <18 μg/l; IDA (iron deficiency anemia) = Hb < 110 g/l and ferritin <18 μg/l; ID (iron deficiency) = ferritin < 18 μg/l. The increasing trend for each stage of iron deficiency is significant (P<0.001). Numbers of children in each ascending level of hookworm epg (egg/g of feces) are 45, 83, 19, and 56. Note children with severe IDA are also counted in categories of IDA and ID, etc. (Adapted from Stoltzfus *et al.* 1996.)

Blood and nutrient loss in hookworm is particularly dangerous for pregnant women and girls and young children, although school age children and adult males also suffer in endemic areas. The loss of >10 mg/g of faecal hemoglobin in Zanzibari children, equivalent to about 2 mg of iron loss per day, more than doubled the children's requirement for dietary iron. At this level of iron loss, 93% of children had IDA and 29% were severely anaemic (Stoltzfus *et al.* 1996). For some women and girls it is almost impossible to meet their daily iron requirements even with good quality iron-fortified diets (Viteri, 1994). IDA is considered responsible for 20% of maternal deaths globally (WHO, 1989). Anaemia increases the risk of prematurity and low birth weight in infants; Seshadri (1997) cites data from Afghanistan, Bangladesh, India, Iran, Nepal, Pakistan, and Sri Lanka to show that the incidence of premature delivery can be 3 times higher in severely anaemic as compared with normal women. One negative influence of IDA on pregnancy outcomes is illustrated by the fact that the prevalence of low birth weight (<2 kg) decreased from 50% to 7% in a study in Nigeria when iron and folate supplements were given (see Viteri, 1994).

Studies have shown that hookworm and iron deficiency can impair growth, appetite, and physical fitness of children and may decrease their intellectual performance as well (Pollitt, 1990; Connolly & Kvalsvig, 1993; Stephenson *et al.* 1993a, 1993b; Lawless *et al.* 1994; Stoltzfus *et al.* 1997, 1998; Seshadri, 1997; Bundy & da Silva, 1998; Guyatt, 2000). Two additional studies in preschool age children in Kenya show that hookworm infection can cause or aggravate anaemia, and that treatment, even of relatively low egg counts, can improve growth. Hookworm has often been considered relatively unimportant in preschoolers because prevalences and egg counts are lower than in older children who have had much more time to acquire significant worm loads (Stephenson, Latham & Ottesen, 2000a). However Brooker and colleagues (1999) found that 28% of 460 preschoolers aged 6-60 months had hookworm, that 76% were anaemic, and that anaemia was significantly more severe in children with hookworm infections > 200 epg. In Bungoma, Manjrekar (1999) reported that treatment of sick, worm-infected two to four year olds with a single dose of mebendazole yielded statistically significant weight and height gains at 6 months follow up. This result was notable because only 12% of children were infected with any helminth, only 6% harbored hookworm, 6% had *Ascaris*, and 1% had *Trichuris*, and egg counts were light.

4. *ASCARIS LUMBRICOIDES*

As of 1990, an estimated 29% of the world's preschool age children (158 million) and 35% of school-age children (320 million) were infected with *A. lumbricoides* (Michael *et al.* 1997). The clinical features and potential nutritional outcomes of the various stages of *Ascaris* infection are shown in Table 3.4.

4.1 Ascaris and Malnutrition in Children

Most *A. lumbricoides* infections are chronic and may significantly impair childhood nutrition, especially in areas where poor growth and ascariasis are common. Bodily growth, absorption of fat, vitamin A and β-carotene, and iodine, and digestion and absorption of protein and lactose are the nutritional parameters most likely to be impaired (Carrera *et al.* 1984; Taren *et al.* 1987; Hadju *et al.* 1997; Jalal *et al.* 1998; see reviews in Stephenson & Holland, 1987; Taren & Crompton, 1989; Thein Hlaing, 1993; O'Lorcain & Holland, 2000; Crompton, 2001). *Ascaris* infection reduces appetite (Hadju *et al.* 1996; 1998). The intestinal pathology documented in children includes villus atrophy and cellular infiltration of the lamina propria (Tripathy *et al.* 1972). Treatment has been shown to lead to both improved appetite and weight gain (see O'Lorcain & Holland, 2000), and numerous studies have shown that anthelminthic treatment can be effective in improving growth rates when given to malnourished children with ascariasis (see Thein Hlaing, 1993; O'Lorcain & Holland, 2000; Crompton, 2001). The increases in appetite and nutrient intake that follow are likely to be the single most important nutritional benefit of anthelminthic treatment for ascariasis. Recent studies have also shown that *Ascaris* infections can affect mental processing in some school children (Connolly & Kvalsvig, 1993; Hadidjaja *et al.* 1998).

Table 3.4. Clinical features and potential nutritional outcomesassociated with *Ascaris lumbricoides* infection

Stage of infection	Event	Clinical Features	Potential nutritional outcome
Larval migration	Migration of larvae through lungs	Pneumonitis, including: asthma cough dyspnea substernal pain	 ?↓ Food intake
		Conjunctivitis	
		Convulsions	↓ Food intake
		Eosinophilia	
		Fever	↑ Nitrogen loss
		Skin rash	?↓ Food intake
Migration, oviposition	Presence of juveniles and patent adult worms in small intestine	Abdominal distension	↓ Food intake
		Abdominal pain	↓ Food intake
		Colic	↓ Food intake
		Nausea	↓ Food intake
		Vomiting	↑ Nutrient loss
		Anal itching	
		Anorexia	↓ Food intake
		Disordered small bowel pattern	? Malabsorption
		D-xylose and lactose malabsorption	↑ Carbohydrate excretion
		Enterocolitis	↑ Nutrient excretion
		Fat malabsorption	↑ Fat excretion
		Intermittent diarrhoea	↑ Nutrient loss
		Jejunal mucosal abnormalities	Malabsorption
		Protein malabsorption	↑ Protein excretion
		Restlessness	
		Vitamin A malabsorption	↑Vitamin A excretion
Complic-ations	Migration or aggreg-ation of adult *Ascaris* in intestine	Intestinal obstruction Intussusception Invasion of bile duct (producing cholangitis obstructive jaundice, gallstones, or liver abscesses) Acute appendicitis Acute pancreatitis Intestinal perforation Peritonitis Upper respiratory tract obstruction. Volvulvus	Life-threatening illnesses which all decrease food intake and may increase nutrient requirements (due to fever) and nutrient losses (due to diarrhoea)

(Adapted from Stephenson & Holland, 1987)

4.2 The Immune Response in Ascariasis

Because *Ascaris* and other infections can lead to nutritional deficiencies, they can lower the immunity that is essential for the maintenance of innate resistance and the genetically constituted immune response that help the body resist parasites (Beisel, 1982; Puri & Chandra, 1985). Intestinal helminths and *A. lumbricoides* in particular stimulate the production of IgE antibody (Jarret & Miller, 1982) (see Chapter 6). Migration of the larvae through the liver and lungs can lead to pneumonitis, which can include asthma, cough, substernal pain, fever, skin rash and eosinophilia (see Crompton, 2001). Regular anthelminthic treatment of *Ascaris* infected asthmatic patients in Venezuela for one year has been shown to decrease the severity of asthma for up to two years (Lynch *et al.* 1997). IgE antibody responses in conjunction with inflammatory processes appeared in one study to be associated with natural immunity to *Ascaris* (McSharry *et al.* 1999). Regarding the mechanisms responsible for predisposition, Holland *et al.* (1992) studied the class I HLA antigen distribution among Nigerian children predisposed to heavy, light or no infection with *Ascaris* and found that those who remained consistently uninfected despite exposure to infection lacked the A30/31 antigen (see Chapter 1). In addition, studies in East Nepal reported that there appeared to be a strong genetic component accounting for 30-50% of the variation in *Ascaris* worm burden among >1200 individuals from a single pedigree in the Jirel population (Willliams-Blangero *et al.* 1999) (see Chapter 10).

4.3 Complications of Intestinal Ascariasis

Children and adults experience acute life-threatening ascariasis, most commonly in the form of intestinal obstruction or biliary complications (see reviews by De Silva *et al.* 1997b and Crompton, 2001). De Silva *et al.* (1997a) estimated that 12 million acute cases occur each year with approximately 10,000 deaths. Complications are much more rare than faltering growth and are most likely associated with higher worm burdens. *Ascaris*-induced intestinal obstruction is the commonest, accounting for 57% of all complications; it is most frequent in children <10 years of age (De Silva *et al.* 1997b). The incidence in published studies was on the order of 0-0.25 cases per year per 1000 population in endemic areas, and was associated with a mean case fatality rate of >5%.

5. *TRICHURIS TRICHIURA*

As of 1990, an estimated 21% of the world's preschool-age children (114 million) and 25% of school-age children (233 million) were thought to harbour *T. trichiura*. The prevalence of *Trichuris* infection may reach 95 % in children in many parts of the world where protein energy malnutrition and anaemias are also prevalent and access to medical care and education is often limited. The clinical signs and potential nutritional outcomes of *Trichuris* infection are shown in Table 3.5.

5.1 Trichuris Dysentery Syndrome (TDS)

The *Trichuris* dysentery syndrome (TDS) associated with heavy *T. trichiura* infection includes chronic dysentery, rectal prolapse, anaemia, poor growth and clubbing of the fingers (Figure 3.3). TDS and lighter but still heavy infections constitute an important public health problem, especially in children. The profound growth stunting seen in TDS can be reversed by repeated treatment for the infection and oral iron (Callendar *et al.* 1992; 1993; 1994; 1998; Cooper *et al.* 1995) (Figure 3.4). However Jamaican studies which treated TDS cases every three to six months with mebendazole and visited them in their homes for four years strongly suggest that the significant developmental and cognitive deficits found are unlikely to disappear unless the positive psychological stimulation in the child's environment is increased (Callendar *et al.* 1998).

The severe stunting seen in TDS is likely a reaction at least in part to a chronic inflammatory response and concomitant decreases in plasma insulin-like growth factor-1, increases in tumor necrosis factor-α, both in the lamina propria of the colonic mucosa and peripheral blood (likely leading to decreased appetite and intake of all nutrients), and a decrease in collagen synthesis [MacDonald *et al.* 1994; Duff, Anderson & Cooper, 1999]. The inflammatory response to the infection produces anaemia, growth retardation and intestinal leakiness which are related to infection intensity (Cooper *et al.* 1992). The deleterious effects of the infection are partly mediated by a specific IgE mediated local anaphylaxis, and increased numbers of mucosal macrophages are thought to contribute to the chronic systemic effects of trichuriasis through their output of cytokines. There is however evidence for the absence of cell-mediated immunopathology (Cooper *et al.* 1992) (see also Chapter 8).

Table 3.5. Clinical features of *Trichuris trichiura* infection and nutritional/functional outcomes

Type of infection	Features	Nutritional and functional outcomes
Supersensitive persons	Non-specific responses such Nervousness Anorexia Urticaria	 ↓ School performance ↓ Food intake ↓ School performance
Light	? Negligible or mild version of moderate	?
Moderate	Allergic symptoms Epigastric and lower abdominal pain Diarrhoea (rarely bloody) Vomiting Flatulence and distention Headache Weight loss/stunting Increased blood and iron loss	↓ Food intake ↓ Food intake ↓ Food offered to child ↑ Nutrient loss ?↓ Food intake ↓ School performance ↓ Food intake Anaemia in malnourished children
Heavy	Bloody diarrhoea with profuse mucus Abdominal pain and tenesmus Weight loss leading to cachexia; severe stunting Severe anaemia Rectal prolapse Raised mucosal and peripheral TNFα levels Decreased collagen synthesis Moderate eosinophilia Clubbing of fingers	↓ Food offered to child; ↑ Blood loss and iron loss ↓ Food intake ↓ Food intake; Severe educational impact ↑ Blood loss and iron loss; Severe educational impact ↓ Appetite; likely cause of cachexia Stunted growth Discrimination at school

Adapted from Holland (1987) and Stephenson, Holland & Cooper (2000). Clinical features compiled from Wolfe (1978); Markell, Voge & John (1986); Beaver *et al.* (1984); Pawlowski (1984); MacDonald *et al.* (1994), Callendar *et al.* (1998), and Duff, Anderson & Cooper (1999).

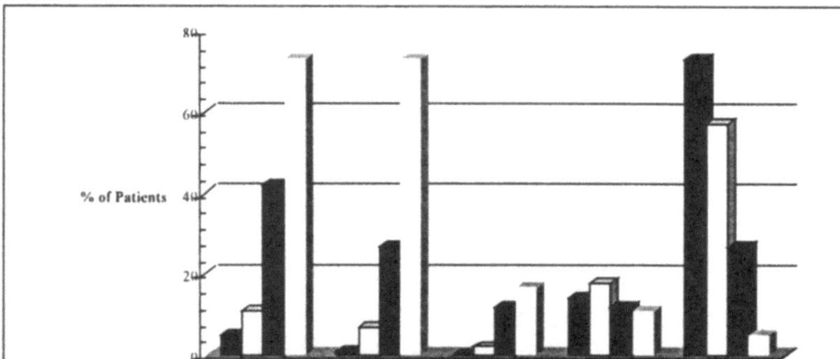

	Diarrhoea	Dysentery	Rectal prolapse	Vomiting	No GI symptoms
■ 0-24 eggs, n=125	5	1	0	14	73
□ 25-99, n=44	11	7	2	18	57
■ 100-299, n=33	42	27	12	12	27
300+, n=18	73	73	17	11	5

Figure 3.3. Relation of symptoms to *T. trichiura* egg counts in 210 patients, Charity Hospital of New Orleans (Source: Stephenson, Holland & Cooper, 2000; reprinted with permission from *Parasitology*. Adapted from Jung & Beaver, 1951.)

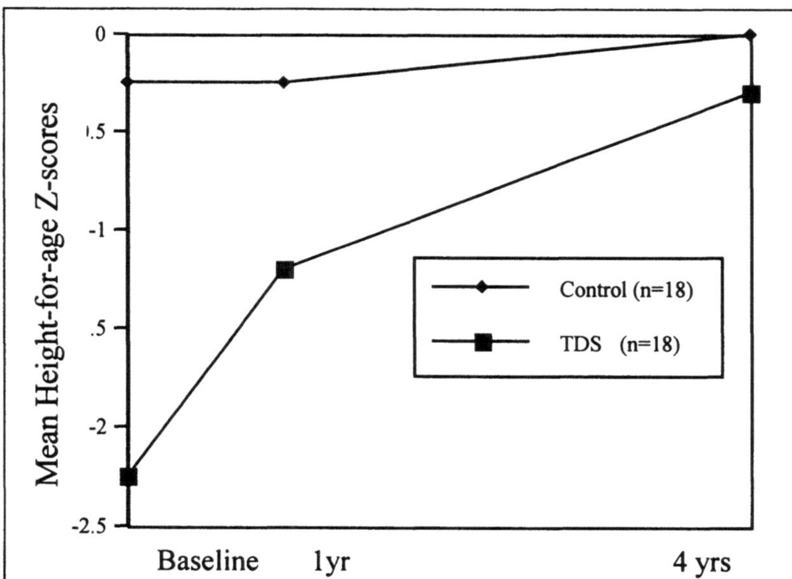

Figure 3.4. Mean height-for-age Z-scores at baseline, 1 yr and 4yr in 18 Jamaican children with *Trichuris* dysentery syndrome given mebendazole 3-6 monthly (and initially, iron supplements) and matched controls. (Source: Stephenson, Holland & Cooper, 2000; Reprinted with permission from *Parasitology*. Adapted from Callendar *et al.* 1998.)

5.2 Growth after Community Treatment for Trichuriasis

A recent important large field study on *Trichuris* and child growth was a randomized, placebo-controlled trial which examined the efficacy and nutritional benefits of combining treatment for intestinal helminths (with albendazole) and lymphatic filariasis (with ivermectin) (Beach *et al.* 1999). The subjects were 853 Haitian school children, 42% of whom harboured *Trichuris*; in addition, 29% had *Ascaris*, 7% had hookworm and 13% exhibited *Wuchereria bancrofti* microfilaraemia. Children were randomly assigned to receive either placebo, albendazole 400 mg, ivermectin 200-400 μg/kg (mean 283 μg/kg), or albendazole + ivermectin and re-examined four months after treatment. The combination of albendazole + ivermectin resulted in significantly higher weight gains in children infected only with *Trichuris* as compared with placebo (0.56 kg more/4 months, $P<0.01$) and significant increases in weight-for-age and weight-for-height Z-scores as well ($P<0.04$, $P<0.03$, respectively; see Figure 3.5). In addition, children infected only with hookworm exhibited a significant increase in height compared with placebo (0.62 cm, $P<0.01$; Figure 3.6). The differences are notable in part because the children were relatively well-nourished and the intensity of infection relatively low. These were positive shifts in growth status in the entire group, underscoring the broad-based community-level benefits of deworming.

5.3 Intestinal Blood Loss in Trichuriasis

The blood loss that can occur in *Trichuris* infection is likely to contribute to anaemia, especially if the child also has hookworm, malaria, and/or has a low intake of dietary iron. The estimated blood loss per worm of 0.005ml per day is only 10-15% of that attributed to a *Necator americanus* worm and 2-3% of that lost due to *Ancylostoma duodenale.* However, *Trichuris* was responsible for a daily blood loss of 0.8 to 8.6 ml in the children studied in Venezuela, vs. only 0.2 to 1.5 ml per day in uninfected childen (Roche *et al.* 1957).

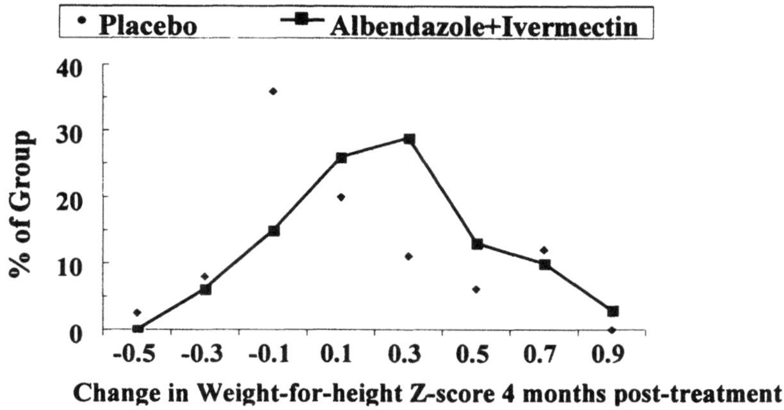

Fig. 3.5. Change in weight-for-height Z-score four months post-treatment in *Trichuris*-infected Haitian children given either 400 mg albendazole + 200-400 µg/kg ivermectin (n = 34) or placebo (n = 36). (Source: Stephenson, Holland & Cooper, 2000; reprinted with permission from *Parasitology*. Adapted from Beach *et al.* 1999).

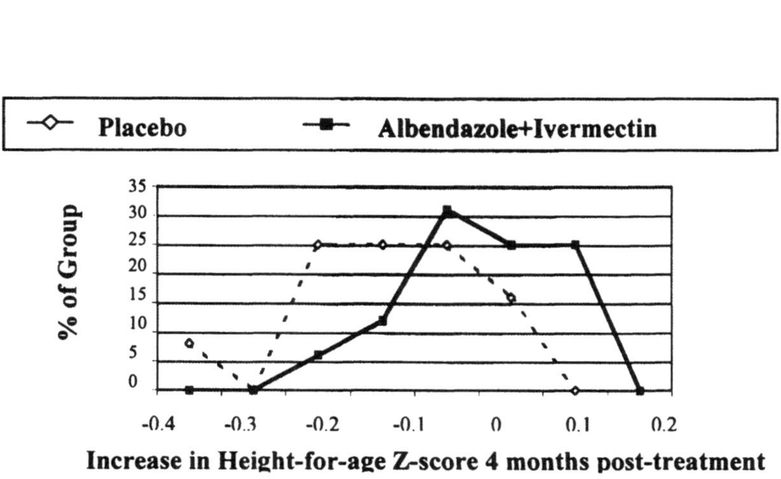

Figure 3.6. Increase in height-for-age Z-score 4 months post-treatment in hookworm-infected Haitian children given either 400 mg albendazole + 200-400 µg/kg ivermectin (n = 17) or placebo (n = 16). (Source: Stephenson, 2001. Reprinted with permission from *Paediatric Drugs*. Adapted from Beach *et al.* 1999).

6. CONCLUSION

Community control of hookworm, *Ascaris* and *Trichuris* is important, especially in cases of heavy infection, which means focusing on children, with special attention to girls, who have increased iron requirements and blood loss due to menstruation, and later, pregnancies and lactation. Detailed discussions of control strategies and implementation of community programs are available, including three recent WHO publications covering (a) the monitoring of drug efficacy in the control of schistosomiasis and intestinal nematodes (WHO, 1999), (b) the monitoring of helminth control programmes with particular reference to school-age children (Montresor *et al.* 1999), and (c) guidelines for the evaluation of soil-transmitted helminthiasis and schistosomiasis at community level (Montresor *et al.* 1998) (also see Chapter 2).

ACKNOWLEDGEMENTS

The author thanks B. Seely for excellent technical help and the graduate students in Savage Hall and the Division of Nutritional Sciences, Cornell University for institutional support.

REFERENCES

ACC/SCN (1992). *Second Report on the World Nutrition Situation: Vol. I: Global and Regional Results.* ACC/SCN, Geneva.

ACC/SCN (2000). *Fourth Report on the World Nutrition Situation.* ACC/SCN, Geneva, March 2000.

BANWELL, J. G. & SCHAD, G. A. (1978). Hookworms. *Clinics in Gastroenterology* **7**, 129-156.

BEACH, M. J., STREIT, T. G., ADDISS, D. G., PROSPERE, R., ROBERTS, J. M. & LAMMIE, P. J. (1999). Assessment of combined ivermectin and albendazole for treatment of intestinal helminth and *Wuchereria bancrofti* infections in Haitian schoolchildren. *American Journal of Tropical Medicine and Hygiene* **60**, 479-486.

BEAVER, P. C., JUNG, R. C. & CUPP, E. W. (1984). Helminths and helminth infections. In *Clinical Parasitology,* pp. 240-245 and pp. 253-261. Lea and Febiger, Philadelphia.

BENTLEY, M. E., STALLINGS, R. Y., FUKUMOTO, M. & ELDER, J. A. (1991). Maternal feeding behavior and child acceptance of food during diarrhea, convalescence, and health in the Central Sierra of Peru. *American Journal of Public Health* **81**, 43-47.

BEISEL, W. R. (1982). Synergism and antagonism of parasitic diseases and malnutrition. *Reviews of Infectious Diseases* **4**, 746-750.

BROOKER, S., PESHU, N., WARN, P. A., MOSOBO, M., GUYATT, H., MARSH, K. & SNOW, R. W. (1999). The epidemiology of hookworm infection and its contribution to anaemia among pre-school children on the Kenya Coast. *Transactions of the Royal Society of Tropical Medicine and Hygiene* **93**, 240-246.

BUNDY, D.A.P. & de SILVA, N. R. (1998). Can we deworm this wormy world? *British Medical Bulletin* **54**, 421-432.

CALLENDAR, J. E., GRANTHAM-MCGREGOR, S., WALKER, S. & COOPER, E. S. (1992). *Trichuris* infection and mental development in children. *Lancet* **339**, 181.

CALLENDAR, J. E., GRANTHAM-MCGREGOR, S., WALKER, S. & COOPER, E. S. (1993). Developmental levels and nutritional status of children with the *Trichuris* dysentery sydrome. *Transactions of the Royal Society of Tropical Medicine and Hygiene* **87**, 528-529.

CALLENDAR, J. E., GRANTHAM-MCGREGOR, S., WALKER, S. & COOPER, E. S. (1994). Treatment effects in *Trichuris* dysentery syndrome. *Acta Paediatrica* **83**, 1182-1187.

CALLENDAR, J. E., WALKER, S., GRANTHAM-MCGREGOR, S. & COOPER, E. S. (1998). Growth and development four years after treatment for the *Trichuris* dysentery syndrome. *Acta Paediatrica* **87**, 1247-1249.

CARRERA, E., NESHEIM, M. C. & CROMPTON, D.W. T. (1984). Lactose maldigestion in *Ascaris*-infected pre-school children. *American Journal of Clinical Nutrition* **39**, 255-264.

CHAN, M.-S. (1997). The global burden of intestinal nematode infections - Fifty years on. *Parasitology Today* **13**, 438-443.

CHAN, M.-S., MEDLEY, G. M., JAMISON, D. & BUNDY, D. A. P. (1994). The evaluation of potential global morbidity attributable to intestinal nematode infections. *Parasitology* **109**, 373-387.

CONNOLLY, K. J. & KVALSVIG, J. D. (1993). Infection, nutrition and cognitive performance in children. *Parasitology* **107** (Suppl) S187-S200.

COOPER, E. (1991). Intestinal parasitoses and the modern description of diseases of poverty. *Transactions of the Royal Society of Tropical Medicine and Hygiene* **85**, 168-170.

COOPER, E. S., DUFF, E. M. W., HOWELL, S. & BUNDY, D. A. (1995). 'Catch-up' growth velocities after treatment for *Trichuris* dysentery syndrome. *Transactions of the Royal Society of Tropical Medicine and Hygiene* **89**, 653.

COOPER, E. S., WHYTE-ALLENG, C. A. M., FINZI-SMITH, J. S. & MACDONALD, T. T. (1992). Intestinal nematode infections in children: the pathophysiological price paid. *Parasitology* **104** (Suppl.) S91-S103.

CROMPTON, D. W. T. (1999). How much human helminthiasis is there in the world? *Journal of Parasitology* **85**, 397-403.

CROMPTON, D. W. T. (2000). The public health importance of hookworm disease. *Parasitology* **121**(Suppl.) S39 – S50.

CROMPTON, D. W. T. (2001). Ascaris and ascariasis. *Advances in Parasitology* **48**, 285-375.

CROMPTON, D. W. T. & STEPHENSON, L. S. (1990). Hookworm infection, nutritional status and productivity. In *Hookworm Disease* (ed. Schad, G. A. & Warren, K. S.), pp. 231-264. Taylor and Francis, Ltd., London & Philadelphia.

DE BENOIST, B. (1999). Anaemia and iron deficiency anaemia: Extent of the public health problem. *INACG Symposium, 12 Mar 1999, Durban, South Africa*. Dept of Nutrition for Health and Development, World Health Organization.

DE SILVA, N. R., CHAN, M. S. & BUNDY, D. A. P. (1997a). Morbidity and mortality due to ascariasis: reestimation and sensitivity of global numbers at risk. *Tropical Medicine and International Health* **2**, 519-528.

DE SILVA, N. R., GUYATT, H. L. & BUNDY, P. (1997b). Morbidity and mortality due to *Ascaris*-induced intestinal obstruction. *Transactions of the Royal Society of Tropical Medicine and Hygiene* **91**, 31-36.

DUFF, E. M., ANDERSON, N. M. & COOPER, E. S. (1999). Plasma insulin-like growth factor-1, type 1 procollagen, and serum tumor necrosis factor alpha in children recovering from *Trichuris* dysentery syndrome. *Pediatrics* **103**, e69.

GUYATT, H. (2000). Do intestinal nematodes affect productivity in adulthood? *Parasitology Today* **16**, 153-158.

HADIDJAJA, P., BONANG, E., SUYARDI, M. A., ABIDIN, S. A., ISMID, I. S. & MARGANO, S. S. (1998). The effect of intervention on nutritional status and cognitive function of primary school children infected with *Ascaris lumbricoides*. *American Journal of Tropical Medicine and Hygiene* **59**, 791-795.

HADJU, V., STEPHENSON, L. S., ABADI, K., MOHAMMED, H. O., BOWMAN, D. D. & PARKER, R. S. (1996). Improvement in appetite and growth in helminth-infected schoolboys three and seven weeks after a single dose of pyrantel pamoate. *Parasitology* **113**, 497-504.

HADJU, V., SATRIONO, ABADI, K. & STEPHENSON, L. S. (1997). Relationships between soil-transmitted helminthiases and growth in urban slum school children in Ujung Pandang Indonesia. *International Journal of Food Science and Nutrition* **48**, 85-93.

HADJU, V., STEPHENSON, L. S., MOHAMMED, H.O., BOWMAN, D. D. & PARKER, R. S. (1998). Improvements of growth, appetite, and physical activity in helminth-infected schoolboys six months after a single dose of albendazole. *Asia Pacific Journal of Clinical Nutrition* **7**, 170-176.

HOLLAND, C.V. (1987). Hookworm infection. In *The Impact of Helminth Infections on Human Nutrition: Schistosomes and Soil-transmitted Helminths*. (Stephenson, L.S. & Holland, C.V.), pp. 128-160. Taylor & Francis, Philadelphia.

HOLLAND, C.V. (1987). Neglected infections - trichuriasis and strongyloidiasis. In *The Impact of Helminth Infections on Human Nutrition: Schistosomes and Soil-transmitted Helminths*. (Stephenson, L.S. & Holland, C.V.), pp. 161-201. Taylor & Francis, Philadelphia.

HOLLAND, C.V. (1989). An assessment of the impact of four intestinal nematode infections on human nutrition. *Clinical Nutrition* **8**, 239-250.

HOLLAND, C. V., CROMPTON, D. W. T., ASAOLU, S. O., CHRICHTON, W. B., TORIMIRO, S. E. A. & WALTERS, D. E. (1992). A possible genetic factor influencing protection from infection with *Ascaris lumbricoides* in Nigerian children. *Journal of Parasitology* **78**, 915-916.

HOTEZ, P.J. & CERAMI, A. (1983). Secretion of a proteolytic anticoagulant by *Ancylostoma duodenale* hookworms. *Journal of Experimental Medicine* **157**, 1594 - 1603.

JALAL, F., NESHEIM, M. C., AGUS, Z., SANJUR, D. & HABICHT, J.-P. (1998). Serum retinal concentrations in children are affected by food sources of β-carotene, fat intake, and anthelmintic drug treatment. *American Journal of Clinical Nutrition* **68**, 623-629.

JARRETT, E. E. & MILLER, H. R. (1982). Production and activities of IgE in helminth infection. In *Immunity and Concomitant Immunity in Infectious Diseases* (ed. Kallos, P.) *Progress in Allergy* **31**, 178-233.

JUNG, R. C. & BEAVER, P. C. (1951). Clinical observations on *Tricocephalus trichiurus* (Whipworm) infestation in children. *Pediatrics* **8**, 548-557.

LAWLESS, J. W., LATHAM, M. C., STEPHENSON, L. S., KINOTI, S. N. & PERTET, A. (1994). Iron supplementation improves appetite and growth in anaemic Kenyan primary school children. *Journal of Nutrition* **124**, 645-654.

LYNCH, N. R., PALENQUE, M., HAGEL, I. & DiPRISCO, M. C. (1997). Clinical improvement of asthma after anthelminthic treatment in a tropical situation. *American Journal of Respiratory and Critical Care Medicine* **156**, 50-54.

MACDONALD, T. T., SPENCER, J., MURCH, S. H., CHOY, M.-Y.,VENUGOPAL, S., BUNDY, D. A. P. & COOPER, E. S. (1994). Immunoepidemiology of intestinal helminth infections. 3. Mucosal macrophages and cytokine production in the colon of children with *Trichuris trichiura* dysentery. *Transactions of the Royal Society of Tropical Medicine and Hygiene* **88**, 265-268.

MANJREKAR, R. R. (1999). *Evaluation of the integrated management of childhood illness guidelines for treatment of intestinal helminth infections in sick children, 2-4 years, Western Kenya.* Master's of Public Health Thesis, Rollins School of Public Health, Emory University, 1999.

MANSFIELD, L. S. & URBAN, J. F. (1996). The pathogenesis of necrotic proliferative colitis in swine is linked to whipworm induced suppression of mucosal immunity to resident bacteria. *Veterinary Immunology and Immunopathology* **50**, 1-17.

MARKELL, E. K., VOGE, M. & JOHN, M. (1986). *Medical Parasitology*, 6th edn. W. B. Saunders, Philadelphia.

MARTINEZ-TORRES, C., OJEDA, A. A., ROCHE, M. & LAYRISSE, M. (1967). Hookworm infection and intestinal blood loss. *Transactions of the Royal Society of Tropical Medicine and Hygiene* **61**, 373-383.

McSHARRY, C., XIA, Y., HOLLAND, C. V. & KENNEDY, M. W. (1999). Natural immunity to *Ascaris lumbricoides* associated with immunoglobulin E antibody to ABA-1 allergen and inflammation indicators in children. *Infection and Immunity* **67**, 1-6.

MICHAEL, E., BUNDY, D. A. P., HALL, A., SAVIOLI, L. & MONTRESOR, A. (1997). This wormy world: Fifty years on - The challenge of controlling common helminthiases of humans today. *Parasitology Today* **13**(11), poster.

MOLLA, A. M., MOLLA, A. M., SARKER, S. A. & RAHAMAN, M. M. (1983). Food intake during and after recovery from diarrhoea in children. In *Diarrhoea and Malnutrition: Interactions, Mechanisms, and Interventions* (ed. Chen, L. D. & Scrimshaw, N. S.), pp. 113-123. Plenum Press, New York.

MONTRESOR, A., CROMPTON, D. W. T., HALL, A. & SAVIOLI, L. (1998). *Guidelines for the evaluation of soil-transmitted helminthiasis and schistosomiasis at community level. A guide for managers of control programmes.* World Health Organization, Geneva; 1998. WHO/CTD/SIP/98.1.

MONTRESOR, A., GYORKOS, T. W., CROMPTON, D. W. T., BUNDY, D. A. P. & SAVIOLI, L. (1999). *Monitoring Helminth Control Programmes. Guidelines for Monitoring the Impact of Control Programmes Aimed at Reducing Morbidity Caused by Soil-transmitted Helminths and Schistosomes, with Particular Reference to School-age Children*. WHO/CDS/CPC/SIP/99.3. Geneva, WHO.

O'LORCAIN, P. & HOLLAND, C. V. (2000). The public health importance of *Ascaris lumbricoides*. *Parasitology* **121** (Suppl.) S51 – S72.

PAWLOWSKI, Z. S. (1984). Trichuriasis. In: *Tropical and Geographic Medicine* (ed. Warren, K. S. & Mahmood, A. A. F.). pp. 380-385. McGraw-Hill, New York.

PAWLOWSKI, Z. S., SCHAD, G.A. & STOTT, G. J. (1991). *Hookworm Infection and Anaemia*. World Health Organisation, Geneva.

POLLITT, E. (1990). *Malnutrition and infection in the classroom*. UNESCO, Paris, France.

PURI, S. & CHANDRA, R. K. (1985). Nutritional regulation of host resistance and predictive value of immunologic tests in assessment of outcome. *Pediatric Clinics of North America* **32**, 499-516.

ROCHE, M. & LAYRISSE, M. (1966). The nature and causes of "hookworm anemia". *American Journal of Tropical Medicine and Hygiene* **15**, 1030-1100.

ROCHE, M., PEREZ-GIMENEZ, M. E., LAYRISSE, M. & DI PRISCO, E. (1957). Study of urinary and faecal excretion of radioactive chromium Cr in man. Its use in the measurement of intestinal blood loss associated with hookworm infection. *Journal of Clinical Investigation* **36**, 1183-1192.

SCRIMSHAW, N. S. & SANGIOVANNI, J. P. (1997). Synergism of nutrition, infection, and immunity: an overview. *American Journal of Clinical Nutrition* **66** (Suppl.) 464S-477S.

SESHADRI, S. (1997). Nutritional anaemia in South Asia. In *Malnutrition in South Asia* (ed. Gillespie, S.), pp 75-124. UNICEF, Kathmandu, Nepal.

STEPHENSON, L. S. (1993). The impact of schistosomiasis on human nutrition. *Parasitology* **107** (Suppl.) S107- S123.

STEPHENSON, L. S. (2001). Benefits of Anthelminthic Treatment in Children. *Paediatric Drugs*. In press 3/01.

STEPHENSON, L. S. & HOLLAND, C. V. (1987). *The Impact of Helminth Infections on Human Nutrition*. Taylor and Francis, London and Philadelphia.

STEPHENSON, L. S, HOLLAND, C. V. & COOPER, E. S. (2000). The public health significance of *Trichuris trichiura*. *Parasitology* **121**(Suppl.), S73 – S96.

STEPHENSON, L. S., LATHAM, M.C., ADAMS, E.J., KINOTI, S.K. & PERTET, A. (1993a). Weight gain of Kenyan school children infected with hookworm, *Trichuris trichiura* and *Ascaris lumbricoides* is improved following once- or twice- yearly treatment with albendazole. *Journal of Nutrition* **123**, 656-665.

STEPHENSON, L. S., LATHAM, M. C., ADAMS, E. J., KINOTI, S. N. & PERTET, A. (1993b). Physical fitness, growth, and appetite of Kenyan schoolboys with hookworm, *Trichuris trichiura* and *Ascaris lumbricoides* infections are improved four months after a single dose of albendazole. *Journal of Nutrition* **123**, 1036-1046.

STEPHENSON, L. S., LATHAM. M. C. & OTTESEN, E. A. (2000a). Malnutrition and parasitic helminth infections. *Parasitology* **121**(Suppl) S23-S38.

STEPHENSON, L. S., LATHAM, M. C. & OTTESEN, E. A. (2000b). Global Malnutrition. *Parasitology* **121**(Suppl.) S5 - S22.

STOLTZFUS, R. J., ALBONICO, M., CHWAYA, H. M., SAVIOLI, L. TIELSCH, J. SCHULZE, K. & YIP, R. (1996). Hemoquant determination of hookworm-related blood loss and its role in iron deficiency in African children. *American Journal of Tropical Medicine and Hygiene* **55**, 399-404.

STOLTZFUS, R. J., ALBONICO, M., CHWAYA, H. M., TIELSCH, J., SCHULZE, K. & SAVIOLI, L. (1998). Effects of the Zanzibar school-based deworming program on iron status of children. *American Journal of Clinical Nutrition* **68**, 179-186.

STOLTZFUS, R.J., CHWAYA, H. M., TIELSCH, J., SCHULZE, K. J., ALBONICO, M. & SAVIOLI, L. (1997). Epidemiology of iron deficiency anaemia in Zanzibari schoolchildren: the importance of hookworms. *American Journal of Clinical Nutrition* **65**, 153-159.

TAREN, D. L., NESHEIM, M. C., CROMPTON, D. W. T., HOLLAND, C. V., BARBEAU, I., RIVERA, G., SANJUR, D., TIFFANY, J. & TUCKER, K. (1987). Contributions of ascariasis to poor nutritional status in children from Chiriqui Province, Republic of Panama. *Parasitology* **95**, 603-613.

TAREN, D. L. & CROMPTON, D. W. T. (1989). Nutrition interactions during parasitism. *Clinical Nutrition* **8**, 227-238.

THEIN HLAING (1993). Ascariasis and childhood malnutrition. *Parasitology* **107** (Suppl.) S125-S136.

TORLESS, H. (1999). *Parasitic infections and anaemia during pregnancy in Sierra Leone.* PhD Dissertation: University of Glasgow.

TRIPATHY, K., DUQUE, E., BOLANOS, O., LOTERO, H. & MAYORAL, L. G. (1972). Malabsorption syndrome in ascariasis. *American Journal of Clinical Nutrition* **25**, 1276-1287.

VITERI, F.E. (1994). The consequences of iron deficiency and anaemia in pregnancy on maternal health, the foetus and the infant. *SCN News* **11**, 14-18.

WILLIAMS-BLANGERO, S., SUBEDI, J., UPADHAYAY, R. P., MANRAL, D. B., RAI, D. R., JHA, B., ROBINSON, E. S. & BLANGERO, J. (1999). Genetic analysis of susceptibility to infection with *Ascaris lumbricoides*. *American Journal of Tropical Medicine & Hygiene* **60**, 921-926.

WOLFE, M. S. (1978). *Oxyuris, Trichostrongylus* and *Trichuris*. *Clinics in Gastroenterology* **7**, 211-217.

WORLD BANK (1993). *World Development Report 1993: Investing in Health.* Oxford University Press, Oxford.

WORLD HEALTH ORGANIZATION (1989). *Report of African Regional Consultation on Control of Anaemia in Pregnancy* (WHO/Brazzaville document AFR/NUT/104), WHO Regional Office for the African Region, Brazzaville, D. R. Congo.

WORLD HEALTH ORGANIZATION (1996). *Report of the WHO Informal Consultation on Hookworm Infection and Anaemia in Girls and Women*, Geneva, 5-7 December. WHO/CDS/IPI/96.1, WHO, Geneva.

WORLD HEALTH ORGANIZATION (1998). *World Health Report 1998: Life in the 21st century. A vision for all.* Report of the Director General. WHO, Geneva.

WORLD HEALTH ORGANIZATION (1999). *Report of the WHO informal consultation on monitoring of drug efficacy in the control of schistosomiasis and intestinal nematodes.* Geneva; WHO, 1999. WHO/CDS/CPC SIP/99.1.

Chapter 4
INTESTINAL NEMATODES AND COGNITIVE DEVELOPMENT

Jane Kvalsvig
School of Anthropology, Psychology and the Centre for Social Work, University of Natal, Durban, South Africa.
e-mail: kvalsvig@nu.ac.za

1. INTRODUCTION

It is no easy matter to assess the changes that investment in parasite control programmes may bring about in the cognitive development of children living in endemic areas. This is particularly so when government health and education policies in the affected countries are themselves in a state of flux, unevenly applied, and subject to fluctuations in economic resources and political will. The problem of assessing the factors that impact on cognitive development is essentially the problem of assessing one dynamic system (the developing child) within another (the developing country).

There is constant negotiation between developed and developing countries as to whether, and how, development funding should be made available, and what affordable and sustainable measures will give the most benefits. School-based parasite control programmes (see Chapter 2) are obvious candidates for support, targeting as they do, a vulnerable sector of the population, the children of the poor. In the case of intense infections the morbidity attributable to geohelminth infections is sufficient reason to advocate treatment, but what of subclinical infections? Do they affect the cognitive development of children to a sufficient extent to warrant the outlay of scarce financial resources?

The difficulties of assessing the impact of parasites on the nutritional status of children are considerable (see Chapter 3), but minor in comparison to the difficulty of measuring the constraints imposed on the development of thinking skills in children by chronic low-level infections. But this is what must be done if we are to assess the damage inflicted by geohelminths and the benefits that might accrue to children if they were free of these organisms.

2. DEVELOPMENTAL PSYCHOLOGY

The purpose of this chapter is to use a developmental psychology perspective for the task of assessing the impact of parasite infections on children in developing countries. Recent trends in child development research per se make it easier to tackle this task.

Developmental psychology has achieved a maturation of its own. In place of a tendency in the science to be intolerant of principles established from a different theoretical perspective, there is a recognition of the need for overarching theories to accommodate data collected from a variety of theoretical perspectives (Horowitz, 2000). Horowitz, in an overview article to mark the beginning of a new millenium, notes a new enthusiasm for models which illuminate dynamic processes. The processes in questions are 'nonlinear, interactive, full of reciprocity between and among levels and variables'. She talks about poverty as 'a dense concentration of disadvantaged circumstances that can swamp development negatively'. Constitutional, social, economic and cultural factors shape development: they interact with one another across the course of development and aggregate to produce different levels of advantage. Extreme poverty such as one finds in a developing country constitutes a swamping factor, placing children at high risk, but children can be protected by special circumstances and measures.

This way of thinking has given rise to a vocabulary of concepts and constructs that enable psychologists to work with large sets of cross-sectional or longitudinal observations, describing and tracing influences on development.

Developmental psychology is naturally concerned with changes over time. Words like *trajectories*, *transactions* and *transitions* afford ways of thinking about behavioural plasticity. Developmental *trajectories* refer to increments over time in a particular developmental domain. With this comes the notion that an infection may alter the altitude peak of skill attained, or slow the velocity of development. The *transactional* nature of development refers to the fact that from moment to moment the child interacts with her environment, bringing about changes in people and objects, and at the same time is herself influenced by those people or objects. Thus happy, healthy, active children may be more sociable, eliciting more responses from caregivers, and allowing more opportunities for social learning. *Transitional periods* refer to periods of rapid qualitative change in behaviour and cognition, such as adolescence or the time of entry into school. These are thought to be times when negative contexts might have a more permanent effect.

Linked to all these is the Piagetian concept of *stages*, where the child is active in the construction of her own mind. Neuroimaging has given support to this, because we now know that the brain responds actively to stimulation, linking new pathways to established connections. In Piagetian terms each stage in the construction of mind is built on the preceding stages, a metaphor with the corollary that the richness in the early construction of mind may make subsequent cognitive development easier. Skill in using symbols, for instance, facilitates other cognitive ventures. In an American study, the early acquisition of reading skills in first grade predicted better verbal ability and knowledge in a wide range of fields 10 years later, indicating the cumulative value across the years of the early skill (Cunningham & Stanovich, 1997).

The risk and resilience literature has brought familiarity with the idea that risks are additive in their effects. Low-birth weight predicts developmental delays in impoverished environments more certainly than it does amongst the well-to-do. There is evidence that geohelminths are more readily acquired by stunted children (Hagel *et al,* 1999).

3. A LONGITUDINAL VIEW

It is obvious that the common geohelminths do not arrive on the first day of school, but that children in an endemic area are at risk from the time when they start to move about independently, and even before that time. In endemic areas many children acquire more than one species of intestinal helminth. Immunologically speaking, two different kinds of host responses are identifiable: the inflammatory response to first-time infections and a more settled chronic response. Psychologically speaking, during the period from birth to six significant skills are acquired in different domains at different times in the lead-up to the rapid cognitive development that takes place at around six.

Developmental milestones do not occur in isolation: cognitive constructs are built on the foundations of what has been previously experienced and learned. In the development of gross motor skills, bipedal locomotion puts the child in contact with more people and objects, allows her to approach and avoid, to explore and to develop a good visual sense of perspective. There are increased energy and muscle building requirements and increased risks of acquiring common parasite infections. The rapid acquisition of language usually follows the development of locomotion, words are used as symbols, and this allows for the further development of social skills: asking for information and seeing the other person's point of view. Underlying the development of these and many other skills is a process of myelination taking place in the brain, allowing qualitatively different processing of information as new areas of the brain become fully functional. There is some evidence that nutritional deficits such as iron-deficiency can slow the developmental process (Stoltzfus *et al*, 2001). For the purposes of assessing the damage done by parasitic infections, the time of acquisition and the duration of infection may determine where the greatest impact may be on cognitive development.

4. A CROSS-SECTIONAL VIEW

Because the current recommendations from the World Health Organisation emphasize the usefulness of school-based control programmes, ministries of education need to be convinced that the time given up to a school-based programme is beneficial in educational terms. For advocacy purposes it is usually important to find out whether the anticipated benefits of improved cognitive processing would further translate into improved school performance on the assumption that improved school performance is more persuasive to policy makers in ministries of education than improved performance on cognitive tests. Table 4.1 sets out the levels of analysis which have bearing on the question of whether there is a causal link between geohelminth infections and poor educational performance or early dropout rates for children in endemic areas.

In any analysis that attempts to link geohelminth infections to educational performance there are clusters of variables which must be accounted for statistically or in the research design. The list of associated factors in Table 4.1 is illustrative rather than exhaustive, and the starting point for research is a testable model of how they might be related to one

Table 4.1. Levels of analysis: clusters of factors which usually occur together, and may provide links connecting geohelminths and poor educational performance.

Health factors	**Geohelminths**; other parasites; other infections; poor nutrition; poor access to health services; other socio-economic factors
Physiology	CNS damage; energy depletion; micronutrient deficiencies; inflammatory responses; pain and discomfort
Behaviour	Minimal activity; little interaction; depressed/withdrawn
Cognition	Poor concentration; impaired memory functions; slow processing; delayed language development; inadequate concept formation
Education	Poor preparation, long distances from school; inadequate schools, a late start & frequent absences; poor participation in classroom **Poor performance/ early dropout**

another. A simple association between geohelminths and educational performance may be explained by underlying socio-economic factors, and the genuine impact of geohelminths on social and cognitive functions may be obscured by any number of school-related factors when educational performance is the outcome measure.

Fortunately some pathways have been explored. At the present time some causal linkages are well accepted and others more or less speculative. An example of the former would be the link between geohelminths and anemia which is quite well worked out and is shown to be dependent on the species of parasite (Stoltzfus *et al*, 2000) and on the intensity of the infection. Hookworm infections are strongly associated with anemia, and *Trichuris trichiura* and *Ascaris lumbricoides* less so. On the other hand the link between inflammatory responses to parasite infections and changes in cognitive performance is plausible at present but not worked out although there is mounting evidence of changes in brain function as a result of infection (Kelley, 2001; Dantzer, 2001). The direction of the link between poor cognitive functioning and poor educational achievement must depend very much on the quality of the schooling being offered. If classrooms are overcrowded and the teaching poor, improved cognitive processing is unlikely to have an effect on school performance. Indeed, there is some evidence that lively-minded children do worse in dull classrooms than their less healthy but more compliant classmates (Olney, *personal communication*).

5. THE EVIDENCE

Intestinal nematodes have at various times occasioned intense interest amongst evolutionary biologists (for example, Dawkins, 1982), public health policy makers (for example Savioli *et al*, 1997; Bundy & De Silva, 1998) and now immunologists. Psychologists have been involved over a very long period (Watkins & Pollitt, 1997) but there was a long gap about the middle of last century and there have been few studies overall relative to the complexity of the issues. In recent years there have been several overview articles (Nokes et al, 1992; Connolly & Kvalsvig, 1993; Watkins & Pollitt, 1997, Connolly, 1998) but still relatively few papers reporting original research.

Although assessing the functional significance of parasite infections in humans is important, it is difficult to design research projects that will test the hypotheses adequately. It has to be said that although there are studies that show cognitive and educational benefits for children after treatment with anthelmintics, the causal evidence is not strong. Even with improved research methodologies such as better cognitive measures and better research designs, the situation has not improved much (Watkins & Pollitt, 1997). Why is this the case, is it because the effect is not there or are there other reasons?

There are difficulties in designing a well-controlled study. The biology of the parasites themselves suggests that they may all have different effects. Poor sanitation favours transmission of all of these common species and polyparasitism is more common in endemic areas than single infections. Thereafter the similarities between them diminish. They are structurally different organisms, feeding differently and causing different kinds of damage to their hosts. Even within one species, the intensity of the infection may evoke quite different host responses: while there is considerable agreement about the damage done by intense infections, mild to moderate infections may be quite well-tolerated; some would even suggest beneficial under some circumstances (Watkins & Pollitt, 1997). First-time infections, even if low-key, may spark off a cytokine-mediated neurobehavioural reaction, whereas later add-on infections may have only small additional effects. What may be tolerated in an otherwise healthy child may be harmful in a malnourished child.

All other things being equal, the design of choice for establishing a link between parasites and cognition would be a randomised placebo-controlled trial, but there are a number of other considerations. It is now difficult to justify withholding or delaying treatment, at least for school-aged children: there is sound evidence that treatment risk is low and there are benefits for the children. The drugs of choice have been so widely used and for so long, that it can be argued that even for very young children and for pregnant women there is sufficient evidence for low treatment risk. On the other hand there are unquestionable health risks if high intensity infections are left untreated: *A. lumbricoides* has been associated with intestinal obstructions (De Silva, Guyatt & Bundy, 1997), *T. trichiura* with severe diarrhoea and rectal prolapse (Bundy & Cooper, 1989) and hookworm species with severe anaemia (Stoltzfus *et al,* 2000). A fair amount of evidence testifies to improvement in anthropometric indicators across the board following treatment, and micronutrient deficiencies may be rectified when deworming is coupled with micronutrient supplementation.

There are practical difficulties in sustaining a randomised placebo-controlled trial. In order to obtain informed consent from parents, medical ethics require that they should made aware of parasite infections as the research issue and of the fact that some children will be treated and some not. Inexpensive, effective and safe treatments are readily available in many countries these days, either across the counter or through clinics, so parents may be able to treat their children if the treatment provided by the researchers does not appear to be working. All of these difficulties suggest that the time has come to consider other options where cognitive benefits are concerned, less hard-headed perhaps, but more informative when dealing with complex adaptive systems.

In terms of measuring psychological or psycho-educational outcomes the time of onset and the duration of the infection may determine the processes and skills which are affected, and the time it may take for rehabilitation to be measurable. Harking back to the ideas of trajectories and transitional periods, socio-cognitive damage may be manifest either in the rate of acquisition of a skill or in the ultimate level of skill attained. An infection acquired at a young age may delay the development of language and social skills. Together with stunting, this may have give the appearance that the child is too young to benefit from school, delaying school entry, and there is some evidence for this kind of indirect effect (Oyewole, 1984). More directly, lack of energy may limit activity, concentration, and perseverence in the face of difficulty, and consequently the level of skill acquired in the language domain, a domain which has obvious connections with later academic skills.

As noted in the introduction, we are dealing with dynamic systems: humans, and especially young humans, are complex adaptive systems. Our research methods and even our research questions assume an 'upward' causality from the biology to the outcome behaviour or test performance but causation moves the other way as well, from cultural interpretation or adaptation or motivation down to the outcome measure. Thus, parents are likely to assimilate new knowledge about what benefits their children and act in their children's best interest rather than the 'general good', in the process ruining a good research design. One can speculate that children may adapt to limited energy or chronic debilitation by concentrating on the demands of the moment, performing well on a cognitive or educational tests while their healthier peers are discharging excess energies and high spirits in play. Another neglected 'downward' causality area concerns emotional state. There are very few descriptive studies, and more observational research may be required to generate hypotheses more in tune with current thinking in the field of developmental psychology.

In spite of all these difficulties there is broad agreement amongst reviewers that cognitive development is likely to be affected. Common helminths are associated with poor cognitive performance (for example Hadidjaja *et al*, 1998, and Sakti *et al* 1999). They are also associated with certain nutritional and micronutrient deficiencies. Both micronutrient deficiencies and parasite infections have been associated with altered behaviour and poor cognition, although the more stringent requirement of a causal connection remains elusive because of the many confounders (Pollitt, 1997; Grantham-McGregor & Ani, 1999).

Dickson *et al* (2000) in a meta-analysis limited to randomised controlled trials reviewed the effects of treatment of intestinal helminth infection on growth and cognitive performance in children and came to the conclusion that routine anthelmintic treatment was not indicated, a conclusion which drew protests from many quarters. Statements like these from scientists confuse the rules of evidence needed for scientific enquiry with those needed for public health policies and have real-life consequences in developing countries.

Different rules of evidence apply to public health policy makers (Shonkoff, 2000). Public health policies in many endemic countries link parasite control to a spectrum of measures to be tackled through school health programmes like school feeding schemes on the assumption that such programmes protect children living in poverty, enabling them to benefit from tuition. Where scientists, like judges in criminal matters, require proof 'beyond reasonable doubt', public health policy-makers have a legal duty to protect, and to point out possible and probable health risks. A recent legal enquiry into the British government measures to protect the public over the bovine spongiform encephalopathy question has highlighted this. Public health policy should operate on a 'balance of probability' principle and there is certainly circumstantial and associative evidence linking parasites with behavioural and cognitive effects. Children, especially those at risk in areas where medical treatment is not easily accessible, merit protection. Parasite effects range from mild discomfort and abdominal pain to death in the case of untreated intestinal obstruction from *A. lumbricoides*. Parasite control programmes *per se* are beneficial to schoolchildren in endemic areas in a variety of ways, including as many do, health education and improved sanitation.

6. THE NEW QUESTIONS

Where does all this lead? The behaviours and cognitive functions under scrutiny are undoubtedly complex and adaptive, and most of the relevant associative connections between factors have been demonstrated. Improved design is not doing much better than the former less stringent methodologies in giving evidence of causal connections. Undoubtedly children in the underdeveloped areas of the world do not perform optimally on either cognitive or educational tests, but there are too few studies and too many variables for us to pin blame convincingly on parasites alone. Time and care are needed to untangle the influences.

Research questions have been based mainly on 'upward' causality from biology, ignoring the well established transactional principles in developmental psychology, whereby the child is not merely a passive recipient but also active in adapting to and coping with a stressor. There have been advances in immunology which have not yet made their way into this area of study. The link between cytokine action and 'sickness behaviour' may give us a productive clue to an explanatory principle, or lead us into yet another set of questions. But the thrill of the chase for scientists should not be allowed to endanger the health and well-being of children in endemic areas.

REFERENCES

BUNDY, D.A.P. & COOPER, E.S. (1989). *Trichuris* and trichuriasis in humans. *Advances in Parasitology* **28**, 107-173.

BUNDY D.A.P. & DE SILVA, N.R. (1998). Can we deworm this wormy world? *The British Medical Journal* **54** (2), 431-432.

CONNOLLY, K.J. & KVALSVIG, J.D. (1993). Infections, nutrition and cognitive performance in children. *Parasitology* **90** (Suppl) 187-S200.

CONNOLLY, K.J. (1998). Mental and behavioral effects of parasitic infection. In *Nutrition, Health and Child Development,* Pan American Health Organisation/ World Bank Scientific Publication No 566.

CUNNINGHAM A.E. & STANOVICH K.E. (1997). Early reading acquisition and its relation to reading experience and ability 10 years later. *Development Psychology* **33**, 943-945.

DANTZER, R. (2001). Cytokine-induced sickness behaviour: where do we stand? *Brain, Behaviour and Immunity* **15**, 7-24.

DAWKINS, R. (1982). *The extended phenotype*. Oxford: Freeman.

DE SILVA, N.R., GUYATT, H.L. & BUNDY, D.A.P. (1997). Morbidity and mortality due to *Ascaris*-induced intestinal obstruction. *Transactions of the Royal Society of Tropical Medicine and Hygiene* **91**, 31-36.

DICKSON, R., AWASTHI, S., WILLIAMSON, P., DEMMELLWEEK, C., & GARNER, P. (2000). Effects of treatment for intestinal helminth infection on growth and cognitive performance in children: systematic review of randomised trials. *British Medical Journal* **320**, 1697-1701.

GRANTHAM-McGREGOR, S.M., & ANI, C.C. (1999). The role of micronutrients in psychomotor and cognitive development. *British Medical Journal* **55**, 511-527.

HADIDJAJA, P., BONANG, E., SUYARDI, M.A., ABIDIN, S.A.N., ISMID, I.S., & MARGONO, S.S. (1998). The effect of intervention methods on nutritional status and cognitive function of primary school children infected with *Ascaris lumbricoides*. The American Journal of Tropical Medicine and Hygiene **59**, 791-795.

HAGEL, I., LYNCH, N.R., DI PRISCO, M.C., PEREZ, M., SANCHEZ, J.E. PEREYRA, B.N., & SOTO DO SANABRIA, I. (1999). Helminthic infection and anthropometric indicators in children from a tropical slum: *Ascaris* reinfection after anthelmintic treatment. *Journal of Tropical Pediatrics* **45**, 215-220.

HOROWITZ, F.D. (2000). Child development and the PITS: Simple questions complex answers, and developmental theory. *Child Development* **71**, 1-10.

KELLEY, K.W. (2001). It's time for psychoneuroimmunology. *Brain, Behaviour and Immunity* **15**,1-6.

NOKES, C., GRANTHAM-MCGREGOR, S.M., SAWYER, A.W., COOPER, E.S. & BUNDY, D.A.P. (1992). Parasitic helminth infection and cognitive function in school children. *Proceedings of the Royal Society, London* **247**, 77-81.

OLNEY, D.K. (2001). The association between iron supplementation and grade repetition in a population. *Personal communication.*

OYEWOLE, A.I. (1984). Home and school: effects of micro-ecology on children's educational achievement. In *Nigerian children: developmental perspectives* (ed. Curran, H.V.) pp156-174. Routledge & Kegan Paul, London.

POLLITT, E. (1997). Iron deficiency and educational deficiency. *Nutrition Reviews* **55**, 133-141.

SAKTI, H., NOKES, C., HERTANTO, W.S., HENDRATNO, S., HALL, A., BUNDY, D.A.P. & SATOTO. (1999). Evidence for an association between hookworm infection and cognitive function in Indonesian school children. *Tropical Medicine and International Health* **4**, 322-334.

SAVIOLI, L., CROMPTON, D.W.T., OTTESON E.A., MONTRESOR, A., & HAYASHI S. (1997). Intestinal worms beware: developments in anthelmintic chemotherapy usage. Par*asitology Today* **13**, 43-44.

SHONKOFF, J.P. (2000). Science, policy and practice: three cultures in search of a shared mission. *Child Development*, **71**, 181-187.

STOLTZFUS, R.J., CHWAYA, H.M., MONTRESOR, A., ALBONICO, M., & SAVIOLI, L TIELSCH, J.M. (2000). Malaria, hookworms and recent fever are related to anemia and iron status indicators in 0-5 yearold Zanzibari children and these relationships change with age. *Journal of Nutrition* **130**, 1724-1733.

STOLTZFUS, R.J., KVALSVIG, J.D., CHWAYA, H.M., MONTRESOR, A., ALBONICO, M., TIELSCH, J.M., SAVIOLI, M.D. & POLLITT, E. (2001). Effects of iron supplementation and anthelminthic treatment on motor and language development of Zanzibari preschool children. *British Medical Journal,* In press.

WATKINS W.E. & POLLITT, E. (1997).'Stupidity or worms': do intestinal worms impair mental performance? *Psychological Bulletin* **121**, 171-191.

Chapter 5
THE ECONOMICS OF WORM CONTROL

Helen Guyatt
Wellcome Trust Research Laboratories-Kenya Medical Research Institute, PO Box 43640, Nairobi, Kenya and Centre for Tropical Medicine, University of Oxford, John Radcliffe Hospital, Oxford OX1 3QU, UK.
e-mail: Hguyatt@wtnairobi.mincom.net

1. INTRODUCTION

Worm infections remain unchecked in much of the developing world. Providing realistic data on the cost of disease and the cost of control is a necessary pre-requisite for moving intestinal nematode control into the operational arena. In the absence of evidence it is unreasonable to expect the policy maker to alter the low priority attached to these chronic parasitic diseases or to expect the health planner to risk limited funds on interventions of unknown cost and efficacy.

This chapter presents some of the evidence on the economic burden of intestinal nematode infections and discusses the affordability of approaches to their control.

2. HOW HARMFUL ARE WORMS?

Acute clinical complications arising from intestinal worm infestation are rare. Although these worms are extremely prevalent, only a small percentage of infected people suffer symptoms such as intestinal obstruction from *Ascaris lumbricoides*, rectal prolapse from *Trichuris trichiura* or severe anaemia from hookworm infection. These symptoms are typically associated with very heavy intensities of infection, presenting in only a few individuals. Most infected individuals habour light-to-moderate infections, which rarely demonstrate overt clinical symptoms, but have important long-term consequences for health. Children are the most at risk group for *Ascaris* and

Trichuris infections. In this vulnerable population, these chronic infections can have a major impact on mental and physical development (see Chapters 3 and 4). The recognition that the chronic effects of helminthiasis on child development are of much greater importance than the acute represented a major change in the perception of worm morbidity, and has been instrumental in putting intestinal nematodes on the international health agenda (Bundy, 1997). There is now convincing evidence that worm infection in children can be associated with impaired physical growth (Stephenson *et al.* 1989; Simeon *et al.* 1995; Stolzfus *et al.* 1998a) and cognitive ability (see Drake *et al.* 2000).

3. WHAT ARE THE OPTIONS FOR CONTROL?

Worms are associated with poverty. Only through economic development, with the concomitant improvements in sanitation, will communities be rid of these parasites. In the meantime, there are available, safe and effective drugs, which in addition to ridding individuals of infection, have also been shown to reverse some of the symptoms of morbidity. Treatment with the benzimidazoles has been shown to improve anaemia status (Stoltzfus *et al.* 1998b) and result in catch-up growth in those stunted (Stephenson *et al.* 1989; Simeon *et al.* 1995). These drugs also have the advantages that they are simple to administer (a single oral dose) and relatively inexpensive (0.03-0.25 US$ per dose). Although mebendazole can be up to 10 times cheaper than albendazole (Stoltzfus *et al.* 1998b), the concerns about its efficacy in treating hookworm infection has lead most control programmes to favour albendazole.

The problem with drug administration as a control measure is that one treatment is not enough. Individuals are continuously exposed to infection and get reinfected after treatment. The need for regular deworming of individuals presents a formidable hurdle in establishing a sustainable control programme for these parasites.

4. IS CONTROL AN EFFICIENT USE OF RESOURCES?

Policy makers in developing countries are faced with a myriad of competing demands for scarce funds. Developing a strong argument for the control of intestinal worms as a priority health issue requires uncontroversial data on the economic and public health importance of the disease. There needs to be some quantified measure of the benefits to society of ridding people of these worms.

The sums on economic loss attributed to intestinal nematodes in livestock, for instance, appear relatively straightforward. Infections may reduce yield by a certain percentage that could be directly translated into monetary loss through market values. Measuring the economic impact of infections in humans is much more difficult as one has to place a monetary value on their poor health. One approach is to treat health as an investment in human capital, contributing to economic output through increased productivity and availability of potential workers. Providing meaningful quantitative estimates of the return on health investment for parasitic diseases has proved difficult, and there are currently no estimates for the intestinal nematodes.

A review of the evidence on the contribution of worm infections to the poor health of children, and the consequences of these on future productivity in the workplace, suggest that the economic impact may be significant (Guyatt, 2000). There is a wealth of evidence that worms can lead to growth stunting in childhood (Stephenson *et al.* 1989; Simeon *et al.* 1995; Stolzfus *et al.* 1998a). Independent studies in adulthood have shown height to be associated with reduced work output and wage-earning capacity, particularly in professions requiring hard physical labour (Spurr *et al.* 1977). For instance, a study in rural Philippines suggested that an adult 15cm taller than average might expect to achieve a 13% increase in wage rates (Haddad & Bouis, 1991).

The effect of stunting on future productivity may work directly through reduced physical strength in adulthood, or indirectly through reduced schooling. Children with low height-for-age have been shown to delay school enrollment (PCD, 1999a), which will have implications for the years of schooling they attain and the age at which they join the workforce. Absenteeism from school has also been shown to be associated with *T. trichiura* infection, with some evidence that this may be causal. For example, studies in Jamaica have shown that the proportion of time absent from school is related to the level of infection (Nokes & Bundy, 1993), and that treatment of moderate whipworm enhances school attendance in the more severely stunted children (Simeon *et al.* 1995). Children who are absent from school are likely to perform poorly at school and drop-out prematurely (Weitzman, 1987).

There is a large literature on the returns to investment in education (Colclough, 1982; Psacharopoulos, 1993), whereby earnings and years of schooling are used to determine the rate of return of one additional year of schooling. Primary education can be shown to yield high returns in developing countries, with these returns declining with the level of schooling and a country's per capitum GNP (Psacharopoulos, 1993). It has been shown, for instance, that giving primary-school leavers four more years of education would increase their earnings by 15-24 % in Kenya and 8-18% in Tanzania (Boissiere, Knight & Sabot, 1985). This work also suggests that the main effects of years of schooling on earnings are indirect, operating through the development of cognitive skills. An effect of worms on cognitive ability is evident, particularly in those with moderate-heavy infections (Nokes *et al.* 1992). Although the evidence of an effect of worms on schooling is suggestive, either through stunting or cognitive impairment, quantitative estimates of the contribution of intestinal nematode infection to years of schooling are not available.

The cost of compromised development in childhood for the productivity of the adult labour force is particularly difficult to assess. The effect of current infection on the work-output of adults is more amenable to estimation. In this case, the focus has been on hookworm infection, where the highest burdens are often found in adults, frequently as a result of occupational exposure. The productivity of anemic (assumed primarily due to hookworm infection) rubber tappers in Java was found to be 19% below that of their non-anaemic colleagues, a difference which was reversible by treatment (Basta *et al.* 1979). Similarly Kenyan road-workers who were <85% weight for height (35% of those studied) were 10% less productive than their normally nourished peers, although it was not conclusively demonstrated that the nutritional deficit was solely a consequence of helminth infection (Brooks *et al.* 1979).

Rather than look directly at productivity, another approach is to estimate working time loss through sickness. However economic loss is more complex than the number of days lost multiplied by average wage rate, as incapacity is not necessarily directly related to production. It will limit potential productivity, but whether this translates to actual loss depends on many factors such as the duration of the incapacity, its correlation with the agricultural cycle and household coping mechanisms (Breiger & Guyer, 1990). It may also be appropriate to look beyond the affected individual to the household or community. An illness in an individual can affect the activities of a whole household with uninfected individuals adopting additional roles and neglecting other tasks. Compensation for an illness of one person could have adverse consequences for all. These indirect effects on production and community development are quite separate from direct effects on individual health, and have yet to be accurately assessed. However, they may be more relevant to policy makers than the human capital approach, since workforce availability *per se* is rarely perceived in developing countries as a limit on production.

Part of the difficulty in getting good data on the economic consequences of worm infection is due to the complexity in the relationships involved, but also in part due to the failure to account for the known epidemiology of the diseases, in particular that morbidity is associated with intense infections. This has been one of the major criticisms of studies investigating the impact of helminths on productivity (Prescott, 1989).

Economic impact assessments have the potential to attract funds for the control of intestinal nematodes, but they are not sufficient to guide the setting of health priorities. Decision makers also need to know how much interventions are going to cost.

5. IS CONTROL AFFORDABLE?

The cost of implementing control measures is a critical component in evaluating the affordability and potential sustainability of any approach. Most of the past and present control programmes have relied heavily on donor financing or the involvement of non-governmental organizations for their sustainability. In most parts of the world, official control programmes for the intestinal nematodes do not exist. An absence of control implies either that the disease is not perceived by health planners to be of high priority compared to other health issues or that control is not affordable.

The average annual per capita expenditure on all forms of health in low-income countries (excluding India and China) is estimated at US $14 (World Bank, 2000) with some countries such as Kenya spending as little as US $3 (World Bank, 2001). Although published prices of orginal formulations of albendazole have decreased slightly from US $0.25 in 1988 (Guyatt, Bundy & Evans, 1993) to estimates of less than US$ 0.20 per dose (PCD, 1999b), a vertical control programme of mass chemotherapy is likely to be too expensive for governments to take on board. Recent cost calculations of vertical approaches of mass albendazole treatment using mobile teams estimate that delivery costs can constitute between 40 and 70% of programme costs (Table 5.1).

Table 5.1. Delivery costs for a single mass treatment with albendazole

Strategy	**Country**	**Delivery cost/ per treatment**		**Source**
		US$	**% of total cost**	
Mobile team to community	Montserrat Nigeria Tanzania Bangladesh	0.51 0.32 [1] 0.21[3] - [5]	67% - [2] - [4] 42%	Guyatt *et al.* (1993) Holland *et al.* (1996) Guyatt *et al.* (1994) Mascie-Taylor *et al.* (1999)
School-based	Ghana Tanzania	0.04 0.03	17% 13%	PCD (1999b) PCD (1999b)
Out-of-school children	Eygpt	0.21	- [5]	Talatt & Evans (2000)

[1] estimated from an exchange rate of 8 Naira / US$ in 1990 (World Bank, 2001) and 4 cycles of treatment over 1 year.
[2] drug delivered was levamisole
[3] mobile team to schools not communities
[4] drug delivered was praziquantel
[5] not available

Assuming a drug price of US $0.20, the total cost per person treated with a single dose of albendazole could be upwards of US $0.50. If the cheaper generics on the market were used, this could be reduced, but the compromise with drug quality is unclear. Some generics are available for as little as US $0.03 per dose, but given the high delivery costs, this could still represent a significant proportion of the average national budget for most developing countries. In Rwanda, it was estimated that a mass treatment programme against intestinal nematodes and schistosomiasis would entail nearly a third of the actual annual drug budget (de Schaepdryver, 1984). Without donor assistance it is unlikely that many countries could sustain this type of expenditure aimed solely at worms.

6. REDUCING THE COSTS

6.1 Targeting the school-aged child

The 1993 World Development Report of the World Bank included school-based anthelmintic delivery in the essential package of public health services it recommended for all developing countries (World Bank, 1993). Subsequently, initiatives such as the Partnership for Child Development (PCD) have emerged to assess the operational feasibility, costs and effectiveness of drug delivery through schools (PCD, 1998; 1999a,b).

The PCD activities of school-based deworming in Tanzania and Ghana suggest a delivery cost for albendazole of US $0.03-0.04 (PCD, 1999b), between 13 and 17% of the total cost (in 1996). Focusing treatment through the school system reduces delivery costs (by more than 10 fold in comparison with mobile teams (Table 5.1)), but has raised concerns about missing school-age children not in school who may be at a higher risk of infection than those in school (Gyorkos *et al.* 1996). Current estimates suggest that around 40% of school-aged children in the least developed countries are not enrolled in primary school (Unicef, 2001), with a further proportion not attending on a regular basis. Reaching out-of-school children with anthelmintics, in this case praziquantel, was the focus of recent work in Eygpt where this was achieved at a financial delivery cost of US $0.21 per child (Talatt & Evans, 2000). This suggests that a child out of school could be treated with albendazole for US$ 0.41 (Table 5.2).

Although the costs per treatment are low, a national deworming programme targeted at school-aged children would represent a major investment in a developing country. For example, in Kenya 69% of school-aged children are attending school (see Table 5.2). A national campaign of mass treatment with albendazole targeting children both in and out of school could cost over 3 million US$. This would represent 4% of the national expenditure on health. If the programme was to be financed from Official Development Aid, then this would need to increase, each year, by 1% (Table 5.2). If a cheaper generic was used instead (eg MedPharm quote CIF for 400mg albendazole to Nairobi at 0.05 US$), it would still involve an investment of over 1.5 million US$ in Kenya every year. Although these very simple sums do not take into account issues such as economics of scale, they provide some order of magnitude to the likely affordability of these programmes in developing countries.

Table 5.2. The annual cost of treating school-aged children in Kenya in 1999

		Value	Source
Population size	Total School-aged IN school[1] School-aged NOT IN school[1]	28,686,607 7,833,668 3,463,423	CBS (2000a) CBS (2000b) CBS (2000b)
Unit cost/child	Albendazole Delivery using school system Delivery to those out-of-school	0.20 US$ 0.03 US$ 0.21 US$	PCD (1999b) PCD (1999b) Talatt & Evans (2000)
Total cost	All school-aged children Cost per capita	3,221,747 US$ 0.11 US$	estimated here estimated here
Resources available	Health expenditure per capita ODA[2] per capita (real)	3 US$ 10 US$	World Bank (2001) World Bank (2001)

[1] Aged 5-19 years
[2] Official Development Aid

6.2 Targeting pregnant women through ante-natal clinics (ANCs)

Hookworm is commonly observed at highest intensities of infection in adults, and poses a particular risk for the anaemic status of pregnant women. A recent randomized controlled trial in Sierra Leone demonstrated an average 6.6 g/L increase in hameoglobin with albendazole treatment of pregnant women (Torlesse & Hodges, 2000). Routine treatment of pregnant women with albendazole could theoretically be integrated into existing ANC services. The integration of control initiatives into existing services can be difficult, requiring a long-term effort. Failure will be guaranteed unless care is taken from the start of the integration process to develop methods and strategies that will afterwards remain feasible and affordable for regular health services. The delivery of any drugs to pregnant women is controversial and mass treatment may not be appropriate. A selective approach of first diagnosing infection before administering treatment is often more expensive, particularly if equipment needs to be purchased or personnel trained (Bundy, 1990).

6.3 User fees

Another approach to identifying cheaper approaches is to investigate other mechanisms of financing. Governments do not necessarily need to provide these services free-of-charge, and there is a potential for cost recovery. There has been surprisingly little work on treatment seeking behaviour, including user-provider interaction and the relative importance of quality of care and cost, and it is difficult to predict how willing and able a population would be to contribute to worm control. A study by Stephenson, Latham & Odouri (1980) suggested people spend a lot of money on anthelmintics, implying they are willing to buy and use them in some situations. More recent work by PCD suggests that parents are willing to pay the costs for anthelminthic treatment of their children, the preferred option being to incorporate these into existing school fees (PCD, in press). The success of any cost recovery programme will depend on the value assigned to deworming by households and the amount of cash at their disposable. Although the benefits of deworming were appreciated by the parents interviewed in Ghana, worms fell low down on their list of health priorities. Annual primary school fees were shown to vary widely (0.06 to > 5 US$ per child), such that the incremental cost of adding albendazole treatment (assuming 0.23 $ per child) could be anywhere between 4 and 400% of current fees paid. This variation in essence also reflects the wide variation in ability to pay. It is clear that the implementation of such a cost recovery scheme would need to carefully consider issues of equity in the ability of parents to pay, with methods put in place to identify and subsidize poorer households.

6.4 Health services savings

The costs of worm control have focussed on the costs to the provider of setting up and running the programme, with no attempt to evaluate the likely savings in the subsequent use of health services brought about by early treatment of the disease. Although these are likely to be difficult to obtain and interpret, not least because the costs associated with a disease will depend on the quality of the health service, they would provide a stronger argument for deworming. Such an analysis would require specific health audits establishing the costs for hospitalization (for example, with intestinal obstruction) and the reduction in outpatient visits. However, the costs associated with disease would not directly reflect the gains obtained with treatment. It is not clear, for instance, whether a single treatment would reduce the risk of subsequent pathology in places where children are continually infected.

7. CONCLUDING REMARKS

Although deworming is a relatively low cost health care intervention, the magnitude of the problem would require a significant and sustained financial investment. Most developing countries where these parasites are a problem would not be able to afford nationwide programmes without some donor support. Attracting investments from overseas requires convincing evidence on the benefits that are likely to accrue. Although there is strong evidence for an impact of worms on health and productivity, it is not available in a tangible format. The future challenge in advocating worm control is to quantify the dollars gained per dollar investment in a way that fully captures the wide-range of benefits that could be obtained from removing these parasitic infections.

ACKNOWLEDGEMENTS

Helen Guyatt is in receipt of a Wellcome Trust Research Career Development Fellowship (#055100).

REFERENCES

BASTA, S.S., SOEKIRMAN, M.S., KARYADI, D. & SCRIMSHAW, N.S. (1979). Iron deficiency anemia and the productivity of adult males in Indonesia. *American Journal of Clinical Nutrition* **32**, 916-925.

BOISSIERE, M., KNIGHT, J.B. & SABOT, R.H. (1985). Earnings, schooling, ability and cognitive skills. *The American Economic Review* **75**, 1016-1030.

BRIEGER, W.R. & GUYER, J. (1990). Farmers' loss due to Guinea worm disease: a pilot study. *Journal of Tropical Medicine and Hygiene* **93**, 106-111.

BROOKS, R.M., LATHAM, M.C. & CROMPTON, D.W. (1979). The relationship of nutrition and health to worker productivity. *East African Medical Journal* **9**, 413-21.

BUNDY, D.A.P. (1990). Control of intestinal nematode infection s by chemotherapy: mass treatment versus diagnostic screening. *Transactions of the Royal Society for Tropical Medicine and Hygiene* **84**, 622-5.

BUNDY, D.A.P. (1997). Health and early child development In *Early Child Development: investing in our Children's Future.* (ed. Young, M.E.), pp.11-38, Elsevier Science.

CENTRAL BUREAU OF STATISTICS (CBS) (2000a). 1999 Population and Housing Census. Volume I. Population distribution by administrative areas and urban centres. Prepared by CBS, Ministry of Finance and Planning, Kenya.

CENTRAL BUREAU OF STATISTICS (CBS) (2000b). 1999 Population and Housing Census. Volume II. Socio-economic profile of the population. Prepared by CBS, Ministry of Finance and Planning, Kenya.

COLCLOUGH, C. (1982). The impact of primary schooling on economic development: a review of the evidence. *World Development* **10**, 167-185.

DE SCHAEPDRYVER, L. (1984). Costs of training and maintenance of expert man-power versus drugs. Policies in the field of helminthic diseases in developing countries. *Social Science and Medicine* **19**, 1113-1116.

DRAKE, L.J., JUKES, M.C.H., STERNBERG, R.J. & BUNDY, D.A.P. (2000). Geohelminth infections (Ascariasis, Trichuriasis and Hookworm): cognitive and developmental impacts. *Seminars in Pediatric Infectious Diseases* **11**, 245-251.

GYORKOS, T.W., CAMARA, B., KOKOSKIN, E., CARABIN, H. & PROUTY, R. (1996). Enquete de prevalence parasitaire chez les infants d'age scolaire en Guinee en 1995. *Cahiers Sante* **6**, 377-381.

GUYATT, H.L. (2000). Do intestinal nematodes affect productivity in adulthood? *Parasitology Today* **16**, 153-158.

GUYATT, H.L., BUNDY, D.A.P., EVANS, D. (1993). A population dynamic approach to the cost-effectiveness analysis of community-based anthelmintic treatment: effects of treatment frequency. *Transactions of the Royal Society for Tropical Medicine and Hygiene* **87**, 570-575.

GUYATT, H.L., EVANS, D., LENGELER, C. & TANNER, M. (1994). Controlling schistosomiasis: the cost-effectiveness of alternative delivery strategies. *Health Policy and Planning* **9**, 385-395.

GYORKOS, T.W., CAMARA, B., KOKOSKIN, E., CARABIN, H. & PROUTY, R. (1996). Enquete de prevalence parasitaire chez les infants d'age scolaire en Guinee en 1995. *Cahiers Sante* **6**, 377-381.

HADDAD, L.J. & BOUIS, H.E. (1991). The impact of nutritional status on agricultural productivity: wage evidence from the Philippines. *Oxford Bulletin of Economics and Statistics* **53**, 45-58.

HOLLAND, C.V., O'SHEA, E., ASAOLU, S.O., TURLEY, O. & CROMPTON, D.W.T. (1996). A cost-effectiveness analysis of anthelminthic intervention for community control of soil-transmitted helminth infection: levamisole and *Ascaris lumbricoides*. *Journal of Parasitology* **82**, 527-530.

MASCIE-TAYLOR, C.G.N., ALAM, M., MONTANARI, R.M., KARIM, R., AHMED, T., KARIM, E. & AKHTAR, S. (1999). A study of the cost-effectiveness of selective health interventions for the control of intestinal parasites in rural Bangladesh. *Journal of Parasitology* **85**, 6-11.

NOKES, C., GRANTHAM-MCGREGOR, S.M., SAWYER, A.W., COOPER, E.S., ROBINSON, B.A. & BUNDY, D.A. (1992). Moderate to heavy infections with *Trichuris trichiura* affect cognitive function in Jamaican school children. *Parasitology* **104**, 539-47.

NOKES, C. & BUNDY, D.A.P. (1993). Compliance and absenteeism in school-children: implications for helminth control. *Transactions of the Royal Society for Tropical Medicine and Hygiene* **87**, 148-152.

PARTNERSHIP FOR CHILD DEVELOPMENT (PCD) (1998). The health of school-age children: experience from school health programs in Ghana and Tanzania. *Transactions of the Royal Society for Tropical Medicine and Hygiene* **92**, 254-261.

PARTNERSHIP FOR CHILD DEVELOPMENT (PCD) (1999a) Short stature and the age of enrolment in primary school: studies in two African countries. *Social Science and Medicine* **48**, 675-682.

PARTNERSHIP FOR CHILD DEVELOPMENT (PCD) (1999b). The cost of large-scale school health programmes which deliver anthelmintics to children in Ghana and Tanzania. *Acta Tropica* **73**, 183-204.

PARTNERSHIP FOR CHILD DEVELOPMENT (PCD) (2001) Community perception of school-based delivery of anthelmintics in Ghana and Tanzania. *Tropical Medicine and International Health*, In press.

PRESCOTT, N. (1989). Economic analysis of schistosomiasis control projects. In *Demography and Vector-Borne Diseases* (ed. Service, M.W.), pp.155-163, CRC Press.

PSACHAROPOULOS, G. (1993). *Returns to Investment in Education : a Global Update.* (Policy Research Working papers in Education and Employment, WPS 1067), World Bank.

SIMEON, D.T., GRANTHAM-MCGREGOR, S.M., CALLENDER, J.E. & WONG, M.S. (1995). Treatment of *Trichuris trichiura* infection improves growth, spelling scores and school attendance in some children. *Journal of Nutrition* **125**, 1875-1883.

SPURR, G.B., BARAC-NIETO, M. & MAKSUD, M.G. (1977). Productivity and maximal oxygen consumption in sugar cane cutters. *American Journal of Clinical Nutrition* **30**, 316-321.

STEPHENSON, L.S., LATHAM, M.C., & ODOURI, M.L. (1980). Costs, prevalence and approaches for control of *Ascaris* infection in Kenya. *Journal of Tropical Pediatrics* **26**, 246-263.

STEPHENSON, L.S., LATHAM, M.C. & KURZ, K.M. (1989). Treatment with a single dose of albendazole improves growth of Kenyan children with hookworm, *Trichuris trichiura* and *Ascaris lumbricoides* infections. *American Journal of Clinical Nutrition* **41**, 78-87.

STOLTZFUS, R.J., ALBONICO, M., TIELSCH, J.M., CHWAYA, H.M. & SAVIOLI, L. (1998a). School-based deworming yields small improvement in growth of Zanzibari school children after one year. *Journal of Nutrition* **128**, 2187-2193.

STOLTZFUS, R.J., ALBONICO, M., CHWAYA, H.M., TIELSCH, J.M., SCHULZE, K.J. & SAVIOLI, L. (1998b). Effects of the Zanzibar school-based deworming program on iron status of children. *American Journal of Clinical Nutrition* **68**, 179-186.

TALAAT, M. & EVANS, D.B. (2000). The costs and coverage of a strategy to control schistosomiasis morbidity in non-enrolled school-age children in Egypt. *Transactions of the Royal Society for Tropical Medicine and Hygiene* **94**, 449-54.

TORLESSE, H. & HODGES, M. (2000). Anthelmintic treatment and haemoglobin concentrations during pregnancy. *Lancet* **356**,1083.

UNICEF (2001). The State of the World's Children 2001. United Nation's Children Fund (UNICEF), New York.

WEITZMAN, M. (1987). Excessive school absences. *Advances in Developmental and Behavioral Pediatrics* **8**, 151-78.

WORLD BANK (1993). *World Development Report 1993: Investing in Health.* Oxford University Press, Oxford.
WORLD BANK (2000). *World Development Indicators.* The World Bank, Washington.
WORLD BANK (2001). *African Development Indicators 2001.* The World Bank, Washington.

Chapter 6
IMMUNE RESPONSES IN HUMANS – *ASCARIS*

Philip J Cooper
Department of Infectious Diseases, St George's Hospital Medical School, Cranmer Terrace, Tooting, London SW17 ORE, UK; and Laboratorio de Investigacion, Hospital Pedro Vicente Maldonado, Pedro Vicente Maldonado, Pichincha Province, Ecuador.
e-mail: pc102d@hotmail.com

1. INTRODUCTION

Although *Ascaris lumbricoides* infections are the most prevalent of all helminth infections of humans, the immune response to human ascariasis remains poorly understood in comparison with other helminthiases such as schistosomiasis and filariasis.

This chapter will review the current state of knowledge of the human immune response to ascariasis. The review will focus particularly on the role of *Ascaris* larvae in stimulating specific immune responses, because adult parasites in the small intestine are not thought to be a major target of host immune responses. The role of protective immunity as a determinant of the epidemiological features of ascariasis will be discussed also, particularly with respect to predisposition to infection (see Chapter 1) and variation in infection intensity with age.

1.1 Clinical pathology of larval ascariasis

Both *A.suum* and *A.lumbricoides* are pathogenic to humans, but there is evidence that human infection with *A.suum* is more likely to cause a larva migrans-like syndrome (Pawlowski, 1978; Maruyama *et al.* 1996) and may only rarely reach sexual maturity (Pawlowski, 1978).

Larval ascariasis may cause damage to the lung during the migration of larvae on their way to the intestine. The majority of the cases of Loeffler's syndrome, characterised by fever, cough, asthma, eosinophilia, and radiological infiltrates of the lungs, have been attributed to larval ascariasis (Keller, Millstrom & Gus, 1932; Loeffler, 1956). Pulmonary ascariasis generally causes a self-limiting illness that resolves within two weeks of onset (Arean & Crandall, 1971). During pulmonary ascariasis, segments of fourth stage larvae have been described in the bronchioles associated with an infiltrate rich in eosinophils (Beaver & Dhanaraj, 1956). It is not clear whether the living, migrating larvae are the stimulus for the development of inflammation or dead and dying larvae are the primary stimulus because in histological sections, *Ascaris* larvae are frequently observed free of inflammatory infiltrates (Arean & Crandall, 1971).

Symptomatic pulmonary ascariasis appears to be rare in endemic areas, and may result from a degree of host tolerance to the parasite as a consequence of uninterrupted contact with *A. lumbricoides* (Spillman, 1975). In contrast, in locations where *Ascaris* infections are seasonal as a result of the failure of eggs to survive throughout the year, symptoms of pulmonary ascariasis may be relatively common. Gelpi & Mustafa (1967) reported outbreaks of eosinophilic pneumonitis associated with *A. lumbricoides* infections occurring every year during and after the short rainy season in Saudi Arabia. Pulmonary ascariasis in Saudi Arabia generally occurs in adults indicating that the full clinical picture of eosinophilic pneumonitis may require repeated sensitisations to *Ascaris* during childhood. The clinical reaction to relatively small inocula of *Ascaris* eggs administered to human volunteers is greater among those with evidence of previous sensitisation (Vogel & Mining, 1942).

2. IMMUNE RESPONSES

2.1 Antibody responses

Human infections with *A. lumbricoides* induce the production of antibodies of all isotypes (IgM, IgG, IgA, and IgE) and IgG subclasses (IgG1-4) (Figures 6.1 and 6.2). The magnitude of the antibody response is likely to be determined by age, infection intensity, history of infection, in addition to individual host genetic differences. In endemic regions where transmission is continuous throughout the year, the pattern and magnitude of antibody production may reflect changes in the relationship between age and infection intensity (Figure 6.2).

The development of a measurable antibody response following *Ascaris* infection is not invariable. Increased IgE and IgM responses were detectable in only two of four subjects accidentally infected with *A.suum* (Phills *et al.* 1972). Experimental infections of four humans with *A.lumbricoides* resulted in the detection of microprecipitating antibodies against *Ascaris* larvae soon after infection that lasted up to 3 to 4 months following inoculation (Lejkina, 1965) and antibody levels had declined to negligible levels by the time of adult sexual maturity. Studies of *Ascaris* serology in children in Northern Europe where *Ascaris* transmission is interrupted during the winter, have shown marked rises in specific antibody levels during the spring, summer, and autumn months coincident with an increase in parasite transmission as measured by increased egg excretion rates (Lejkina, 1965).

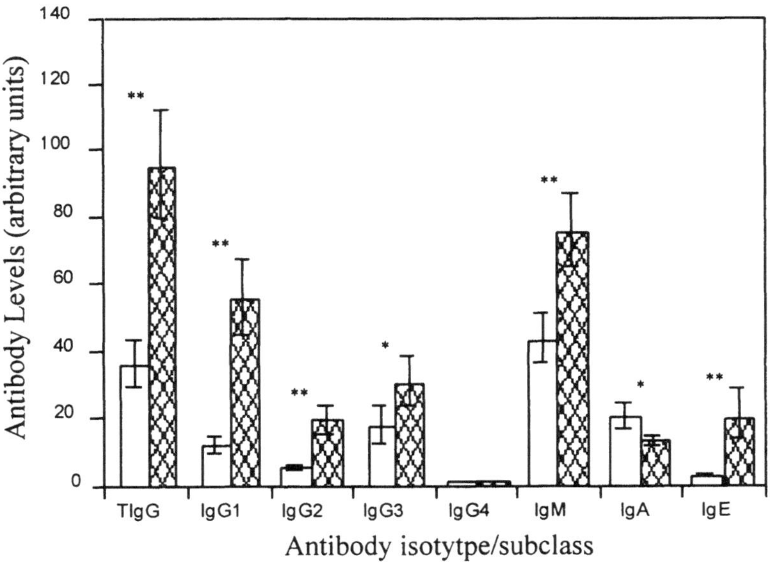

Figure 6.1. Levels of *A.lumbricoides*-specific antibodies from groups of *A.lumbricoides*-infected (hatched columns) (n=73) and uninfected (open columns) (n=40) individuals living in Manabi Province, Ecuador. Infected subjects were from endemic rural communities where infection intensities are moderate (geometric mean 6,728 epg (range 1,278-61,200)) and uninfected subjects were from a nearby town. Uninfected subjects had immunological evidence of exposure to *A.lumbricoides* as indicated by the presence of specific antibodies and a measurable cellular response to adult and larval-stage antigens. Shown are geometric mean antibody levels (arbitrary units) and 95% confidence intervals . Adapted from Cooper *et al.* (2000). *- p<0.05, **P<0.001.

Investigations conducted in an area of high transmission and pronounced age-convexity in India suggest that age-related changes in antibody responses to *Ascaris* larval antigens mirror age-intensity patterns rather than being protective (Haswell-Elkins *et al.* 1989; 1992). Semi-quantitative analysis of antibody banding profiles to L3/L4 larval antigen preparations using Western blot indicated that age banding profiles broadly reflected infection intensity: children aged between five and nine years showed both the strongest banding patterns and heaviest infection intensities (Haswell-Elkins *et al.* 1989). Immunoprecipitation of *Ascaris* L3/L4 antigens by sera identified 12 major antigens (Haswell-Elkins *et al.* 1992). Although sera from most subjects were reactive with *Ascaris* larval antigens, there was considerable variation in individual recognition profiles. However, banding intensity scores using sera from children collected four months after treatment corresponded broadly with infection intensity suggesting that antibody responses to larval excretory/secretory antigens develop during exposure to migrating larvae and occur at a level proportional to the number of larvae that develop into adult worms.

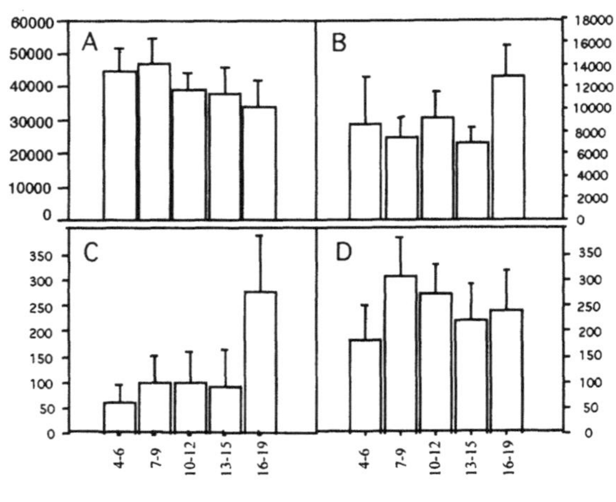

Figure 6.2. Changes in levels of antibodies with age in children and young adults aged 4 to 19 years (n=92) in an endemic community in Esmeraldas Province in Ecuador. Shown are mean (+sem) *A.lumbricoides* egg counts (per gram of stool) (A), and levels of total IgE in IU/mL (B), and *A.lumbricoides*-specific levels of IgG4 (C), and IgG 1(D). Levels of specific IgG4 and IgG are in arbitrary units. Data are from Cooper *et al.* (unpublished).

Few studies have examined levels of the different antibody isotypes in endemic populations (Palmer *et al.* 1995; Cooper *et al.* 2000). These studies indicate also that antibody levels are reflective of parasite burdens. Stool egg counts in a group of infected young adults were strongly correlated with levels of *Ascaris*-specific IgG, IgG1 and IgG2 antibodies (Cooper *et al.* 2000). Further, specific levels of IgG1 parallel age-dependent changes in infection intensity in children and young adults (Figure 6.2).

2.2 IgE and immediate hypersensitivity

The most consistent findings in populations endemic for this geohelminth are high circulating levels of total IgE and *Ascaris*-specific IgE (Johansson, Melbin & Vahlquist, 1968; Jarrett & Miller, 1982; Hagel *et al.* 1993a; Lynch *et al.* 1993a) (Figure 6.2). The high levels of IgE that are observed in infections with *A.lumbricoides* are probably stimulated by the tissue migratory stage of the life cycle (Radermucker *et al.* 1974) because levels of specific IgE associated with infections with non-invasive helminths (e.g. *Trichuris trichiura*) are either not detectable or are low level (Orren & Dowdle, 1975).

Children living in endemic areas often have total IgE levels in excess of 10,000 IU/mL (Figure 6.2), and such high levels are often attributed to infections with ascariasis. The production of polyclonal IgE may be induced by a combination of direct mitogenic effects on B cells of *Ascaris* allergens (Lee & Xia, 1995) and the non-specific induction of IgE secretion in an immune environment associated with the production of large amounts of IL-4 (King *et al.* 1993; Cooper *et al.* 2000).

Ascaris adult and larvae contain large quantities of potent allergens (O'Donnell & Mitchell, 1978; Coles, 1985; Fraser *et al.* 1993). Larvae also secrete allergenic substances (Kennedy & Qureshi, 1986) and probably are the primary stimulus of IgE production in infected individuals. In common with other tissue invasive helminth infections, immediate hypersensitivity (IH) reactions appear to be relatively rare in endemic populations. Typical allergic reactions such as allergic rashes, angioneurotic oedema, bronchospasm, or even acute anaphylaxis (Odunjo *et al.* 1970), have been reported rarely and such allergic phenomena are more typical of acute infections resulting from experimental/malicious (Koino, 1922; Vogel & Mining, 1942; Lejkina, 1965; Phills *et al.* 1972), discontinuous (Alkan *et al.*

1952; Loeffler, 1956), or massive exposures (Beaver & Dhanaraj, 1958; Barlow *et al.* 1961). Allergic reactions to *Ascaris* proteins occur frequently in laboratory workers exposed to *Ascaris* parasites – these reactions include typical IH phenomena (Andrews, 1962; Coles, 1985), and become less severe with prolonged contact (Coles, 1985).

The production of large amounts of polyclonal IgE in ascariasis may modulate IH reactions by inhibition of the activity of mast cells by saturation of Fc϶R1 receptors (Bazaral *et al.* 1973; Godfrey & Gradidge, 1976; Lynch *et al.* 1987; Hagel *et al.* 1993b). Further, immediate hypersensitivity reactions may be modulated by parasite-specific IgG4 'blocking' antibodies (Butterworth *et al.* 1989; Hussain *et al.* 1992) – the levels of specific IgG4 increase with age in a high transmission area of Ecuador (Figure 6.2).

2.3 Cellular responses

The cellular immune response to *Ascaris* antigens is characterised by a highly polarised Th2 cytokine response (Cooper *et al.* 2000). Peripheral blood mononuclear cells (PBMCs) from young adults with moderate infection intensities from an endemic region of Ecuador proliferated to both adult (Figure 6.3) and larval stage antigens and secreted significant amounts of interleukin (IL)-5. High frequencies of IL-4 and IL-5- secreting PBMCs were observed after stimulation with *Ascaris* antigens and the ratios of Th2 (IL-4 and IL-5) to Th1 (IFN-γ) expressing PBMCs was significantly greater in the same group compared to uninfected controls (Cooper *et al.* 2000). The polarized Th2 response would explain the high circulating levels of IgE (Palmer *et al.* 1995) (IL-4-mediated) and peripheral eosinophilia (Gelpi & Mustafa, 1967) (IL-5-mediated) that are characteristic features of *A.lumbricoides* infection.

The relative importance of Th2 cytokines in protective immunity against *A.lumbricoides* infection is not clear. A number of animal models have demonstrated the importance of Th2-mediated mechanisms in parasite expulsion (Finkelman *et al.* 1997). Protective immune responses may operate at several points during the parasite life cycle in *A.lumbricoides* infection. Parasite killing in the intestinal lamina propria, in the liver, or in the lungs, may occur via mast cell-mediated mechanisms with the recruitment of cytotoxic effector cells such as activated eosinophils, neutrophils and macrophages. The migrating larvae may become mechanically trapped in the tissues or be immobilised by a combination of antibody and cellular mechanisms with larval destruction occurring via antibody-dependent cellular cytotoxicity (Butterworth, 1984) and other cellular pathways such as the release of nitrous oxide and reactive oxide products by activated macrophages (Taylor *et al.* 1996; Wynn & Hoffmann, 2000).

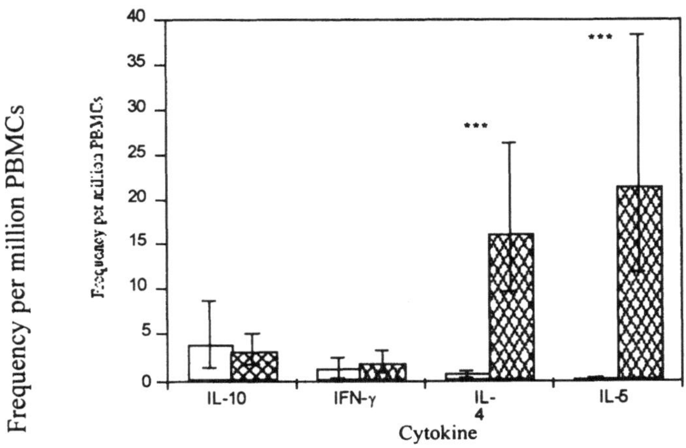

Figure 6.3. Frequencies of peripheral blood mononuclear cells (PBMC) secreting interleukin (IL)-10, interferon (IFN)-γ, IL-4, or IL-5 in *A.lumbricoides*-infected (hatched columns) (n=73) and uninfected (open columns) (n=40) subjects. Cultures were performed in the presence or absence of *A.lumbricoides* adult antigen, and cellular frequencies were assessed by ELISPOT. Findings are expressed as the frequency of cytokine-positive cells per million PBMC. Geometric means and 95% confidence intervals are shown. ***P<0.001. Adapted from Cooper *et al.* (2000).

There is a large published literature of the mechanisms of destruction of helminth larvae, particularly schistosome and filarial parasites (Butterworth, 1984). The mechanisms acting against migrating schistosomula (Butterworth, 1984) and *O. volvulus* microfilariae (Johnson *et al.* 1991; Taylor *et al.* 1996) may be similar to those occurring during larval ascariasis. Both Th1 (e.g. IFN-γ) and Th2 (e.g. IL-5) cytokines are thought to be important in the development of protective immunity during infections with both schistosome (Brunet *et al.* 1998; Correa-Oliveira *et al.* 2000) and filarial parasites (Elson *et al.* 1995; Soboslay *et al.* 1999; Doetze *et al.* 2000), particularly early in the course of infection (Cooper *et al.* 2001a) and in 'putatively immune' individuals who do not develop clinical or parasitologic evidence of infection despite apparent exposure (Correa-Oliveira *et al.* 2000; Elson *et al.* 1995; Soboslay *et al.* 1999; Doetze *et al.* 2000). The role of Th1/Th2 cytokines in protective immunity against ascariasis has not been investigated specifically.

There is some indirect evidence that potent Th1 responses may be important in protective immunity against geohelminth parasites as suggested by the apparently protective effects of a previous BCG scar against hookworm infection (Barreto *et al.* 2000). A combination of Th1 and Th2 responses may be important in controlling *Ascaris* infections during different stages of the parasite life cycle: a mixed Th1/Th2 response may be important in the killing of migrating larvae,

while Th2 responses such as increased mucus production and mast cell-dependent mechanisms may be important in the expulsion of juvenile adults (Finkelman *et al.* 1997).

As discussed already, clinical and epidemiological observations suggest that continuous exposure to *Ascaris* results in some degree of tolerance of the pathological effects of larval invasion (Beaver & Dhanraj, 1958; Lejkina, 1965; Spillman, 1975). The apparent lack of symptomatology associated with larval invasion may be a consequence of more effective protective immune responses preventing larval penetration of the intestinal mucosa or follow downregulation of the immune response to invading larvae. Observations from other tissue invasive helminth infections would support the development of cytokine-driven immune downregulatory mechanisms following persistent exposure and chronicity of infection that may serve to suppress inflammation and prevent host pathology. Such mechanisms may act in addition to those postulated to modulate IH reactions described above, and may be mediated by immunosuppressive cytokines such as IL-10 and TGF-β1 (Soboslay *et al.* 1999; Doetze *et al.* 2000; Montenegro *et al.* 2000; Cooper *et al.* 2001a). However, there is little evidence of increased IL-10 expression during human ascariasis, and addition of an IL-10 neutralising antibody to PBMC cultures stimulated with *Ascaris* antigens did not result in enhanced secretion of either Th1 or Th2 cytokines (Cooper *et al.* 2000). Increased mRNA expression of TGF-β1 was observed by RT-PCR (Cooper *et al.* unpublished data), but the significance of this remains to be determined.

2.4 Immune responses after anthelmintic treatment

Treatment with anthelmintic drugs appears to have relatively little impact on *Ascaris*-specific antibody and cellular responses. A small study of *A.lumbricoides*-infected young adults who were randomised to receive either albendazole or placebo revealed no significant declines in levels of specific antibodies over a six week posttreatment observation period (Cooper *et al.* unpublished data). In contrast, a 22-month follow-up study in Venezuela demonstrated an increase in levels of both total and specific IgE despite monthly treatments with pyrantel that resulted in high and sustained cure rates (Lynch *et al.* 1993b; Hagel *et al.* 1993a). The production of Th1 (IL-2, IFN-γ) and Th2 (IL-5) cytokines by PBMCs stimulated with *Ascaris* adult and larval stage antigens did not alter significantly after anthelmintic treatment (Figure 6.4).

The results of such studies indicate that antibody and cellular responses are probably determined by the intensity of transmission and invasion with larval stages of *Ascaris*. Chemotherapy, at least in the short term, has little impact on exposure to viable eggs in the environment, and uninterrupted immune stimulation would explain the lack of impact on parameters of both the humoral and cellular immune responses directed towards *Ascaris* antigens.

3. EVIDENCE FOR PROTECTIVE IMMUNITY IN HUMANS

Protective immunity to *A.lumbricoides* is likely to follow exposure to larval stages of the parasite (Lekjina, 1965), develop slowly in a manner relating to accumulated experience of infection, and manifests as a reduction in the number of adult parasites. As immune responses are directed primarily against invasive larvae, protective immunity is probably most active against larval stages of the parasite. Newly infecting larvae may face progressively stronger immune attack until the rate of establishment of new parasites is balanced or exceeded by the rate of attrition of adults (Smithers & Terry, 1976).

An important epidemiological feature of ascariasis is apparent predisposition to infection (see Chapter 1), and several studies have investigated immune correlates of predisposition to *A.lumbricoides* infection (Hagel *et al.* 1993; Palmer *et al.* 1995; McSharry *et al.* 1995).

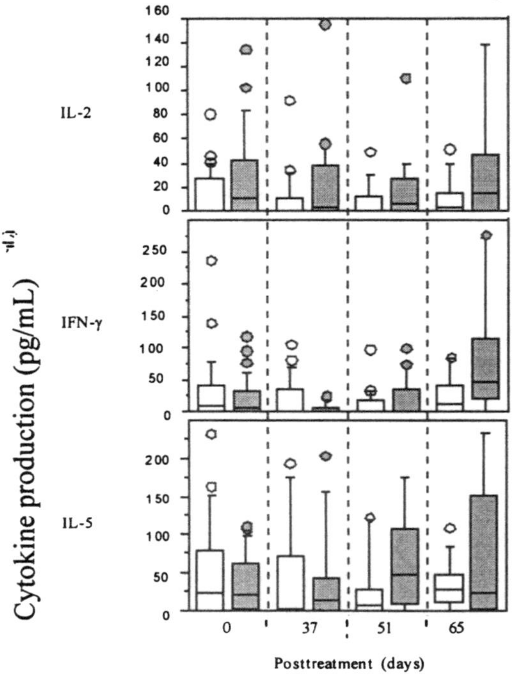

Figure 6.4. Production of cytokines, IL-2, IFN-γ, and IL-5 by PBMCs stimulated with L2/L3 *Ascaris* antigen from 28 *A.lumbricoides*-infected subjects. Shown are box plots of net cytokine production (pg/mL) before treatmen_t, and at various time after treatment with albendazole (shaded columns) (n=15) or placebo (n=13) (clear columns). There were no significant inter- or intragroup differences over the study period for any of the 3 cytokines. Box plots represent median values (central line), interquartile range (box margins), 95% confidence intervals (bars), and outlying values (circles). Adapted from Cooper *et al.* (2001b).

Children predisposed to reinfection following anthelmintic treatment were shown to have higher pre-treatment levels of total IgE and lower levels of *Ascaris*-specific IgE compared to those who did not become reinfected (Hagel *et al.* 1993a). Previous studies of the same cohort had demonstrated evidence of mast cell saturation in this group (Lynch *et al.* 1987; Hagel *et al.* 1993b), and the low ratios of specific to total IgE in the group of reinfected children were suggested as evidence in support of the role of mast cell saturation in preventing mast cell-driven killing responses against the parasite.

A study of children in an endemic community in Nigeria demonstrated that children with light parasite burdens upon reinfection after anthelmintic treatment had higher levels of IgE specific to the major allergen of Ascaris, ABA-1 (McSharry *et al.* 1999). Further, the same children had high serum levels of the inflammatory markers ferritin, C-reactive protein, and eosinophil cationic protein indicating ongoing acute inflammatory processes, while children predisposed to heavy reinfections had little evidence of inflammatory activity (McSharry *et al.* 1999). However, a study of children predisposed to either heavy or light infections in urban Bangladesh where transmission of *A.lumbricoides* is very intense, were unable to demonstrate a protective role for IgE or any other antibody subclass (Palmer *et al.* 1995), and actually showed that levels of IgG1, IgG4, and IgE were higher in the heavy infection group.

Human helminthiases, including ascariasis, are associated with potent Th2 activation (Cooper *et al.* 2000), and there are numerous effector pathways by which Th2 cytokines may act (Finkelman *et al.* 1997). High levels of specific IgE may merely serve as a marker of generalized Th2 activation rather than be the actual mechanism by which protective immunity occurs.

4. CONCLUSION

The human immune response against *Ascaris* parasites is characterised by prominent antibody and cellular responses that are directed primarily against the larval stages of infection. *Ascaris* infection stimulates the secretion of all antibody isotypes, although high circulating levels of total IgE and parasite-specific IgE are the most characteristic features. In endemic regions, the cellular response to *A.lumbricoides* infection is characterised by a polarized Th2 response with the prominent production of both IL-4 and IL-5. Evidence for the development of a protective immune response in human ascariasis is derived from observations of a decline in infection intensity with age in many endemic regions and by evidence for predisposition to either light or heavy infections. Protective immune responses may involve IgE-mediated mechanisms directed against invading larvae, although elevated IgE responses may merely be a marker of enhanced Th2 activation. There are differences in the clinical response to ascariasis between those suffering acute (one-off or seasonal exposure) or chronic (continuous transmission throughout the year) and these clinical and epidemiological observations may reflect profound differences in the immune response to the parasite, as has been demonstrated for other tissue-invasive helminthiases.

REFERENCES

ALKAN, W.J., FREUDENTHAL, M.L. & STEINITZ, E. (1952). An epidemic of pulmonary disease with blood eosinophilia (Eosinophilia disease). *Transactions of the Royal Society of Tropical Medicine and Hygiene* **46**, 666.

ANDREWS J.M. (1962). Parasitism and allergy. *Journal of Parasitology* **48**, 3.

AREAN, V.M. & CRANDALL, C.A. (1971). Ascariasis. In *Pathology of Protozoal and Helminthic Diseases*. (ed. Marcial-Rojas, R.A.), pp. 769-807. Williams & Wilkins, New York.

BARLOW, J.B., POCOCK, W.A. & TABATZNICK, B.A. (1961). An epidemic of 'acute eosinophilic pneumonia' following 'beer drinking' and probably due to infestation with *Ascaris lumbricoides*. *South African Medical Journal* **35**, 390.

BARRETO, M.L., RODRIGUES, L.C., SILVA, R.C., ASSIS, A.M., REIS, M.G., SANTOS, L.A. & BLANTON, R.E. (2000). Lower hookworm incidence, prevalence, and intensity of infection in children with a Bacillus Calmette-Guerin vaccination scar. *Journal of Infectious Diseases* **182**, 1800-1803.

BAZARAL, M., ORGEL, H.A. & HAMBURDER, R.N. (1973). The influence of serum IgE levels of selected recipients, including patients with allergy, helminthiasis and tuberculosis, on the apparent P-K titre of a reaginic serum. *Clinical and Experimental Immunology* **14**, 117-125.

BEAVER, P.C. & DANARAJ, T.J. (1958). Pulmonary ascariasis resembling eosinophilic lung. Autopsy report with description of larvae in bronchioles. *American Journal of Tropical Medicine and Hygiene* **7**, 100-111.

BRUNET, L.R., DUNNE, D.W. & PEARCE, E.J. (1998). Cytokine interaction and immune responses during *Schistosoma mansoni* infection. *Parasitology Today* **14**, 422-427.

BUTTERWORTH, A.E. (1984). Cell-mediated damage to helminths. *Advances in Parasitology* **23**, 143-235.

BUTTERWORTH, A.E., CORBETT, E.L., DUNNE, D.W., FULFORD, A.J.C., KIMANI, G., GACHUHI, R.K., KLUMP, R., MBUGUA, G., OUMA, J.H., ARAP SIONGOK, T.K. & STURROCK, R.F. (1989). Immunity and morbidity in human schistosomiasis. In *Frontiers of Infectious Diseases: New Strategies in Parasitology*. (ed. McAdam, K.P.W.), pp. 193-210. Churchill Livingstone, Edinburgh.

COLES GC. Allergy and immunopathology of ascariasis. (1985). In *Ascariasis and its public health significance*. (eds. Crompton, D.W.T., Nesheim, M.C., Pawlowski, Z.S.), pp. 167-184. Taylor & Francis, London.

COOPER, P.J., CHICO, M.E., SANDOVAL, C., ESPINEL, I., GUEVARA, A., KENNEDY, M.W., URBAN, J.F. Jr., GRIFFIN, G.E. & NUTMAN, T.B. (2000). Human infection with *Ascaris lumbricoides* is associated with a polarized cytokine response. *Journal of Infectious Diseases* **182**, 1207-1213.

COOPER, P.J., MANCERO, T., ESPINEL, M., SANDOVAL, C., LOVATO, R., GUDERIAN, R.H. & NUTMAN, T.B. (2001a). Early human infection with *Onchocerca volvulus* is associated with an enhanced parasite-specific cellular immune response. *Journal of Infectious Diseases* **183**, In press.

COOPER, P.J., CHICO, M., SANDOVAL, C., ESPINEL, I., GUEVARA, A., LEVINE, M.M., GRIFFIN, G.E. & NUTMAN, T.B. (2001b). Human infection with *Ascaris lumbricoides* is associated with suppression of the interleukin-2 response to recombinant cholera toxin B subunit following vaccination with the live oral cholera vaccine CVD 103-HgR. *Infection and Immunity* **69**, 1574-1580.

CORREA-OLIVEIRA, R., CALDAS, I.R. & GAZZINELLI, G. (2000). Natural versus drug-induced resistance in *Schistosoma mansoni* infection. *Parasitology Today* **16**, 397-399.

DOETZE, A., SATOGUINA, J., BURCHARD, G., RAU, T., FLEISCHER, B. & HOERAUF, A. (2000). Antigen-specific cellular hyporesponsiveness in a chronic human helminth infection is mediated by Th3/Tr1-type cytokines IL-10 and TGF-β but not by a Th1 to Th2 shift. *International Immunology* **12**, 623-30.

ELSON, L.H., CALVOPINA, H.M., PAREDES, W., ARAUJO, E., BRADLEY, J.E., GUDERIAN, R.H. & NUTMAN, T.B. (1995). Immunity to onchocerciasis: putative immune persons produce a Th1-like response to *Onchocerca volvulus*. *Journal of Infectious Diseases* **171**, 652-8.

FINKELMAN, F.D., SHEA-DONOHUE, T. & GOLDHILL, J., SULLIVAN, C.A., MORRIS, S.C., MADDEN, K.B., GAUSE, W.C. AND URBAN Jr, J.F. (1997). Cytokine regulation of host defence against parasitic gastrointestinal nematodes: lessons from studies with rodent models. *Annual Reviews of Immunology* **15**, 505-533.

FRASER, E.M., CHRISTIE, J.F. & KENNEDY, M.W. (1993). Heterogeneity amongst infected children in IgE antibody repertoire to the antigens of parasitic nematode *Ascaris*. *International Archives of Allergy and Immunology* **100**, 283-286.

GELPI, A.P. & MUSTAFA, A. (1967). Seasonal pneumonitis with eosinophilia. *American Journal of Tropical Medicine and Hygiene* **16**, 646-657.

GODFREY, R.D. & GRADIDGE, C.F. (1976). Allergic sensitisation of human lung fragments prevented by saturation of IgE binding sites. *Nature* **259**, 484-486.

HAGEL, I., LYNCH, N.R., DI PRISCO, M.C., ROJAS, E., PEREZ, M. & ALVARAREZ, N. (1993a). *Ascaris* reinfection of slum children: relation with the IgE response. *Clinical and Experimental Immunology* **94**, 80-83.

HAGEL, I., LYNCH, N.R., PEREZ, M., DI PRISCO, M.C., LOPEZ, R. & ROJAS, E. (1993b). Modulation of the allergic reactivity of slum children by helminthic infection. *Parasite Immunology* **15**, 311-315.

HASWELL-ELKINS, M.R., KENNEDY, M.W., MAIZELS, R.M., ELKINS, D.B. & ANDERSON, R.M. (1989). The antibody recognition profiles of humans naturally infected with *Ascaris lumbricoides*. *Parasite Immunology* **11**, 615-627.

HASWELL-ELKINS, M.R., LEONARD, H., KENNEDY, M.W., ELKINS, D.B. & MAIZELS, R.M. (1992). Immunoepidemiology of *Ascaris lumbricoides*: relationships between antibody specificities, exposure and infection in a human community. *Parasitology* **104**, 153-159.

HUSSAIN, R., POINDEXTER, R.W. & OTTESEN, E.A. (1992). Control of allergic reactivity in human filariasis: predominant localization of blocking antibodies to the IgG4 subclass. *Journal of Immunology* **148**, 2731-2737.

JARRETT, E.E. & MILLER, H.P. (1982). Production and activities of IgE in helminth infection. *Progress in Allergy* **32**, 178-233.

JOHANSSON, S.G., MELBIN, T. & VAHLQUIST, B. (1968). Immunoglobulin levels in Ethiopian preschool children with special reference to high concentrations of immunoglobulin E (IgND). *Lancet* **i**, 1118-1121.

JOHNSON, E.H., LUSTIGMAN, S., BROTMAN, B., BROWNE, J. & PRINCE, A.M. (1991). *Onchocerca volvulus*: in vitro killing of microfilariae by neutrophils and eosinophils from experimentally infected chimpanzees. *Tropical Medicine and Parasitology* **42**, 351-355.

KELLER, A.E., MILLSTROM, H.T. & GAS, R.S. (1932). The lungs of children with ascariasis. *Journal of the American Medical Association* **99**, 1249.

KENNEDY, M.W. & QURESHI, F. (1986). Stage-specific secreted antigens of the parasitic larval stages of the nematode *Ascaris*. *Immunology* **58**, 515-522.

KING, C.L., MAHANTY, S., KUMARASWAMI, V., ABRAMS, J.S., REGUNATHAN, J., JAYARAMAN, K., OTTESEN, E.A. & NUTMAN, T.B. (1993). Cytokine control of parasite-specific anergy in human lymphatic filariasis. Preferential induction of a regulatory T helper type 2 lymphocyte subset. *Journal of Clinical Investigation* **92**, 1667-1673

KOINO, S. Infection experiments with ascariasis in human body with special reference to the symptoms of *Ascaris* pneumonia. A preliminary report. (1922). *Japanese Medical World* **3**, 30.

LEE, T.D.G. & XIA, C.Y. (1995). IgE regulation by nematodes: the body fluid of *Ascaris* contains a B-cell mitogen. *Journal of Allergy and Clinical Immunology* **95**, 1246-1254.

LEJKINA, E.S. (1965). Research on ascariasis immunity and immunodiagnosis. *Bulletin of the World Health Organisation* **32**, 699-708.

LOEFFLER, P. (1956). Transient lung infiltrations with blood eosinophilia. *Archives of International Allergy* **8**, 54-59.

LYNCH, N.R., LOPEZ, R.I., DI PRISCO, M.C., HAGEL, I., MEDOUZE, L. & VIANA, G., ORTEGA, C. & PRATO, G. (1987). Allergic reactivity and socio-economic level in a tropical environment. *Clinical Allergy* **17**, 199-207.

LYNCH, N.R., HAGEL, I., VARGAS, M., PEREZ, M., LOPEZ, R.I., GARCIA, N.M., DI PRISCO, M.C. & ARTHUR, I.H. (1993a). Effect of age and helminthic infection on IgE levels in slum children. *Journal of Investigative Allergy and Clinical Immunology* **3**, 96-99.

LYNCH, N.R., HAGEL, I., PEREZ, M., DI PRISCO, M.C., LOPEZ, R. & ALVAREZ, N. (1993B). EFFECT OF ANTHELMINTIC TREATMENT ON THE ALLERGIC REACTIVITY OF CHILDREN IN A TROPICAL SLUM. *Journal of Allergy and Clinical Immunology* **92**, 404-411.

MARUYAMA, H., MAWA, Y., NODA, S., MIMORI, T. & CHOI, W.Y. (1996). An outbreak of visceral larva migrans due to *Ascaris suum* in Kyushu, Japan. *Lancet* **347**, 1766-1777.

McSHARRY, C., XIA, Y., HOLLAND, C.V. & KENNEDY, M.W. (1999). Natural immunity to *Ascaris lumbricoides* associated with immunoglobulin E antibody to ABA-1 allergen and inflammation indicators in children. *Infection and Immunity* **67**, 484-489.

MONTENEGRO, S.M., MIRAND, P., MAHANTY, S., ABATH, F.G., TEIXEIRA, K.M., COUTINHO, E.M., BRINKMAN, J., GONCALVES, I., DOMINGUES, L.A., DOMINGUES, A.L., SHER, A. & WYNN, T.A. (2000). Cytokine production in acute versus chronic human *Schistosomiasis mansoni*: the cross-regulatory role of interferon gamma and interleukin-10 in the response of peripheral blood mononuclear cells and splenocytes to parasite antigens. *Journal of Infectious Diseases* **179**, 1502-1514.

O'DONNELL, I.J. & MITCHELL, G.F. (1978). An investigation of the allergens of *Ascaris lumbricoides* using a radioallergosorbent test (RAST) and sera of naturally infected humans: comparison with an allergen for mice identified by a passive cutaneous anaphylaxis test. *Australian Journal of Biological Science* **31**, 459-487.

ODUNJO, E.O. (1970). Helminthic anaphylactic syndrome (HAS) in children. *Pathology and Microbiology* **35**, 220.

ORREN, A. & DOWDLE, E.B. (1975). Effects of allergy, intestinal helminth infestation and sex on serum IgE concentrations and immediate hypersensitivity in three ethnic groups. *International Archives of Allergy and Applied Immunology* **49**, 814-830.

PALMER, D.R., HALL, A., HAQUE, R. & ANWAR, K.S. (1995). Antibody isotype responses to antigens of *Ascaris lumbricoides* in a case-control study of persistently heavily infected Bangladeshi children. *Parasitology* **111**, 385-393.

PAWLOWSKI, Z.S. (1978). Ascariasis. *Clinics in Gastroenterology* **7**, 157-178.

PHILLS, J.A., HARROLD, A.J., WHITEMAN, G.V. & PERLMUTTER, L. (1972). Pulmonary infiltrates, asthma, and eosinophilia due to *Ascaris suum* infestation in man. *New England Journal of Medicine* **286**, 965-970.

RADERMUCKER, M., BEKHTI, A., PONCELET, E. & SALMON, J. (1974). Serum IgE levels in protozoal and helminthic infections. *International Archives of Allergy* **47**, 285-295.

SMITHERS, S.R. & TERRY, R.J. (1976). The immunology of schistosomiasis. *Advances in Parasitology* **14**, 399-422.

SOBOSLAY, P.T., LUDER, C.G.K., RIESCH, S., GEIGER, S.M., BANLA, M., BATCHASSI, E. & STADLER, A. (1999). Regulatory effects of Th1-type (IFN-γ, IL-12) and Th2-type (IL-10, IL-13) cytokines on parasite-specific cellular responsiveness in *Onchocerca volvulus*-infected humans and exposed endemic controls. *Immunology* **97**, 219-225.

SPILLMAN, R.K. (1975). Pulmonary ascariasis in tropical communities. *American Journal of Tropical Medicine and Hygiene* **24**, 791-800.

TAYLOR, M.J., CROSS, H.F., MOHAMMED, A.A., TREES, A.J. & BIANCO, A.E. (1996). Susceptibility of *Brugia malayi* and *Onchocerca lienalis* microfilariae to nitric oxide and hydrogen peroxide in cell-free culture and from IFN-γ-activated macrophages. *Parasitology* **112**, 315-322.

VOGEL, H. & MINING, W. (1942). Contributions to clinical knowledge of lung ascariasis and to the question of transient eosinophilic lung infiltrates. *Beitrage zur Klinik der Tuberkulose* **98**, 620-654.

WYNN, T.A. & HOFFMANN, K.F. (2000). Defining a schistosomiasis vaccination strategy – is it really Th1 versus Th2? *Parasitology Today* **16**, 497-501.

Chapter 7
IMMUNITY AND IMMUNE RESPONSES TO *ASCARIS SUUM* IN PIGS

Gregers Jungersen
Danish Veterinary Laboratory, Bülowsvej 27, DK-1790 Copenhagen V, Denmark
e-mail: gju@svs.dk

1. INTRODUCTION

Traditionally, *Ascaris* infections in pigs are attributed to *A. suum* while human infections to *A. lumbricoides*. The parasites are ubiquitous, with more than 1 billion people infected, predominantly in the third-world (Chan *et al.* 1994; Crompton, 1989) and most likely with an even more widespread distribution in the pig (Kennedy, 1988; Roepstorff & Nansen 1994). However, eggs from human *Ascaris* infections have been shown to infect pigs (Galvin, 1968) and *vice versa* (Takata, 1951; Maruyama *et al.* 1996) and there is no morphological distinction between the two species. Whether or not *A. suum* and *A. lumbricoides* are separate species, or strains belonging to one species, is therefore still an unresolved question, but unequivocally *A. suum* and *A. lumbricoides* have a strong affinity for their respective hosts (Anderson *et al.* 1993; Anderson, 1995).

Human ascariasis poses a significant health problem, and under certain conditions is accompanied by reduced growth (see Chapter 3) and sometimes requires surgical intervention in children (Crompton, 2001). In contrast, the health effects of *A. suum* on naturally infected pigs are usually subclinical, and economic losses directly attributable to reduced growth parameters following *A. suum* infection in the modern pig industry have been difficult to establish. However, the often profound lesions in the liver of pigs experiencing larval migration result in a high level of condemnation of livers at slaughter. Thus, in spite of the gross appearance of expelled worms on the pen floor, *A. suum* infection in pigs constitutes more of an economical than an animal health issue. As a result of this, much of the research into porcine immunity against *Ascaris* has focused on the possibility of stimulating a pre-hepatic immunity rather than protecting animals from acquiring intestinal infections. However, studies on immunity against *A. suum* in pigs can contribute substantially to a better understanding of human ascariasis as many features of migration and development of protective immunity appear to be generally similar.

2. *ASCARIS SUUM* LIFE-CYCLE

The migration of *A. suum* larvae in the natural host has been described by Douvres *et al.* (1984), Murrell *et al.* (1997), Roepstorff *et al.* (1997) and others. A schematic representation of the life cycle is shown in Figure 7.1. Briefly, the eggs with the infective larvae are picked up from the environment and ingested. The eggs hatch in the intestine and the larvae penetrate the wall of caecum and upper large intestine, then, most likely via the venous blood stream, they move to the liver by 6-12 hours after ingestion. By convention, the *Ascaris* hatchling is designated the second-stage (L2) (Douvres *et al.* 1969), but, recent data have confirmed older reports that the hatchling is an early third-stage (L3) larvae covered by the L2 cuticle (Maung, 1978; Geenen *et al.* 1999). In the liver the larvae shed the cuticle and advance to the lung where they develop further, penetrate the alveolar space and migrate up the trachea to be swallowed once again. This hepato-tracheal migration is accomplished by day 10 after egg uptake, although there are indications that the time course can vary with pig strain (Urban, *personal communication*) Once back in the small intestine the larvae moult to L4, where after the majority or all of the parasites are expelled from the gut (self cure) from 14-21 days after infection. A few larvae may remain in the small intestine for the rest of their lives, developing to the L5 stage and sexual maturity, which is reached when the worms are 6-10 weeks old. The worms do not attach to the mucosal surface, but move freely in the ingesta and feed on the intestinal contents. Estimates of daily *Ascaris* female egg production generally are in the range of 200,000 eggs (Brown & Cort, 1927; Sinniah, 1982) although up to two million eggs per day has been estimated (Olsen *et al.* 1958). Inside the egg a larva develop to the infective stage over a period of at least three to five weeks, depending on environmental conditions (Seamster,1950).

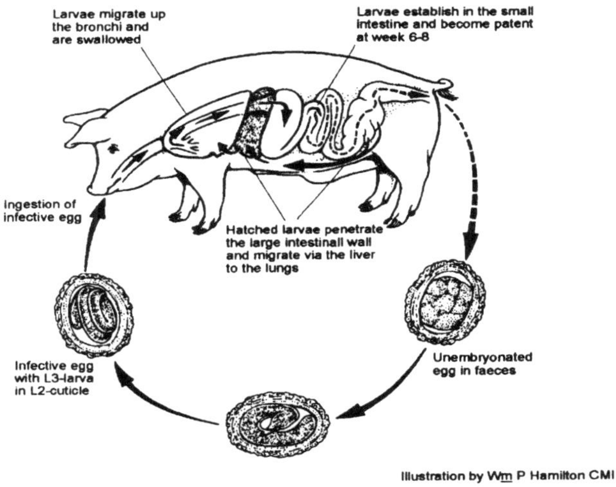

Figure 7.1. The *Ascaris suum* life-cycle is distinctly divided into a short systemic hepato-tracheal migratory phase and a resident intestinal luminal phase.

3. IMMUNOLOGIC AND IMMUNO-PATHOLOGIC RESPONSE TO *A. SUUM* INFECTIONS

Although the immunologic response to *A. suum* infections has been studied since the beginning of the century, many investigators used unnatural hosts such as rabbits, mice and guinea pigs with incomplete migration and development of larvae. The early studies on the immune responses to *A. suum* infection in pigs and mice have been reviewed by Eriksen (1981).

3.1 Changes in blood parameters

Pigs exposed to *A. suum* develop a sustained serum antibody response to larval excretory-secretory antigens of both IgG and IgA isotype (Lind *et al.* 1993). As IgE in the pig has not been conclusively identified (Roe *et al.* 1993), and as reagents for the detection of IgE responses in pigs therefore have not been developed, there is no direct information for this kind of response. However, biological activities normally related to IgE production such as parasite antigen-specific passive cutaneous anaphylaxis (Roe *et al.* 1993; Urban, Jr. *et al.* 1988) and degranulation of intestinal mucosal mast cells (Ashraf *et al.* 1988) are evident in pigs infected with *A. suum.*

The larval migration also induces a dose-dependent blood eosinophilia, in primary and secondary infections peaking at day 14 and 10 post-infection (p.i.) respectively. Eosinophil levels return to normal levels around 20-30 days p.i. irrespective the of development or presence of adult worms (Ronéus, 1971; Eriksen *et al.* 1980; Rhodes *et al.* 1982; Jungersen *et al.* 1999a). The eosinophilic response is mounted in both newborn piglets and young growing pigs, while maternal antibodies abrogate serum responses to inoculations in three day old piglets (Eriksen *et al.* 1980). Using monoclonal antibodies and flow cytometry, Lunney *et al.* (1986) have demonstrated a transient increase in peripheral blood macrophages and MHC-II expression per cell in pigs naturally exposed to *A. suum* eggs, while experimentally trickle infected pigs only showed very moderate increases in macrophage numbers. In none of the groups could changes in the number of circulating T helper ($CD4^+$) or cytotoxic T ($CD8^+$) lymphocytes be detected. A transient and short-lived lymphocyte blastogenesis response of peripheral blood lymphocytes, occurring earlier than the antibody response, has also been demonstrated (Urban & Tromba, 1982; Rhodes *et al.* 1982; Barta *et al.* 1986). Antibody secreting cells specific to larval antigens appeared in peripheral blood one week following inoculation with infective eggs as measured by the ELISPOT technique. A 10-fold increase in circulating *A. suum* specific antibody secreting cells was observed with a memory response compared to that of primary infections (Jungersen *et al.* 1999a).

3.2 Lesions of the liver, lung and small intestine

The liver lesions following migration of larvae, the classic white spots or milk spots, have been studied in great detail by Ronéus (1966), Copeman (1971) and Eriksen (1981), who, in addition, used specific fixation and staining for mast cell visualisation. In primary infections the early histological changes were of a non-specific response to mechanical damage with the first indications of larval migration apparent 12 hours after oral dosing of eggs. The lesions increased for the first four days, developing from a focus of septal and peri-septal haemorrhage and necrosis resulting from larval rupture of a portal vessel and larval migration into the parenchyma. Cellular infiltration with neutrophils, eosinophils and mononuclear cells followed, along with hyperplasia of the interlobular connective tissue, bile ducts and arteries. A second wave of eosinophils was seen 10-15 days after the primary inoculation with a general distribution along inter-lobular septa not associated with areas of previous larval damage. In secondary infections far greater numbers of eosinophils were seen than in primary infections, and not merely localized to foci of larval migration but with a general distribution resembling the second wave of primary infection as early as three to five days after inoculation. The kinetics and generalized septal distribution of eosinophils in secondary infections is equivalent to that of a hypersensitivity reaction (Copeman, 1971). Beyond the white spots formed by the lesions described above, another type of white spot is formed by lympho-nodular aggregates with a cellular distribution similar to that found in cortex of lymph nodes (Perez *et al.* 2001).

Eriksen (1981) further described the pathological lesions from migrating larvae in the lung and small intestine. As early as three days after primary inoculation eosinophil infiltration of the alveolar septa preceded the arrival of larvae in the lung. At day 7, migrating larvae in the tissue caused oedema and severe haemorrhages, and the eosinophil infiltration was extensive with some mononuclear cells and neutrophils. By day 14 after inoculation the alveolar and perilobular septa were thickened from mononuclear cell infiltration, while eosinophil infiltration was decreasing. In secondary inoculations oedema and emphysema was again accompanied by marked eosinophilic bronchitis. Furthermore, larvae were surrounded by eosinophils and mononuclear cells and there were pronounced proliferation of peribronchiolar lymphoid tissue. In the small intestine 14 and 21 days after inoculation, a dense population of eosinophils was present in the full length of the villi. There were also increased numbers of plasma cells and mucus secretion. When larvae, in histological sections, were present in the intestinal lumen they appeared to cause no damage. Mesenterical lymph nodes appeared depleted of lymphocytes and with eosinophil infiltration. *A. suum* infections also induces increased numbers of IgA and IgM antibody secreting cells in the jejunal *lamina propria* (Marbella & Gaafar, 1989), and specific antibodies detectable in isolated intestinal loop washings (Rhodes *et al.* 1978). The weight of the small intestine is significantly increased after repeated inoculations with eggs, irrespective of the presence of adult worms (Rhodes *et al.* 1982). It has not been investigated whether any of this hypertrophy can be attributed to the hatching and penetration of L3 larvae. However, Stephenson *et al.* (1980) found significant increase in intestinal weight, mainly due to hypertrophy of the *tunica muscularis*, after experimental infection with orally transferred larvae. There was a positive correlation between the number of established intestinal worms and muscular hypertrophy.

During larval migration in the liver, lung or small intestine, granule content in the mast cells decreases, especially in the vicinity of larvae and their tracks. When the larvae leave the tissue a considerable increase in the number of mast cells occurs (Eriksen, 1981). Lymph nodes along the route of migratory larvae respond with significantly increased numbers of $CD4^+$ T cells as well as B cells producing antibodies corresponding to antigens of the larval stage present in the draining area (Jungersen *et al.* 2001). In the intestine, isolated mucosal mast cells release histamine in response to larval antigens as early as 18 days after uptake of infective eggs while mast cells from naive pigs do not respond to antigen stimulation (Ashraf *et al.* 1988).

3.3 Induction of immunity

The typical immune response to helminth parasites is that of a Th2 cell type response with eosinophilia and mucosal mastocytosis (Baker *et al.* 1994) and a cytokine profile dominated by IL-4, IL-5, IL-6, IL-9, IL-10 and IL-13 (Jankovic & Sher, 1996). Recently, such a polarization was confirmed for human *A. lumbricoides* infection (Cooper *et al.* 2000). Although this has not been investigated in detail for *A. suum*, it is reasonable to expect that it is also the case in porcine ascariasis, with massive eosinophilia and eosinophil and mast cell proliferation in tissues exposed to larval migration.

Pre-hepatic (intestinal) protective immunity to *A. suum* (as measured by absence of liver white spots and a 99% reduction in the number of larvae isolated from lungs of pigs after a 10,000 egg challenge inoculation) has been demonstrated after chronic natural and repeated (daily for 16 weeks) experimental exposure (Urban, Jr. *et al.* 1988; Urban, Jr., 1986). Eriksen *et al.* (1992b) reported acquisition of pre-hepatic immunity as early as six weeks after the beginning of a twice-weekly dosing regimen. However, the majority of other studies on protective immunity in pigs have failed to induce pre-hepatic protection. In contrast, increased liver pathology has accompanied development of immunity following repeated experimental inoculations (Jungersen *et al.* 1999a), immunisation with different antigen preparations (Urban, Jr. & Romanowski, 1985; Hill *et al.* 1994) and in piglets from a naturally exposed herd with acquired protective immunity as early as five to six weeks of age (Eriksen *et al.* 1992a).

A significant acquisition of resistance with age has also been reported, although age-related resistance played a minor role in regulation of parasite infections (Eriksen *et al.* 1992a). Furthermore, Gaafar *et al.* (1973) demonstrated non-specific resistance to larval migration in pigs which had been previously infected with the transmissible gastroenteritis virus (TGEV).

Typically, challenge infections have been administered from zero to two weeks after the last immunization, where considerable inflammatory changes following previous larval migration are present and could provide a nonspecific contribution to the observed protection. However, a protective immune response 10 weeks after three infective egg immunisations has been shown to result in a 60 % and more than 90 % reduction in recovery of larvae from lungs and small intestine, respectively (Jungersen *et al.* 1999a). This protective effect was unaffected by the presence of adult worms in the small intestine. In spite of the numerous studies on immunity against *A. suum* infections it has not been determined how the protective immunity is expressed. In naturally infected children it has been shown that the magnitude of the serological response reflects the intensity of the infection rather than the individual's protective capacity (Palmer *et al.* 1995). However, natural resistance to *Ascaris* may be associated with increased IgE levels to specific antigens like ABA-1 and with higher levels of innate inflammation indicators (McSharry *et al.* 1999). In pigs, conflicting results have been obtained as to the possibility of transferring resistance to larval migration with sera, lymphocyte lysate, or colostrum from hyperimmune pigs to naive pigs (Kelley & Nayak, 1965b; Kelley & Nayak, 1965a; Rhodes *et al.* 1986). In one study, the development of immunity to reinfection after repeated drug abbreviated inoculations did not correlate with the strength of peripheral blood lymphocyte blastogenic response to *A. suum* antigens (Barta *et al.* 1986). In contrast, immunity to *A. suum* after vaccination has been correlated with the increase in lymphocyte response to specific antigens in another study (Urban & Tromba, 1982). All the studies so far have, however, used crude antigen preparations, often from different stages of the parasite, which makes it difficult to compare the results of different studies and to establish the contribution of individual antigens to the development of protective immunity.

The eosinophilia commonly observed in blood and tissues as a prominent feature of helminth infections (Butterworth & Thorne, 1993) is well documented in studies of both primary and secondary *Ascaris* infections (Ronéus 1966; Jungersen *et al.* 1999a; Copeman, 1971; Eriksen *et al.* 1980; Rhodes *et al.* 1982). The kinetics and generalized septal distribution of eosinophils in the liver in secondary infections is equivalent to that of an IgE and mast cell-mediated hypersensitivity reaction (Copeman, 1971). It is likely that, in a secondary infection, activated eosinophils are involved in specific antibody-dependent cellular cytotoxicity (ADCC) reactions against the migrating larvae (Butterworth & Thorne, 1993; Rainbird *et al.* 1998) as part of the specific host defence. However, the role of eosinophils in protection against parasitic infections is still controversial and incompletely understood (Behm & Ovington, 2000; Onah & Nawa, 2000). Although this has not been studied for *Ascaris* in the pig, evidence from *in vivo* activated eosinophils of the anterior eye chamber of guinea-pigs suggests that eosinophil interaction with *A. suum* L2 larvae *in vitro* is dependent on soluble factors present in aspirates of the infected eyes (Rockey *et al.* 1983). In addition, the nature of the intestinal pre-hepatic protective immunity following long-term exposure must be of immediate-type hypersensitivity because *A. suum* larvae hatch and begin to migrate from the large intestinal lumen only few hours after ingestion (Urban, Jr., 1986; Murrell *et al.* 1997).

Although the development of protective immunity, as evidenced by significantly reduced larval recovery after challenge inoculation, is well-substantiated, there is no evidence that pigs are thereby protected from acquisition of patent infections. On the contrary, Stankiewicz *et al.* (1992) showed development of patent infections in all pigs that had previously been immunised with eggs and subsequently treated with an anthelmintic before a single challenge inoculation. Pigs under a similar immunisation schedule had previously been shown to develop acquired resistance to migrating larvae (Stankiewicz *et al.* 1990). Furthermore, it has been shown that elimination of adult worms from the intestine of chronic naturally infected pigs does not reduce the immunity to larval migration (Urban, Jr. *et al.* 1988) and that the effect of anthelmintic treatment is only transitory when pigs are reinfected continuously (Nilsson, 1982). The discrepancy between apparent immunity and the acquisition of patent infections is further substantiated by the difficulties in establishing statistical differences in prevalence of *A. suum* between herds with and without anthelmintic cover (Roepstorff & Nansen, 1994).

In conclusion, it would be reasonable to state that mechanisms of immunity to *A. suum* in pigs are rarely completely protective and involve humoral antibodies, cell mediated responses, secretory antibody responses in the gut, non-specific resistance mechanisms in the gut, and probably also in the liver and lungs.

3.4 Antigens of *A. suum*

Stage-specific changes in *A. suum* antigens of developing larvae, egg-hatching fluid, cultured larvae and both male and female adults is well documented (Justus & Ivey, 1969; Williams & Soulsby, 1970; Fetterer & Urban, 1988; Jungersen *et al.* 2001). Soulsby (1957) reported that when larvae were placed in immune sera, immune precipitates formed around the mouth and the excretory pore, indicating that reacting antigens were mainly of excretory-secretory origin. The same precipitates were seen in *A. suum* larvae attempting to migrate in an immune guinea pig host. Others report that it is probably the glycocalyx overlying the cuticle of the parasite that is directly involved in immunological interactions with the host. The glycocalyx is of a secretory origin with high carbohydrate contents and is assumed to consist of glycoproteins, proteoglycans and glycolipids, of which the antigenic entities are predominantly phosphocholine (Dennis *et al.* 1995). Commonly, a well-characterized 14.6 kDa protein termed ABA-1 (Christie *et al.* 1993), being the major protein component of ABF and present in antigen preparations of all stages of Ascaris (Kennedy & Qureshi, 1986), is considered a major allergen of *Ascaris* (Kennedy *et al.* 1987). In pigs, however, recognition of natural and recombinant

ABA-1, as shown in Figure 7.2, seems to be highly restricted in both natural and experimental infections (Jungersen *et al.* 1999a). Likewise, Recognition of ABA-1 in naturally infected humans varies considerably (Haswell-Elkins *et al.* 1989; McSharry *et al.* 1999; Kennedy *et al.* 1990) and has been shown to be MHC-restricted in experimentally infected rodents (Christie *et al.* 1990).

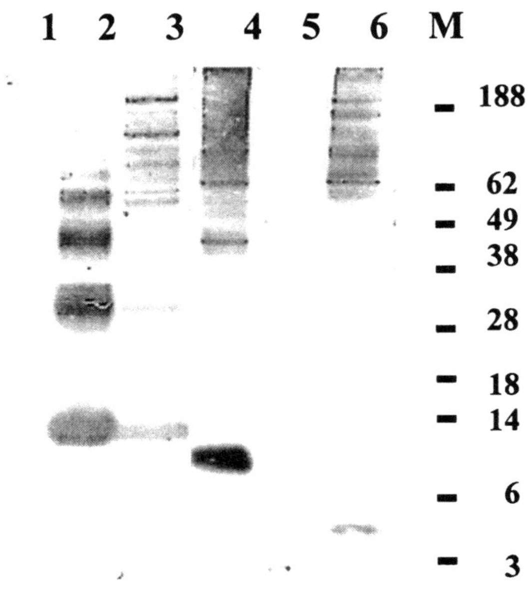

Figure 7.2. Immunoblot showing lack of antibody recognition of ABA-1 in pigs. Novex 10 % NuPAGE Bis-Tris gel was loaded as follows: Lanes 1 and 4: Extract of newly hatched *A. suum* larvae; Lanes two and 5: recombinant ABA-1; Lanes 3 and 6: ABF, M: Novex See Blue marker. Following blotting, lanes 1-3 were reacted with serum from a rabbit hyperimmunized with recombinant ABA-1 and lanes 4-6 with serum 14 days post challenge infection from an immunized pig experimentally infected with adult worms as described in Jungersen *et al.* (1999a). The prominent band at the 10 kDa position of the larval antigen (Lane 4) may be an early antigen or an IgE binding molecule (Jungersen *et al.* 2001). M. W. Kennedy kindly donated recombinant ABA-1 and rabbit anti-ABA-1 serum.

There have been described many other antigens of *Ascaris*, but, as immunity has not been conclusively linked to any of these, no further discussion is made here. It is interesting, however, that *A. suum* high molecular weight extracts has been shown to impair Th1- and Th2-dependent cell functions in mice, probably through induction of IL-4 and IL-10 (Faquim-Mauro & Macedo, 1998; Macedo *et al.* 1998).

4. EXPERIMENTAL *A. SUUM* INFECTIONS AND THEIR OUTCOME

Experimental *A. suum* infections are characterized by a high degree of unpredictability with regard to the outcome of a patent infection (Andersen *et al.* 1973; Roepstorff *et al.* 1997; Jørgensen *et al.* 1975; Eriksen *et al.* 1980; Eriksen *et al.* 1992b; Urban, Jr. *et al.* 1989; Yang *et al.* 1990; Stankiewicz *et al.* 1992). Typically, the population is overdispersed with only a few pigs harbouring the majority of worms; others will have only light worm loads, and most pigs have no infection (Eriksen *et al.* 1992b; Boes *et al.* 1998).

4.1 The self-cure expulsion of larvae

The key factor responsible for the unpredictability of whether adult worms develop from a single inoculation with infective eggs appears to be the self-cure reaction, whereby the majority, if not all, of pre-adult intestinal worms are lost from the gut soon after their arrival. Following a single dose inoculation this takes place from 14-twol days after infection in a process where the immature worms are shifted distally in the small intestine(Roepstorff *et al.* 1997). Although this has not been investigated in detail, it is tempting to speculate that the few worms remaining in the proximal parts of the small intestine are those that may survive the critical phase, while worms in the distal parts of the small intestine are destined for expulsion. The number of surviving worms is apparently not related to inoculation dose as comparable numbers of surviving intestinal worms were observed irrespective of inoculation with 100, 1000 or 10000 eggs (Roepstorff *et al.* 1997). It is one of the open questions regarding immunity and regulation of the load of infection in pigs, which mechanism underlies this self-cure reaction. Neither has it been investigated whether similar events take place in secondary infections, where the number of worms re-entering the intestine following hepato-tracheal migration is significantly reduced compared to primary inoculations (Jungersen *et al.* 1999a). It may be important to consider that, while some injurious effects from migrating larvae on the liver and lungs of the host may be expected, the host would most certainly die of intestinal obstruction if all larvae of a primary infection continued development into adulthood. Such a scenario would, from an *Ascaris* reproductive point of view, be more damaging to the parasite's population than to the host species, and this elimination of immature worms may therefore an integral part of *Ascaris* evolutionary success.

There are at least three possible events that could be responsible for the expulsion of immature worms from the gut following primary inoculations: (1) a specific immune-mediated reaction, (2) a density-dependent non-specific stimulation of innate intestinal host responses, or (3) a density-dependent self-reduction of the larval population that is unrelated to host responses. The basis of an immune mediated expulsion is supported by the findings of Ashraf *et al.* (1988) that porcine intestinal mast cells respond with histamine release at 18 (but not 14) days post initiation of egg inoculations. These dynamics coincide with the time of intestinal expulsion of worms. In addition, immune mediated expulsion of intestinal worms in other host-parasite systems is well-documented (Stewart, 1953; Onah & Nawa, 2000). Following intravenous inoculation with *in vitro* hatched larvae a high number of pigs developed patent infections (Jungersen *et al.* 1999b) indicating that circumvention of the liver-phase may reduce intestinal expulsion. However, significant numbers of worms were expelled even following intravenous inoculations, just as difficulties in establishing intestinal infections with oral or surgical transfer of day-10 larvae to naive pigs has been observed (Jungersen *et al.* 1996). Thus, a relatively short 7-10 day priming period in the small intestine from the return of larvae following migration (day 9-11) to specific immune expulsion would be required to explain a specific immune recognition and expulsion. Evidence against an immune mediated expulsion is circumstantial at best, but is supported by the gradual shifting of viable worms distally in the small intestine, and the fact that comparable numbers of worms survive irrespective of inoculation dose size. Furthermore, the evolutionary aspect that elimination of immature worms is essential for *Ascaris* propagation may suggest an active role for the parasites in the regulation of infection load. A self-limitation will only rarely be needed in secondary infections, where the host has developed some degree of protective immunity, and only a minor fraction of hatched larvae will complete the hepato-tracheal migration and develop into L4 larvae by day 14 p.i. (Jungersen *et al.* 1999a). This is in agreement with the fact that even in highly immune sows, patent *Ascaris* infections recur.

4.2 Experimental infections by transfer of larvae or adult worms

To overcome the problem of elimination of larvae following experimental infection, and the resulting unpredictability of parasite load, there are several reports on oral transfer of *A. suum* larvae in the literature. However, in many experiments the methods have not been evaluated for larval survival and establishment before the effects of natural self-cure would confound the results. The first report is probably that of Buckley (1931) who dosed himself with a piece of bread infested with 20 larvae collected from the lung of a pig recently infected with a massive dose of *A. suum* eggs. Simultaneously, a green monkey and two pigs were infected with much larger numbers of larvae. While neither he nor the monkey showed any sign of adult infection (judged by faecal egg excretion?) both pigs were found to harbour large numbers of adult worms three months after the transfer. Later Stephenson *et al.* (1977) orally transferred larvae encapsulated in gelatine to pigs with reasonable success. These larvae were recovered from rabbits 15 days after infection, which, in this unnatural host, is the time when larvae begin to appear in large numbers in the intestine, i.e. comparable to day 10 in pigs. Stewart and Rowell (1986) orally transferred 366 larvae recovered from the lungs of pigs seven days post inoculation to other pigs. They recovered up to 9 % of the transferred larvae 17 days after transfer, which is too long after transfer to discriminate the effects of passing through the stomach and self-cure on survival of the larvae. Jungersen *et al.* (1996) transferred larvae collected from the small intestine at day 10 p.i. and re-introduced them into naive pigs by gastric lavage or injection into the upper small intestine during laparotomy. Only the surgically transferred larvae had survival rates (day 7 post transfer) comparable to that of a single inoculation with eggs, indicating that the second passage of larvae through the stomach was detrimental. Other studies have shown difficulties establishing larvae recovered from lungs at day 7 p.i. or intestine at day 14 p.i., at which time the intestinal moult to L4 should have been completed (Jungersen, 1998).

More success has been obtained with the technique for oral transfer of adult worms to pigs whereby infections with more or less known numbers and sex of adult worms have been attained (Jungersen *et al.* 1996). This method has been used to study the egg production of female worms in the presence or absence of adult male worms (Jungersen *et al.* 1997), and to study the effects of intestinal adult worms on the host resistance to reinfection (Jungersen *et al.* 1999a). Of a total of 310 female and 148 male worms transferred to 48 pigs, 46% of the females and 42% of the males were recovered from the small intestine five to eight weeks after transfer. Female worms transferred to previously parasite-naive pigs ceased producing eggs two to three weeks after transfer. However, these females readily resumed excretion of fertilised eggs a few days after oral transfer of adult male worms. The presence of adult *Ascaris* in the small intestine was found to be without influence on the host response against migrating larvae and had no effect on the survival of larvae from a single challenge inoculation. This was found irrespective of whether the challenged pig host was parasite-naive or had been immunized previous to the transfer of adult worms (Jungersen *et al.* 1999a). With the development of a technique for the direct establishment of adult worms it is now possible to design studies on the importance and the mechanisms involved in specific and non-specific host reactions to *A. suum*. Such studies could contribute to new information on *Ascaris* immunology. Due to the distinct division of the *A. suum* life cycle into a systemic migratory phase and a resident intestinal luminal phase, a combination of natural infection and the oral worm transfer technique provides a powerful model for studies on the immune mechanisms in the local small intestinal environment. The increasing international interest in the pig immune system has caused an increasing number of porcine immunological reagents to become available, designed to investigate various cellular and humoral responses to both specific and non-specific antigenic stimulation. The use of these new immunological tools in studies with experimental *Ascaris* infections can help the discovery of new aspects of parasite immunology, and might also illuminate basic immunological features such as mucosal immune regulation and oral tolerance.

ACKNOWLEDGEMENTS

Dr. Lis Eriksen and Dr. Darwin Murrell of the Danish Centre for Experimental Parasitology are thanked for their revision and comments on the manuscript.

REFERENCES

ANDERSEN, S., JØRGENSEN, R. J., NANSEN, P. & NIELSEN, K. (1973). Experimental *Ascaris suum* infection in piglets. *Acta Pathologica et Microbiologica Scandinavica Sect.B* **81**, 650-656.

ANDERSON, T. J. C. (1995). *Ascaris* infections in humans from North America: molecular evidence for cross-infection. *Parasitology* **110**, 215-219.

ANDERSON, T. J. C., ROMERO-ABAL, M. E. & JAENIKE, J. (1993). Genetic structure and epidemiology of *Ascaris* populations: patterns of host affiliation in Guatemala. *Parasitology* **107**, 319-334.

ASHRAF, M., URBAN, J. F., Jr., LEE, T. D. & LEE, C. M. (1988). Characterization of isolated porcine intestinal mucosal mast cells following infection with *Ascaris suum*. *Veterinary Parasitology* **29**, 143-158.

BAKER, D. G., BRYANT, J. D., URBAN, J. F., Jr. & LUNNEY, J. K. (1994). Swine immunity to selected parasites. *Veterinary Immunology and Immunopathology* **43**, 127-133.

BARTA, O., STEWART, T. B., SHAFFER, L. M., HUANG, L.-J. & SIMMONS, L. A. (1986). *Ascaris suum* infection in pigs sensitizes lymphocytes but suppresses their responsiveness to phytomitogens. *Veterinary Parasitology* **21**, 25-36.

BEHM, C. A. & OVINGTON, K. S. (2000). The role of eosinophils in parasitic helminth infections: insights from genetically modified mice. *Parasitology Today* **16**, 202-209.

BOES, J., MEDLEY, G. F., ERIKSEN, L., ROEPSTORFF, A. & NANSEN, P. (1998). Distribution of *Ascaris suum* in experimental and naturally infected pigs and comparison with *Ascaris lumbricoides* infections in humans. *Parasitology* **117**, 589-596.

BROWN, H. W. & CORT, W. M. (1927). The egg production of *Ascaris lumbricoides*. *Journal of Parasitology* **14**, 88-90.

BUCKLEY, J. J. C. (1931). An observation on human resistance to infection with *Ascaris* from the pig. *Journal of Helminthology* **9**, 45-46.

BUTTERWORTH, A. E. & THORNE, K. J. I. (1993). Eosinophils and Parasitic Diseases. In: *Immunopharmacology of Eosinophils*, (ed. Smith,H. & Cook,R.M.), pp. 119-150. London: Academic Press.

CHAN, M. S., MEDLEY, G. F., JAMISON, D. & BUNDY, D. A. P. (1994). The evaluation of potential global morbidity attributable to intestinal nematode infections. *Parasitology* **109**, 373-387.

CHRISTIE, J. F., DUNBAR, B., DAVIDSON, I. & KENNEDY, M. W. (1990). N-terminal amino acid sequence identity between a major allergen of *Ascaris lumbricoides* and *Ascaris suum*, and MHC-restricted IgE responses to it. *Immunology* **69**, 596-602.

CHRISTIE, J. F., DUNBAR, B. & KENNEDY, M. W. (1993). The ABA-1 allergen of the nematode *Ascaris suum*: epitope stability, mass spectrometry, and N-terminal sequence comparison with its homologue in *Toxocara canis*. *Clinical and Experimental Immunology* **92**, 125-132.

COOPER, P. J., CHICO, M. E., SANDOVAL, C., ESPINEL, I., GUEVARA, A., KENNEDY, M. W., URBAN JR, J. F., GRIFFIN, G. E. & NUTMAN, T. B. (2000). Human infection with Ascaris lumbricoides is associated with a polarized cytokine response. *Journal of Infectious Diseases* **182**, 1207-1213.

COPEMAN, D. B. (1971). Immunopathological response of pigs in ascariasis. In: *Pathology of Parasitic Diseases*, (ed. Gaafar,S.M.), pp. 135-144. Lafayette: Purdue Univ Press.

CROMPTON, D. W. T. (1989). Prevalence of ascariasis. In: *Ascariasis and its prevention and control*, (ed. Crompton,D.W.T., Nesheim,M.C. & Pawlowski, Z.S.), pp. 45-70. London: Taylor & Francis.

CROMPTON, D. W. T. (2001). Ascaris and ascariasis. *Advances in Parasitology* **48**, 285-375.

DENNIS, R. D., BAUMEISTER, S., SMUDA, C., LOCHNIT, C., WAIDER, T. & GEYER, E. (1995). Initiation of chemical studies on the immunoreactive glycolipids of adult *Ascaris suum*. *Parasitology* **110**, 611-623.

DOUVRES, F. W., TROMBA, F. G. & MALAKATIS, G. M. (1969). Morphogenesis and migration of *Ascaris suum* larvae developing to fourth stage in swine. *Journal of Parasitology* **55**, 689-712.

ERIKSEN, L. (1981). *Host parasite relations in Ascaris suum infection in pigs and mice*, Royal Veterinary and Agricultural University. Commisioned by Carl F. Mortensen, Copenhagen.

ERIKSEN, L., ANDERSEN, S., NIELSEN, K., PEDERSEN, A. & NIELSEN, J. (1980). Experimental *Ascaris suum* infection in pigs. Serological response, eosinophilia in peripheral blood, occurrence of white spots in the liver and worm recovery from the intestine. *Nordisk Veterinærmedicin* **32**, 233-242.

ERIKSEN, L., LIND, P., NANSEN, P., ROEPSTORFF, A. & URBAN, J. F. (1992a). Resistance to *Ascaris suum* in parasite naïve and naturally exposed growers, finishers and sows. *Veterinary Parasitology* **41**, 137-149.

ERIKSEN, L., NANSEN, P., ROEPSTORFF, A., LIND, P. & NILSSON, O. (1992b). Response to repeated inoculations with *Ascaris suum* eggs in pigs during the fattening period. I. Studies on worm population kinetics. *Parasitological Research* **78**, 241-246.

FAQUIM-MAURO, E. L. & MACEDO, M. S. (1998). The immunosuppressive activity of Ascaris suum is due to high molecular weight components. *Clinical and Experimental Immunology* **114**, 245-251.

FETTERER, R. H. & URBAN, J. F. (1988). Developmental changes in cuticular proteins of *Ascaris suum*. *Comparative Biochemistry and Physiology B* **90**, 321-327.

GAAFAR, S. M., DUGAS, S. & SYMENSMA, R. (1973). Resistance of pigs recovered from transmissible gastroenteritis against infection with *Ascaris suum*. *American Journal of Veterinary Research* **34**, 793-795.

GALVIN, T. J. (1968). Development of human and pig Ascaris in the pig and rabbit. *Journal of Parasitology* **54**, 1085-1091.

GEENEN, P. L., BRESCIANI, J., BOES, J., PEDERSEN, A., ERIKSEN, L., FAGERHOLM, H. P. & NANSEN, P. (1999). The morphogenesis of Ascaris suum to the infective third-stage larvae within the egg. *Journal of Parasitology* **85**, 616-622.

HASWELL-ELKINS, M. R., KENNEDY, M. W., MAIZELS, R. M., ELKINS, D. B. & ANDERSON, R. M. (1989). The antibody recognition profiles of humans infected with *Ascaris lumbricoides*. *Parasite Immunology* **11**, 615-627.

HILL, D. E., FETTERER, R. H., ROMANOWSKI, R. D. & URBAN, J. F., Jr. (1994). The effect of immunization of pigs with *Ascaris suum* cuticle components on the development of resistance to parenteral migration during a challenge infection. *Veterinary Immunology and Immunopathology* **42**, 161-169.

JANKOVIC, D. & SHER, A. (1996). Initiation and regulation of CD4+ T-cell function in host-parasite models. In: *Th1 and Th2 cells in health and disease*, Vol. 63 (ed. Romagnani,S.), pp. 51-65. Basel: Karger.

JUNGERSEN, G. (1998). Experimental *Ascaris suum* infections. Manipulated infections and immune responses to larval migration in pigs. Royal Veterinary and Agricultural University, Copenhagen, Denmark. pp.1-127. PhD Thesis.

JUNGERSEN, G., ERIKSEN, L., NIELSEN, C. G., ROEPSTORFF, A. & NANSEN, P. (1996). Experimental transfer of *Ascaris suum* from donor pigs to helminth naive pigs. *Journal of Parasitology* **82**, 752-756.

JUNGERSEN, G., ERIKSEN, L., NANSEN, P. & FAGERHOLM, H.-P. (1997). Sex-manipulated *Ascaris suum* infections in pigs: implications for reproduction. *Parasitology* **115**, 439-442.

JUNGERSEN, G., ERIKSEN, L., ROEPSTORFF, A., LIND, P., MEEUSEN, E. N., RASMUSSEN, T. & NANSEN, P. (1999a). Experimental *Ascaris suum* infection in the pig: protective memory response after three immunizations and effect of intestinal adult worm population. *Parasite Immunology* **21**, 619-630.

JUNGERSEN, G., FAGERHOLM, H. P., NANSEN, P. & ERIKSEN, L. (1999b). Development of patent *Ascaris suum* infections in pigs following intravenous administration of larvae hatched *in vitro*. *Parasitology* **119**, 503-508.

JUNGERSEN, G., ERIKSEN, L., NANSEN, P., LIND, P., RASMUSSEN, T. & MEEUSEN, E. N. T. (2001). Regional immune responses with stage-specific antigen recognition profiles develop in lymph nodes of pigs following *Ascaris suum* larval migration. *Parasite Immunology* **23**, 185-194.

JUSTUS, D. E. & IVEY, M. H. (1969). *Ascaris suum:* Immunoelectrophoretic analysis of antigens in developmental stages. *Experimental Parasitology* **26**, 290-298.

JØRGENSEN, R. J., NANSEN, P., NIELSEN, K., ERIKSEN, L. & ANDERSEN, S. (1975). Experimental *Ascaris suum* infection in the pig. Population kinetics following low and high levels of primary infection in piglets. *Veterinary Parasitology* **1**, 151-157.

KELLEY, G. W. & NAYAK, D. P. (1965a). Passive immunity to *Ascaris suum* transferred in colostrum from sows to their offspring. *American Journal of Veterinary Research* **26**, 948-950.

KELLEY, G. W. & NAYAK, D. P. (1965b). Passive immunity to migrating *Ascaris suum* transmitted by parenterally administered immune serum or immune globulins. *Cornell Veterinarian* **55**, 607-612.

KENNEDY, M. W. & QURESHI, F. (1986). Stage-specific secreted antigens of the parasitic larval stages of the nematode *Ascaris*. *Immunology* **58**, 515-522.
KENNEDY, M. W., QURESHI, F., HASWELL-ELKINS, M. R. & ELKINS, D. B. (1987). Homology and heterology between the secreted antigens of the parasite larval stages of *Ascaris lumbricoides* and *Ascaris suum*. *Clinical and Experimental Immunology* **67**, 20-30.
KENNEDY, M. W., TOMLINSON, L. A., FRASER, E. M. & CHRISTIE, J. F. (1990). The specificity of the antibody response to internal antigens of *Ascaris*: heterogeneity in infected humans, and MHC (H-two) control of the repertoire in mice. *Clinical and Experimental Immunology* **80**, 219-224.
KENNEDY, T. J. (1988). Prevalence of swine parasites in major hog producing areas in the United States. *Agri-Practice* **9**, 25-32.
LIND, P., ERIKSEN, L., NANSEN, P., NILSSON, O. & ROEPSTORFF, A. (1993). Response to repeated inoculations with *Ascaris suum* eggs in pigs during the fattening period. II. Specific IgA, IgG, and IgM antibodies determined by enzyme-linked immunosorbent assay. *Parasitological Research* **79**, 240-244.
LUNNEY, J. K., URBAN, J. F., Jr. & JOHNSON, L. A. (1986). Protective immunity to *Ascaris suum:* analysis of swine peripheral blood subsets using monoclonal antibodies and flow cytometry. *Veterinary Parasitology* **20**, 117-131.
MACEDO, M. S., FAQUIM-MAURO, E., FERREIRA, A. P. & ABRAHAMSOHN, I. A. (1998). Immunomodulation induced by *Ascaris suum* extract in mice: effect of anti-interleukin-4 and anti-interleukin-10 antibodies. *ScandinavianJournal of Immunology* **47**, 10-18.
MARBELLA, C. O. & GAAFAR, S. M. (1989). Production and distribution of immunoglobulinbearing cells in the intestine of young pigs infected with *Ascaris suum* . *Veterinary Parasitology* **34** , 63-70.
MARUYAMA, H., NAWA, Y., NODA, S., MIMORI, T. & CHOI, W.-Y. (1996). An outbreak of visceral larva migrans due to *Ascaris suum* in Kyushu, Japan. *Lancet* **347**, 1766-1767.
MAUNG, M. (1978). The occurrence of the second moult of *Ascaris lumbicoides* and *Ascaris suum*. *International Journal for Parasitology* **8**, 371-378.
MCSHARRY, C., XIA, Y., HOLLAND, C. V. & KENNEDY, M. W. (1999). Natural immunity to *Ascaris lumbricoides* associated with immunoglobulin E antibody to ABA-1 allergen and inflammation indicators in children. *Infection and Immunity* **67**, 484-489.
MURRELL, K. D., ERIKSEN, L., NANSEN, P., SLOTVED, H. C. & RASMUSSEN, T. (1997). *Ascaris suum:* revision of its early migratory path and implication for human ascariasis. *Journal of Parasitology* **83**, 255-260.

NILSSON, O. (1982). Ascariasis in the pig. An epizootological and clinical study. *Acta Veterinaria Scandinavia Supplementum* **79**, 1-108.

OLSEN, L. S., KELLEY, G. W. & SEN, H. G. (1958). Longevity and egg-production of *Ascaris suum. Transactions of the American Microscopy Society* **77**, 380-383.

ONAH, D. N. & NAWA, Y. (2000). Mucosal immunity against parasitic gastrointestinal nematodes. *Korean Journal of Parasitology* **38**, 209-236.

PALMER, D. R., HALL, A., HAQUE, R. & ANWAR, K. S. (1995). Antibody isotype responses to antigens of *Ascaris lumbricoides* in a case-control study of persistently heavily infected Bangladeshi children. *Parasitology* **111**, 385-393.

PEREZ, J., GARCIA, P. M., MOZOS, E., BAUTISTA, M. J. & CARRASCO, L. (2001). Immunohistochemical characterization of hepatic lesions associated with migrating larvae of *Ascaris suum* in pigs. *Journal of Comparative Pathology.* **124**, 200-206.

RAINBIRD, M. A., MACMILLAN, D. & MEEUSEN, E. N. (1998). Eosinophil-mediated killing of *Haemonchus contortus* larvae: effect of eosinophil activation and role of antibody, complement and interleukin-5. *Parasite Immunology* **20**, 93-103.

RHODES, M. B., MCCULLOUGH, R. A., MEBUS, C. A. & KLUCAS, C. A. (1978). *Ascaris suum:* specific antibodies in isolated intestinal loop washings from immunized swine. *Experimental Parasitology* **45**, 255-262.

RHODES, M. B., KERALIS, M. B. & STAUDINGER, L. A. (1982). Immune responses of swine to oral inoculation with embryonated eggs of *Ascaris suum. American Journal of Veterinary Research* **43**, 1604-1607.

RHODES, M. B., KERALIS, M. B., STAUDINGER, L. A. & BAKER, P. K. (1986). Immunity of swine to *Ascaris suum. Veterinary Parasitology* **22**, 87-94.

ROCKEY, J. H., JOHN, T., DONNELLY, J. J., MCKENZIE, D. F., STROMBERG, B. E. & SOULSBY, E. J. (1983). In vitro interaction of eosinophils from ascarid-infected eyes with *Ascaris suum* and *Toxocara canis* larvae. *Investigative Opthalmology & Visual Science* **24**, 1346-1357.

ROE, J. M., PATEL, D. & MORGAN, K. L. (1993). Isolation of porcine IgE, and preparation of polyclonal antisera. *Veterinary Immunology and Immunopathology* **37**, 83-97.

ROEPSTORFF, A., ERIKSEN, L., SLOTVED, H. C. & NANSEN, P. (1997). Experimental *Ascaris suum* infection in the pig: worm population kinetics following single inoculations with three doses of infective eggs. *Parasitology* **115**, 443-452.

ROEPSTORFF, A. & NANSEN, P. (1994). Epidemiology and control of helminth infections in pigs under intensive and non-intensive production systems. *Veterinary Parasitology* **54**, 69-85.

RONÉUS, O. (1966). Studies on the aetiology and pathogenesis of white spots in the liver of pigs. *Acta Veterinaria Scandinavia* **7 supplementum 16**, 7-112.

RONÉUS, O. (1971). Studies on the inter-relationship between the number of orally administered *Ascaris suum* eggs, blood eosinophilia and the number of adult intestinal ascarids. In: (ed. Gaafar,S.M.), pp. 339-343. Purdue University.

SEAMSTER, A. P. (1950). Developmental studies concerning the eggs of *Ascaris lumbricoides* var. *suum. The American Midland Naturalist* **43**, 450-470.

SINNIAH, B. (1982). Daily egg production of *Ascaris lumbricoides:* The distribution of eggs in the feces and the variability of egg counts. *Parasitology* **84**, 167-175.

SOULSBY, E. J. L. (1957). Some immunological phenomena in parasitic infections. *Veterinary Record* **69**, 1129-1139.

STANKIEWICZ, M., JESKA, E. L. & FROE, D. L. (1990). Acquired resistance to migrating larvae of *Ascaris suum* in young pigs by repeated drug-abbreviated infections. *Journal of Parasitology* **76**, 383-388.

STANKIEWICZ, M., JONAS, W. & FROE, D. L. (1992). Patent infections of *Ascaris suum* in pigs: effect of previous exposure to multiple, high doses of eggs and various treatment regimes. *International Journal for Parasitology* **22**, 597-601.

STEPHENSON, L. S., GEORGI, J. R. & CLEVELAND, D. J. (1977). Infection of weanling pigs with known numbers of *Ascaris suum* fourth stage larvae. *Cornell Veterinarian* **67**, 92-102.

STEPHENSON, L. S., POND, W. G., NESHEIM, M. C., KROOK, L. P. & CROMPTON, D. W. T. (1980). *Ascaris suum*: Nutrient absorption, growth, and intestinal pathology in young pigs experimentally infected with 15-day old larvae. *Experimental Parasitology* **49**, 15-25.

STEWART, D. F. (1953). Studies on resistance of sheep to infestation with *Haemonchus contortus* and *Trichostrongylus* spp. and on the immunological reactions of sheep exposed to infestation: V. The nature of the "self-cure" phenomenon. *Australian Journal of Agricultural Research* **4**, 100-117.

STEWART, T. B. & ROWELL, T. J. (1986). Susceptibility of fourth-stage *Ascaris suum* larvae to fenbendazole and to host response in the pig. *American Journal of Veterinary Research* **47**, 1671-1673.

TAKATA, I. (1951). Experimental infection of man with *Ascaris* of man and the pig. *Kitasato Archives of Experimental Medicine* **23**, 49-59.

URBAN, J. F., Jr. (1986). The epidemiology and control of swine parasites. Immunity and vaccines. *Vet Clin.North Am.Food Anim Pract.* **2**, 765-778.

URBAN, J. F., Jr. & DOUVRES, F. W. (1984). Culture requirements of *Ascaris suum* larvae using a stationary multi-well system: Increased survival, development and growth with cholesterol. *Veterinary Parasitology* **14**, 33-42.

URBAN, J. F., Jr. & ROMANOWSKI, R. D. (1985). *Ascaris suum:* Protective immunity in pigs immunized with products from eggs and larvae. *Experimental Parasitology* **60**, 245-254.

URBAN, J. F., Jr & TROMBA, F. G. (1982). Development of immune responsiveness to *Ascaris suum* antigens in pigs vaccinated with ultraviolet-attenuated eggs. *Veterinary Immunology and Immunopathology* **3**, 399-409.

URBAN, J. F., Jr., ALIZADEH, H. & ROMANOWSKI, R. D. (1988). *Ascaris suum:* Development of intestinal immunity to infective second-stage larvae in swine. *Experimental Parasitology* **66**, 66-77.

URBAN, J. F., Jr., ROMANOWSKI, R. D. & STEELE, N. C. (1989). Influence of helminth parasite exposure and strategic application of anthelmintics on the development of immunity and growth of swine. *Journal of Animal Science* **67**, 1668-1677.

WILLIAMS, J. F. & SOULSBY, E. J. L. (1970). Antigenic analysis of developmental stages of *Ascaris suum* I. Comparison of eggs, larvae and adults. *Experimental Parasitology* **27**, 150-162.

YANG, S., GAAFAR, S. M. & BOTTOMS, G. D. (1990). Effects of multiple dose infections with *Ascaris suum* on blood gastrointestinal hormone levels in pigs. *Veterinary Parasitology* **37**, 31-44.

Chapter 8
IMMUNE RESPONSES IN HUMANS – *TRICHURIS TRICHIURA*

Helen Faulkner and Janette E. Bradley
School of Life and Environmental Sciences, University of Nottingham UK
e-mail: Helen.Faulkner@nottingham.ac.uk

1. INTRODUCTION

There are thought to be over 1000 million people in the world today who have ingested embryonated *Trichuris trichiura* eggs resulting in infection (Chan, 1996). *Trichuris* eggs hatch in the intestinal tract whereupon the emergent larvae migrate to the caecal crypts and burrow into the epithelium, thus occupying an intracellular niche. Once the posterior end breaks free into the lumen, fertilization can occur allowing eggs to void with the faeces. This pre-patent period takes approximately 60 days and the adult life span is estimated to be 3 years (Bundy & Cooper, 1989). Considering the scale of this gastrointestinal infection it is surprising how little we know about the immune response to it. Perhaps there has been a tendency to overlook the disease because it does not cause sudden serious debilitating symptoms. Trichuriasis, or whipworm infection, is largely asymptomatic: the size of the worm burden determines the severity of the clinical symptoms (see Chapter 3) and relatively few individuals within a community are heavily infected (Anderson & Medley, 1985, Cooper & Bundy, 1987). However, there are several reasons why we should be interested in the immune response to this rather insidious intestinal nematode.

2. THE IMPORTANCE OF TRICHURIASIS

Trichuriasis represents a major public health problem of global significance. Although a large percentage of infected people harbour light infections, which may go unnoticed, the cost to those with heavier infections is extremely high. When worm loads begin to exceed 50 worms, abdominal discomfort and frequent and watery stools become evident. With larger worm burdens the illness becomes so severe it is often assigned the name *Trichuris* dysentery syndrome (TDS) (Ramsey, 1962). In these cases a heavily infected person can suffer from profuse diarrhoea, rectal prolapse, finger-clubbing, anaemia and growth retardation (for a collation of relevant work see Bundy & Cooper, 1989). The latter two symptoms are particularly devastating in young children because there is a strong correlation between them and cognitive development and it is children who tend to suffer the heaviest infections (see Figure 8.1, Simeon & Grantham-McGregor, 1990; Bundy *et al.* 1987). In fact moderate to heavy infections have been shown to impair learning ability in a large group of school children (Nokes *et al.* 1992; Nokes & Bundy, 1994) (also see Chapter 4). Clearly the consequences of intense infection present both a significant health and economic impact on a community (see Chapter 5).

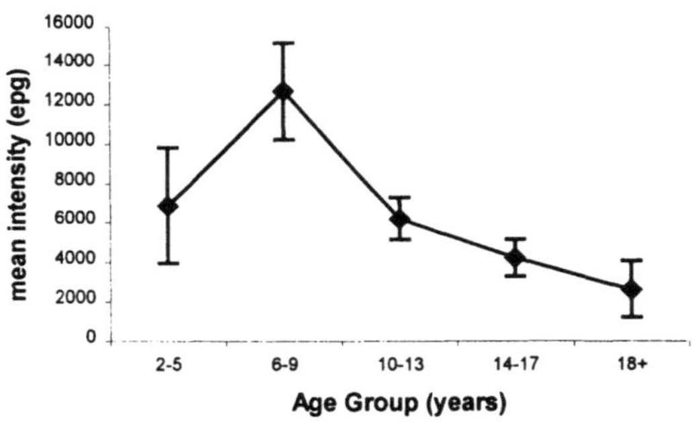

Figure 8.1. Intensity of *T. trichiura* infection by age as assessed by eggs per gram. Ayéné, Cameroon, 2000.

With an estimated 46 million of those infected suffering some level of associated morbidity there is a need to understand immunity in order to develop better transmission control strategies and ultimately develop a vaccine (Montresor *et al.* 1999). At present, in the absence of such a vaccine, treatment is given in the form of a multiple dose course of a benzimidazole carbamate (Rossignol & Maisonneuve, 1984). This effectively expels all worms from their intestinal niche, but within 6-9 months of living in the same endemic area a person often becomes re-infected (Bundy, 1988). The goal of controlling intestinal helminth infection worldwide by the selective targeting of treating school-aged children is becoming a reality (see Chapter

2). However, as eggs may remain viable for long periods and a single female worm can release up to 20,000 eggs per day there is a strong argument for preventing rather than simply treating infection (Bundy & Cooper 1989). Undoubtedly improvements in sanitation would do this, but until the necessary investment in infrastructure becomes available vaccine development represents a viable option.

Despite the need to study immunity because *T. trichiura* is an important helminth species or because of a genuine desire to reduce sufferance there is a further important reason. Numerous reports, from both field studies and the laboratory setting, suggest that helminth infections can modify the response to secondary infections or vaccination (Curry *et al.* 1995; Rousseau *et al.* 1997; Sabin *et al.*1996; Cooper *et al.* 1999, 2001,). This may have deleterious consequences. For example, with regard to the current HIV and tuberculosis epidemics, disease progression and susceptibility to infection is more rapid in areas where helminths are prevalent (Bentwich *et al.* 1999) (see Chapter 16). One explanation for this could be the influence of pre-existing helminths on the cellular immune response to mycobacterial and HIV antigens (Pearlman *et al.* 1993; Stewart *et al.* 1999; Elias *et al.* 2001). Helminth infections have long been associated with strong Th2 type responses that can down-regulate the production of Th1 type cytokines (Finkelman *et al.* 1991; Sher & Coffman, 1992; Urban *et al.* 1992). Indeed this very fact has recently been exploited in patients with inflammatory bowel disease by giving them a non-patent *Trichuris* infection with the aim of alleviating their Th1 mediated immunopathology (Shirakawa *et al.* 1997). A better understanding of the immune response to *Trichuris* and other intestinal helminths will help us elucidate whether co-infections or immune-mediated conditions are likely to be exacerbated or abbreviated. This has obvious implications for vaccine development because promotion of a specific type of response may prove to be a hindrance in terms of these other diseases. When a third of the world's population lives with a life-long exposure to helminth parasites the immunological response, to any other infectious agent, must be considered in the context of having intestinal worms.

3. THE MOUSE MODEL: *TRICHURIS MURIS*

Our knowledge of immunity to intestinal helminths has been greatly aided by a number of laboratory models of infection. *Trichuris muris* in the mouse represents an excellent model for human trichuriasis. The parasite naturally infects mice, undergoes a comparable life cycle and is morphologically and antigenically similar. The ready availability of immunological reagents and inbred or transgenic mouse strains have allowed studies to be performed that have resulted in a detailed understanding of the mechanisms controlling resistance and susceptibility to infection. It is possible to write a review on just this subject, indeed several have been written already (Grencis, 1997; Else & Finkelman, 1998; Artis & Grencis 2001), so only a brief overview will be presented here.

In the laboratory model system of *Trichuris* inbred strains of mice are found which display a mixture of phenotypes ranging from wholly resistant to completely susceptible (Else & Wakelin, 1988). The categorisation of CD4+ T helper cell subsets has helped understand the regulatory mechanisms involved in this resistance and susceptibility. A critical role for CD4+ T-cells in host immunity has now been long established (For review see Artis & Grencis 2001). When resistant and susceptible strains are examined they are found to have developed a Th2 or a Th1 type of response depending on whether they have launched a protective or non-protective immune response respectively (Else & Grencis, 1991, Else; Hültner & Grencis, 1992). Manipulation of cytokines *in vivo*, either by using specific anti-cytokine antibodies or using mice with deletions in cytokines or cytokine receptor genes, has allowed the relative contributions of individuals cytokines to resistance to be analysed. Cytokines produced by Th2 cells such as IL-4 (Else *et al.* 1994), IL-13 (Bancroft *et al.* 2000) and IL-9 (Faulkner *et al.* 1998) have all been shown to be important in worm expulsion.

Susceptibility to infection is associated with a strong Th1 type response characterised by high levels of IFN-γ and IL-12. This susceptible phenotype can be manipulated by the depletion of IFN-γ, or IFN-γ receptor, causing infected mice to produce Th2 cytokines and expel their worms (Else *et al.* 1994). Conversely, resistant mice can be made susceptible by the early administration of IL-12, which promotes Th1 type cytokine production (Bancroft *et al.* 1997). Chronic infection therefore seems to be due to the development of an inappropriate immune response. It is possible that the parasite is able to modulate the host response in a Th1 direction in order promote its survival. This was indicated by the work of Else, Wakelin & Roach, 1989 where it was shown that, by drug abbreviating sub-threshold infections in resistant mice, modulation was dependent on the survival of larvae beyond 21 days post infection.

As yet there is very little known about the actual effector mechanisms operating. Antibody has been shown not to be essential in resistance to infection because CD4+ T cell transfers into susceptible SCID mice can induce worm expulsion in the absence of antibody (Else & Grencis, 1996). However recently a role for B cells has been identified following the discovery that μMT mice are susceptible to infection (Blackwell & Else, 2001). Reconstitution of these mice with B cells, as well as parasite specific IgG1, resulted in resistance to infection. B cells and antibody may therefore contribute to parasite loss in certain circumstances. The classical cellular hallmarks of helminth infection, known to be controlled by Th2 cytokines, are the mast cell and the eosinophil and these have long been hypothesised to play a major role in immunity to helminths. But, against *T. muris* neither cell appears to have an important role. The removal of mast cells *in vivo* using an anti-stem cell factor receptor antibody (99% effective) did not effect on the expulsion of the nematode (Betts & Else, 1999). Equally, removal of eosinophils using anti-IL-5 monoclonal antibodies did not affect resistance to the worm. Consequently, as yet there is no known definitive effector mechanism against *T. muris* in the mouse. This leaves the intriguing possibility that the Th2 cytokines clearly associated with resistance to infection are having direct effects on cells of the gastrointestinal tract.

4. IMMUNITY TO *TRICHURIS TRICHIURA*

Although in the *T. muris* mouse model system there is good evidence to suggest that protective responses are controlled by Th2 type cytokines the situation is far from clear in human infections. Indeed despite evidence of a vigorous immune response it is not certain that protective immune responses operate at all in chronic helminth infections such as *T. trichiura*. Convex cross sectional age-intensity profiles suggest that patterns of infection are determined by age-related differences in both exposure and the ability to acquire immunity. The relative contribution of these two parameters is a matter of much debate. Bundy & Medley (1992) hold the view that the establishment rate of infection can entirely explain the patterns of infection observed. They found little evidence that expulsion or death of the parasite by any cause played a significant role. Nevertheless the ecology versus immunity question has not been resolved (see Chapter 1) and there has been very little by way of immunological studies in human populations to address this issue. It is likely that exposure, immunity, and the genetic backgrounds of host and parasite are all important in defining the outcome of infection. We do not aim in this review to address all of the questions but will summarise the findings on immune responses to *T. trichiura* in humans and address the question of whether there is any evidence that aspects of the immune response confer protection.

4.1 B Cell Responses

When measuring antibody responses to *T. trichiura* several factors should be considered. Firstly, whilst it is comparatively easy to assess antibody levels peripherally it is difficult to determine those present at the site of infection. Consequently, the overwhelming body of data describes serum isotype levels that may not accurately represent the levels and types found within the gut. This is particularly true of IgA, which is present in a dimeric secretory form only in mucosal areas. Whether the form found in the general circulation is representative of intestinal IgA is unknown. However, encouraging data on salivary IgA, which is predominantly in the secretory form and therefore likely to reflect mucosal levels, shows a similar age-dependency to serum IgA (Needham & Lillywhite, 1994). In addition, positive correlations have been found between infection intensity and certain IgG subclasses suggesting that serum antibody can be informative regarding the current infection status of a community (Bundy *et al.* 1991). Measuring serum antibody levels can therefore give us some indication of local B cell responses but should be considered in the context that all gut-associated antigens may not enter the systemic circulation. This is particularly relevant for *T. trichiura* because unlike other intestinal nematodes, such as *Ascaris lumbricoides*, it does not leave its intestinal niche.

A second consideration when examining specific antibody is the choice of parasite antigen. It is common practice to obtain whole adult worms, make them into a crude antigen preparation and analyse the antibodies capable of binding. Although useful, in recent years there has been a move away from single stage somatic extracts towards different life cycle stages and more defined antigens. Key responses can remain hidden unless this approach is adopted. For example, in an area where *A. lumbricoides* is endemic an association was found between susceptibility and ABA-1 specific IgE, a recombinant *Ascaris* allergen (McSharry *et al.* 1999). No association was found between IgE directed toward a more general *Ascaris* preparation. With regard to *T. trichiura* such work is still in its infancy. Immunoblot analyses of adult *T. trichiura* extracts have revealed several prominent antigens (Needham *et al.* 1993). Although there is a high degree of individual heterogeneity, the number and intensity of the bands recognised largely reflects the intensity of infection at the population level. Interestingly, when the 16-17 KDa and 90 KDa antigens were selected for analysis a key difference emerged; levels of specific IgA persisted into adulthood and did not decline with infection intensity as seen for IgG1. The IgA response to these antigens within a community is therefore high when infection levels are low. The molecular characterisation of these and other antigens is now needed, particularly those that are excreted/secreted and consequently more accessible to B and T cells (see Lillywhite *et al.* 1995). Hopefully such work will lead to antigens emerging that may play a role in immunity and represent suitable vaccine candidates.

A further consideration is the fact that intestinal helminth infections rarely arise in isolation. For example, a person infected with trichuriasis is often likely to harbour or have harboured *A. lumbricoides* or a hookworm species. Considering the degree of cross-reactivity between nematodes and the inability to obtain a complete infection history for an individual there are obvious problems when analysing a species-specific response. In trichuriasis a successful technique, for removing antibodies capable of recognising *A. lumbricoides* has been adopted in several studies because these two species are often co-endemic (Lillywhite *et al* 1991; Bundy *et al.* 1991; Needham *et al.* 1992). This is extremely beneficial in terms of negating the problem of single versus dual infections. However, epitopes shared between these nematodes may be important and if so the response to them will be missed. There are therefore arguments for and against 'blanket' whole parasite specific antibody removal from sera. Again there is a need to examine antibodies specific for single species or shared nematode antigens.

What is currently known about antibody responses in trichuriasis has largely come from studies conducted in St Lucia in the Caribbean where there are varying levels of endemicity. The first description of parasite-specific serum isotypic responses, in which IgM, IgA, IgE, IgG1, IgG2 and IgG4 were all detected, revealed their diversity. Only the IgG3 response was minimal (Lillywhite *et al.* 1991). Furthermore, because *A. lumbricoides* was prevalent within the study group *Ascaris*-specific antibodies were experimentally depleted. Interestingly, the IgG and IgE responses were predominantly *T. trichiura* specific whereas greater degrees of cross-reactivity were found for the IgM and IgA subclasses. Whether any of these specific isotypes are involved in immunity can only be postulated.

As can be seen from a typical age-intensity profile there are defined phases of infection; an ascending phase in early childhood, a peak and a descending phase leading into a stable plateau in adulthood (Figure 8.1). In order to address the significance of parasite-specific antibody levels these can be related to worm burden. If systemic antibody levels are simply a reflection of burden then the two are likely to correlate positively and reflect the age-dependent intensity profile. However, if antibody levels are high when infection intensity is low this may suggest an association with immunity or even a protective effector mechanism. A subsequent study therefore examined an age-stratified group and found all isotypes largely paralleled infection levels (Bundy *et al.* 1991). In contrast, when individual age classes were examined, approximating to the different infection phases, IgA levels were found to negatively correlate with infection intensity in adults. This finding therefore suggests that if age-acquired resistance does exist IgA might be involved.

Another means of relating antibody levels to worm burden is to examine serum antibody in areas of high and low *T. trichiura* transmission. One such study revealed greater levels in the high infection intensity area and an age-dependency reflecting the intensity profile (Needham *et al.* 1992). The exception was IgA where levels persisted rather than declined through adulthood. However, this cross-sectional approach cannot separate age from length of exposure and consequently there is a need to follow individuals over a time period, monitor their infection status and relate changes in the levels of the antibody response. Such a longitudinal approach has been adopted and corroborated the previous investigations; IgG1, IgG2, IgG4 and IgE levels decreased significantly from the time the children were aged five to their tenth year, in common with the decrease in infection intensity (Needham *et al.* 1994). In contrast IgA levels did not decline significantly but remained constant over this time course.

Taken together, the above data suggest that the amount of anti-worm IgG is a reflection of the current infection intensity at the population level whereas IgA persists at a high level in adulthood even when worm numbers have declined. Interestingly, a move away from the Caribbean foci has now revealed a possible link between IgE, rather than IgA, and immunity. According to mathematical models of helminth transmission dynamics, age-acquired immunity is likely to operate in areas of hyperendemicity where there is a pronounced convex age-intensity profile (Anderson, 1986; Anderson & May, 1985).

Therefore a recent study was conducted in an area of Cameroon where, according to WHO guidelines, 26% of the population were classified as being heavily infected (> 10 000 epg) (Montresor *et al.* 1999). Here the investigators found a positive correlation between serum IgE and age and a negative association between IgE and infection intensity (Faulkner *et al.*

manuscript in preparation). This is interesting because IgE has previously been associated with immunity to *Necator americanus* and in immunity to re-infection with the helminths *Schistosoma mansoni* and *S. haematobium* (Pritchard *et al.* 1995; Hagan *et al.* 1991; Rihet *et al.* 1991; Dunne *et al.* 1992). In schistosomiasis IgG4 has been postulated to compete for the same binding sites as IgE and to block protective immunity operating in children (Hagan *et al.* 1991). Further studies are needed to assess whether this is the case in trichuriasis and to determine the relative contributions of IgA and IgE, particularly in terms of re-infection. No doubt the specificities of these antibodies will be important. In fact they may prove to be markers of exposure in adulthood rather than immunity, which is of interest because they are known to be markers in animals of a protective Th2 type response. An assessment of T cell responses in infected people would indicate whether this is also the case in humans.

4.2 T Cell Responses

In mice, it is clear that there is a resistant phenotype, where in response to a single infection of larvae, mice are able to expel worms and are resistant to challenge infection. The situation in humans is not so simple for although older children and adults have fewer worms than young children, they are not necessarily resistant. They may be more appropriately defined as chronically susceptible. Such is the defined nature of the polarization in mice it is necessary to pose the question, can comparisons be made between these laboratory investigations in a model system and human field studies?

People do display a range of infection intensities. Certainly within a community there are people who are more heavily infected than others and whom following treatment become re-infected to a similar extent as before (Bundy *et al.* 1988). However, it is difficult to find individuals living in an endemic area who are presumably ingesting eggs but remain uninfected for a long period of time. Perhaps this is due to the fact that people tend to ingest repeated small doses. There is evidence to suggest, in laboratory models of infection, that a small antigen dose promotes Th1 cell development and consequently susceptibility to trichuriasis (Bretscher *et al.* 1992; Bancroft *et al.* 1994). Furthermore, repeated small infective doses of *Trichuris* given to mice result in cumulatively higher worm burdens until expulsion is initiated and are accompanied by a mixed cytokine response (Bancroft *et al.* 2001). Therefore in the human condition different grades of intensity are likely to exist rather than a simple positive or negative outcome. A clear-cut polarization of the specific T cell response may simply not occur.

There are very few studies where the cytokines produced in response to gastrointestinal nematodes have been evaluated. This may be due to the difficulties in performing such assays in patients living in developing countries but it probably also reflects the relative lack of interest in studying parasites where there is little mortality. A recent study by Cooper *et al.* 2000 (see Chapter 6) showed that the cytokines produced in *Ascaris* infection were polarised towards a Th2 type response, in comparison to an uninfected control group. The same patient group was also infected with *T. trichiura* and cytokines produced in recall assays of peripheral blood mononuclear cells to this parasite were evaluated. No responses were found, which may have been due to the extremely low levels of infection.

Recently a study was undertaken where a comprehensive survey of cytokines produced by whole blood cultures was evaluated in a cross sectional age profile study of children aged between four and 15 years of age infected with *T. trichiura.* Interestingly only a small proportion (5-17%) of the study group produced *Trichuris*-specific IL-4, IL-9 and IL-13 whereas a larger proportion produced IL-10, TNF-α, and IFN-γ. No correlations were observed between any cytokine and intensity of infection but *Trichuris*-stimulated IL-10 production decreased with age, whereas when both TNF-α, and IFN-γ increased (Faulkner *et al*, manuscript in preparation). This suggests a switch with age (or exposure) to a more chronic susceptible phenotype with a mixed cytokine response. Studies are currently underway to examine responses in a study after drug treatment. This will allow correlations of immune responses with resistance to reinfection.

4.3 Trichuris in the Intestine

Trichuris species all have a life cycle occurring entirely in the gut and the adults are embedded in the epithelia (Figure 8.2). We therefore have to consider what mechanisms of resistance can operate in this very specialised environment. Although it has been shown in *T. muris* infection that peripheral cytokine responses are reflective of those occurring in mesenteric lymph nodes (Taylor *et al.* 2000) it remains a possibility that there are very local responses that mediate the expulsion of worms.

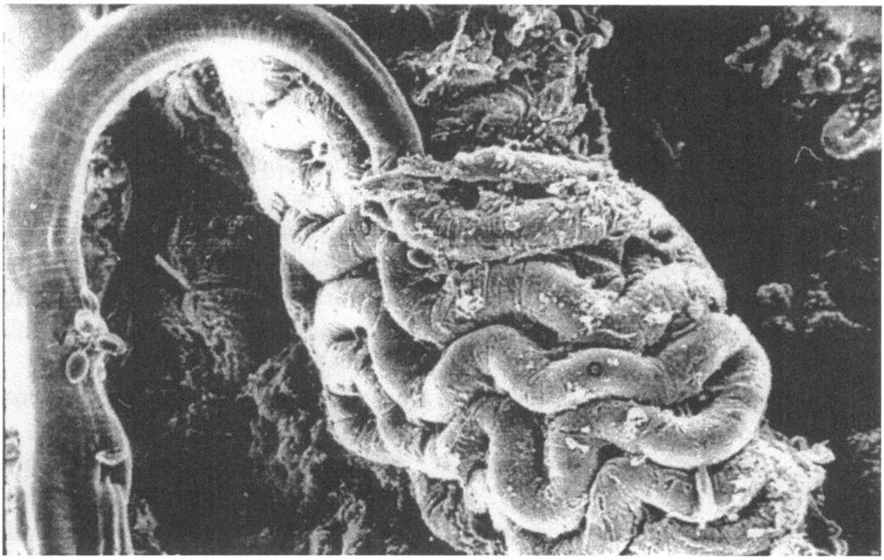

Figure 8.2. Adult *T. muris* embedded within the caecal epithelium where v = vulva and eggs and O = oesophagus. Photograph courtesy of Telfryn Jenkins.

Defining the cellular and immune responses to *Trichuris* in the human intestine is problematic and detailed studies on the patterns of reaction in individuals with unremarkable *Trichuris* infection has never been undertaken. Studies carried out in patients with severe disease or TDS have shown a mild to moderate mucosal inflammation (MacDonald *et al.* 1991). Analysis of the results has also proven difficult because "normal" individuals in the tropics have increased numbers of histiocytes, lymphocytes, and plasma cells compared to European controls (Jenkins 1988). The studies that have been performed have shown a remarkable absence of immunopathology despite heavy infections with the parasite (MacDonald *et al.* 1991, 1994; Cooper *et al.* 1990). There is seemingly no evidence of activation of T cells but there is evidence of a local mastocytosis with high levels of histamine being produced in the mucosa (Cooper *et al.* 1991; Cooper *et al.* 1992). There was also a 10-fold increase of cells with surface IgE; although these cells were not identified it is likely that many were mucosal mast cells. This evidence suggests that the inflammation in TDS could be a local anaphylactic response to *T. trichiura* mediated by parasite specific IgE. Non-specific immune mechanisms may also have an important role in worm expulsion and the pathogenesis of trichuriasis. Increased numbers of macrophages and cells containing TNF-α were observed in caecal biopsies from children with TDS (MacDonald *et al.* 1994). It is postulated that this may be the source of the elevated serum levels of this cytokine found in children with TDS, but the leakiness of the gut in this syndrome will allow activation of macrophages from other sources.

These human studies have associated certain specific-and non-specific immune parameters with intestinal pathology but no associations with protective immunity have been made. It has long been suggested that pathology and protection are co-dependant aspects of responses against intestinal nematodes: the immune response directed towards the parasite causes pathology which in turn results in the expulsion of the worm (Larsh, 1975).

There is an apparent paradox, however, because in mice worm expulsion is clearly a Th2 cytokine controlled mechanism, but the sort of pathology associated with intestinal nematode infections is usually attributable to Th1 cytokines: in *T. muris* infection host intestinal epithelial cell hyperproliferation has been shown to be regulated by IFN-γ (Artis *et al.* 1999a). Furthermore TNF-α, which is usually considered to be a Th1 type cytokine that can be down regulated by IL-4, is known to be associated with various intestinal pathogeneses. This cytokine has been shown to be critical in the expulsion of *T. muris* because TNF-α KO mice, with the background of a normally resistant phenotype, are unable to expel worms (Artis *et al.* 1999b). Interestingly, these mice also failed to mount a Th2 type of response suggesting that TNF-α has a role in regulating Th2 cytokine mediated responses at mucosal sites. Certainly, some of the changes seen in the intestinal epithelium during helminth infection are likely to be under the control of cytokines. Whether these changes can cause the expulsion of worms has yet to be defined. There is known to be a goblet cell hyperplasia in *T.muris* infection (W. I. Khan & R. K. Grencis, *unpublished observations*) and there is evidence in other nematode infections that the mucins they secrete may have a role in expulsion (see Garside *et al.* 2000; Lawrence *et al.* 2001). Also, the administration of IL-4 to SCID mice can cause worm loss in the absence of an adaptive immune response (K. J. Else, *unpublished observations*). Consequently, a greater understanding of the interplay between cytokine mediated intestinal pathology and effector function in response to helminth infections is needed (see Lawrence *et al.* 2001).

5. CONCLUSIONS

In the mouse model of trichuriasis, there is a considerable base of knowledge defining the immunological control of resistance and susceptibility to infection, yet the mechanisms of worm expulsion remain elusive. Our knowledge on the immune responses in the human infection is very limited, in part because it has been neglected due to a lack of consideration of its importance, but also due to difficulties inherent in human population studies. They are logistically difficult because long-term follow up studies are impossible due to funding and ethical considerations and cross-sectional population studies can at best be correlative. However, it is becoming obvious that gastrointestinal nematode infections may be much more important than the direct symptoms that they cause, as they may have profound effects on the outcome of other infections and may reduce vaccination efficacy. We therefore need to understand in greater detail the immunological responses induced by these parasites. The interactions between *Trichuris* infection, intestinal pathology and the mechanisms of worm explusion are complex and as yet it is far from clear how pathology and resistance to infection are controlled and whether they are associated. In order to be able to vaccinate against this parasite without inducing severe pathology it is essential for these mechanisms to be understood.

REFERENCES

ANDERSON, R. M. (1986). The population dynamics and epidemiology of intestinal nematode infections. *Transactions of the Royal Society of Tropical Medicine and Hygiene* **80**, 686-696.

ANDERSON, R. M. & MAY, R. M. (1985). Helminth infections of humans: mathematical models, population dynamics and control. *Advances in Parasitology* **24**, 1-101.

ANDERSON, R. M. & MEDLEY, G. F. (1985). Community control of helminth infections of man by mass and selective immunotherapy. *Parasitology* **90**, 629-660.

ARTIS, D. & GRENCIS, R. K. (2001). T helper cytokine responses during intestinal nematode infection: Induction, regulation and effector function. In *Parasitic Nematodes: Molecular Biology, Biochemistry and Immunology* (ed. Kennedy, M.W. & Harnett, W.), pp. 311-371. CABI publishing.

ARTIS, D., POTTEN, C. S., ELSE, K. J., FINKELMAN, F. D. & GRENCIS, R. K. (1999a). *Trichuris muris*: host intestinal epithelial cell proliferation during chronic infection is regulated by interferon-γ. *Experimental Parasitology* **92**, 144-153.

ARTIS, D., HUMPHREYS, N. E., BANCROFT, A. J., ROTHWELL, N. J., POTTEN, C. S. & GRENCIS, R. K. (1999b). Tumor necrosis factor α is a critical component of Interleukin-13-mediated protective T helper cell type 2 responses during helminth infection. *Journal of Experimental Medicine* **190,** 953-962.

BANCROFT, A. J., ARTIS, D., DONALDSON, D. D., SYPEK, J. P. & GRENCIS, R. K. (2000). Gastrointestinal nematode expulsion in IL-4 knockout mice is IL-13 dependent. *European Journal of Immunology* **30,** 2083-2091.

BANCROFT, A. J., ELSE, K. J. & GRENCIS, R. K. (1994). Low level infection of *Trichuris muris* significantly affects the polarisation of the CD4 reponse. *European Journal of Immunology* **24,** 3113-3118.

BANCROFT, A. J., ELSE, K. J., HUMPHREYS, N. E. & GRENCIS, R. K. (2001). The effect of challenge and trickle *Trichuris muris* infections on the polarisation of the immune response. *International Journal of Parasitology* In press.

BANCROFT, A. J., ELSE, K. J., SYPEK, J. & GRENCIS, R. K. (1997). IL-12 promotes a chronic intestinal nematode infection. *European Journal of Immunology* **27,** 866-870.

BENTWICH, Z., KALINKOVICH, A., WEISMAN, Z., BORKOW, G., BEYERS, N. & BEYERS, A. D. (1999). Can eradication of helminthic infections change the face of AIDS and tuberculosis? *Immunology Today* **20,** 485-487.

BETTS, K. & ELSE, K. J. (1999). Mast cells, eosinophils, and antibody-mediated-cellular-cytotoxicity are not critical in resistance to *Trichuris muris*. *Parasite Immunology* **21,** 45-52.

BRETSCHER, P. A., WEI, G., MENON, J. N. & BIELEFELDT-OHMANN, H. (1992). Establishment of stable, cell-mediated immunity that makes "susceptible" mice resistant to *Leishmania major*. *Science* **257,** 539-542.

BLACKWELL, N. M. & ELSE, K.J. (2001). B cells and antibodies are required for resistance to the parasitic gastrointestinal nematode parasite *Trichuris muris*. *Infection and Immunity* **69,** 3860-3868.

BUNDY, D. A. P. (1988). Population ecology of intestinal helminth infections in human communities. *Philosophical Transactions of the Royal Society of London* **B321,** 405-420.

BUNDY, D. A. P. & COOPER, E. S. (1989). *Trichuris* and trichuriasis in humans. *Advances in Parasitology* **28,** 107-173.

BUNDY, D. A. P. & MEDLEY, G. F. (1992). Immuno-epidemiology of human helminthiasis: ecological and immunological determinants of worm burden. *Parasitology* **104,** S105-S119.

BUNDY, D. A. P., COOPER, E. S., THOMPSON, D. E., ANDERSON, R. M. & DIDIER, J. M. (1987). Age-related prevalence and intensity to *Trichuris trichiura* infection in a St. Lucian community. *Transactions of the Royal Society of Tropical Medicine and Hygiene* **81,** 85-94.

BUNDY, D. A. P., COOPER, E. S., THOMPSON, D. E., DIDIER, J. M. & SIMMONS, I. (1988). Effect of age and initial infection status on the rate of reinfection with *Trichuris trichiura* after treatment. *Parasitology* **97,** 469-476.

BUNDY, D. A. P., LILLYWHITE, J. E., DIDIER, J. M., SIMMONS, I. & BIANCO, A. E. (1991). Age-dependency of infection status and serum antibody levels in human whipworm (*Trichuris trichiura*) infection. *Parasite Immunology* **13,** 629-638.

CHAN, M-S. (1996). The global burden of intestinal nematode infections-fifty years on. *Parasitology Today* **16,** 71-77.

COOPER, E. S. & BUNDY, D. A. P. (1987). Trichuriasis. *Baillières Clinical and Tropical Medicine and Communicable Diseases* **2,** 629-643.

COOPER, E. S., SPENCER, J., MURCH, S., VENUGOPAL, S., HANCHARD, B., BUNDY, D. A. P. & MACDONALD, T. T. (1990). Mucosal macrophages and plasma cachectin (TNF) in *Trichuris* colitis. *Bulletin de la Societe Francaise de Parasitologie* (Suppl 2) 347-351.

COOPER, E. S., SPENCER, J., WHYTE, C. A. M., CROMWELL, O., VENUGOPAL, S., WHITNEY, P., BUNDY, D. A. P., HAYNES, B. & MACDONALD, T. T. (1991). Immediate hypersensitivity in the colon of children with chronic *Trichuris trichiura* dysentery. *Lancet* **338**, 1104-1107.

COOPER, E. S., WHYTE-ALLENG, C. A. M., FINZI-SMITH, J. S. & MACDONALD, T. T. (1992). Intestinal nematode infections in children: the pathophysiological price paid. *Parasitology* **104** (Suppl.) S91-S103.

COOPER, P. J., CHICO, M. E., SANDOVAL, C., ESPINAL, I., GUEVARA, A., KENNEDY, M. W., URBAN, J. F., GRIFFIN, G. E. & NUTMAN, T. B. (2000). Human infection with *Ascaris lumbricoides* is associated with a polarised cytokine response. *Journal of Infectious Diseases* **182**, 1207-1213.

COOPER, P. J., CHICO, M., SANDOVAL, C., ESPINAL, I., GUEVARA, A., LEVINE, M. M., GRIFFIN, G. E. & NUTMAN, T. B. (2001). Human Infection with *Ascaris lumbricoides* is associated with suppression of the interleukin-2 response to recombinant cholera toxin B subunit following vaccination with the live oral cholera vaccine CVD 103-HgR. *Infection and Immunity* **69**, 1574-1580.

COOPER, P. J., ESPINAL, I., WEISEMAN, M., PAREDES, W., ESPINAL, M., GUDERIAN, R. H. & NUTMAN, T. B. (1999). Human Onchocerciasis and tetanus vaccination: impact on postvaccination antitetanus antibody response. *Infection and Immunity* **67**, 5951-5957.

CURRY, A. J., ELSE, K. J., JONES, F., BANCROFT, A., GRENCIS, R. K. & DUNNE, D. W. (1995). Evidence that cytokine-mediated immune interactions induced by *Schistosoma mansoni* alter disease outcome in mice concurrently infected with *Trichuris muris*. *Journal of Experimental Medicine* **181**, 769-774.

DUNNE, D. W., BUTTERWORTH, A. E., FULFORD, A. J. C., KARIUKI, H. C., LANGLEY, J. G., OUMA, J. H., CAPRON, A., PIERCE, R. J. & STURROCK, R. F. (1992). Immunity after treatment of human schistosomiasis: association between IgE antibodies to adult worm antigens and resistance to reinfection. *European Journal of Immunology* **22**, 1483-1494.

ELSE, K. J. & FINKELMAN, F. D. (1998). Intestinal parasites, cytokines and effector mechanisms. *International Journal of Parasitology* **28**, 1145-1158.

ELSE, K. J. & GRENCIS, R. K. (1991). Cellular immune responses to the murine nematode parasite *Trichuris muris*. I. Differential cytokine production during acute or chronic infection. *Immunology* **72**, 508-513.

ELSE, K . J. & GRENCIS, R. K. (1996). Antibody-independent effector mechanisms in resistance to the intestinal nematode parasite *Trichuris muris*. *Infection and Immunity* **64**, 2950-2954.

ELSE, K, J & WAKELIN, D. (1988). The effects of H-2 and non-H-2 genes on the expulsion of the nematode *Trichuris muris* from inbred and congenic mice. *Parasitology* **96**, 543-550.

ELSE, K. J., HULTNER, L. & GRENCIS, R. K. (1992). Modulation of cytokine production and response phenotypes in murine trichuriasis. *Parasite Immunology* **14**, 441-449.

ELSE, K. J., WAKELIN, D. & ROACH, T. I. A. (1989). Host predisposition to trichuriasis: the mouse-*T.muris* model. *Parasitology*. **98**, 275-282.

ELSE, K. J., FINKELMAN, F. D., MALISZEWSKI, C. R. & GRENCIS, R. K. (1994). Cytokine mediated regulation of chronic intestinal infection. *Journal of Experimental Medicine* **179**, 347-351.

ELIAS, D., WOLDAY, D., AKUFFO, H., PETROS, B., BRONNER, U. & BRITTON, S. (2001). Effect of deworming on human T cell responses to mycobacterial antigens in helminth-exposed individuals before and after bacilli calmette-Guerin (BCG) vaccination. *Clinical Experimental Immunology* **123**, 219-225.

FINKELMAN, F. D., PEARCE, E. J., URBAN, J. F. & SHER, A. (1991). Regulation and biological function of helminth-induced cytokine responses. *Immunoparasitology Today* **12/7**: A62-A66.

FAULKNER, H., RENAULD, J. C., VAN SNICK, J. & GRENCIS, R. K. (1998). Interleukin-9 enhances resistance to the intestinal nematode *Trichuris muris*. *Infection and Immunity* **66,** 3832-3840.

GARSIDE, P., KENNEDY, M. W., WAKELIN, D. & LAWRENCE, C. E. (2000). Immunopathology of intestinal helminth infection. *Parasite Immunology* **22,** 605-612.

GRENCIS, R. K. (1997). Th2-mediated host protective immunity to intestinal nematode parasites. *Philosophical Transactions of the Royal Society of London (B)* **29,** 1377-1384.

HAGAN, P., BLUMENTHAL, U. J., DUNN, D., SIMPSON, A. J. & WILKINS, H. A. (1991). Human IgE, IgG4 and resistance to reinfection with *Schistosoma haematobium*. *Nature* **349**, 243-245.

JENKINS, D. (1988). Computing and histopathology of intestinal inflammation. In *Computers in Gastroeneterology*, Vicary, F. R. (editor). London: Springer Verlag. 193-204.

LARSH, J. E. (1975). Allergic inflammation as a hypothesis for the expulsion of worms from tissues: a review. *Experimental Parasitology* **37,** 251-266.

LAWRENCE, C. E., KENNEDY, M. W. & GARSIDE, P. (2001). Gut Immunopathology in Helminth infections-paradigm lost? In *Parasitic Nematodes: Molecular Biology, Biochemistry and Immunology* (ed Kennedy, M.W. & Harnett, W.), pp.373-397. CABI publishing.

LILLYWHITE, J. E., BUNDY, D. A. P., DIDIER, J. M., COOPER, E. S. & BIANCO, A. E. (1991). Humoral immune responses in human infection with the whipworm *Trichuris trichiura*. *Parasite Immunology* **13**, 491-507.

LILLYWHITE, J. E., COOPER, E. S., NEEDHAM, C. S., VENUGOPAL, S., BUNDY, D. A. P. & BIANCO, A. E. (1995). Identification and characterization of excreted/secreted products of *Trichuris trichiura*. *Parasite Immunology* **17**, 47-54.

MACDONALD, T. T., CHOY, M-Y., SPENCER, J., RICHMAN, P. I., DISS, T., HANCHARD, B., VENGOPAL, S., BUNDY, D. A. P. & COOPER, E. S. (1991). Histopathology and immunohistochemistry of the caecum in children with the *Trichuris* dysentery syndrome. *Journal of Clinical Pathology* **44**, 194-199.

MACDONALD, T. T., SPENCER, J., MURCH, S. H., CHOY, M. –Y., VENUGOPAL, S., BUNDY, D. A. P. & COOPER, E. S. (1994). Immunoepidemiology of intestinal helminth infections. 3. Mucosal macrophages and cytokine production in the colon of children with *Trichuris trichiura* dysentery. *Transactions of the Royal Society of Tropical Medicine and Hygiene* **88**, 265-268.

MCSHARRY, C., XIA, Y., HOLLAND, C. V. & KENNEDY, M. W. (1999). Natural Immunity to *Ascaris lumbricoides* associated with Immunoglobulin E antibody to ABA-1 allergen and inflammation indicators in children. *Infection and Immunity* **67**, 484-489.

MONTRESOR, A., GYORKOS, T. W., CROMPTON, D. W. T., BUNDY, D. A. P. & SAVIOLI, L. (1999). *Monitoring Helminth Control Programmes. Guidelines for Monitoring the Impact of Control ProgrammesAimed at Reducing the Morbidity Caused by Soil-Transmitted Helminths and Schistosomes, With Particular reference to School-Age of Children* WHO/CDS/CPC/SIP/99.3 Geneva, WHO.

NEEDHAM, C. S. & LILLYWHITE, J. E. (1994). Immunoepidemiology of intestinal helminthic infections. 2. Immunological correlates with patterns of *Trichuris* infection. *Transactions of the Royal Society of Tropical Medicine and Hygiene* **88**, 262-264.

NEEDHAM, C. S., BUNDY, D. A. P., LILLYWHITE, J. E., DIDIER, J. M., SIMMONS, I. & BIANCO, A. E. (1992). The relationship between *Trichuris trichiura* transmission intensity and the age-profiles of parasite-specific antibody isotypes in two endemic communities. *Parasitology* **105**, 273-283.

NEEDHAM, C. S., LILLYWHITE, J. E., DIDIER, J. M., BIANCO, A. E. & BUNDY, D. A. P. (1993). Age-dependency of serum isotype responses and antigen recognition in human whipworm (*Trichuris trichiura*) infection. *Parasite Immunology* **15**, 683-692.

NEEDHAM, C. S., LILLYWHITE, J. E., DIDIER, J. M., BIANCO, A. E. & BUNDY, D. A. P. (1994). Temporal changes in *Trichuris trichiura* infection intensity and serum isotype responses in children. *Parasitology* **109**, 197-200.

NOKES, C. & BUNDY, D. A. P. (1994) Does helminth infection affect mental processing and educational achievement? *Parasitology Today* **10.** 14-18.

NOKES, C., GRANTHAM-MCGREGOR, S. M., SAWYER, A. W., COOPER, E. S., ROBINSON, B. A. & BUNDY, D. A. P. (1992). Moderate to heavy infections of *Trichuris trichiura* affect cognitive function in Jamaican school children. *Parasitology* **104**, 539-547.

PEARLMAN, E., KAZURA, J. W., HAZLETT, F. E. & BOOM, W. H. (1993). Modulation of murine cytokine responses to mycobacterial antigens by helminth-induced T helper 2 cell responses. *Journal of Immunology*. **151,** 4857-4864.

PRITCHARD, D. I., QUINNELL, R. J. & WALSH, E. A. (1995). Immunity in humans to *Necator americanus*: IgE, parasite weight and fecundity. *Parasite Immunology* **17**, 71-75.

RAMSEY, F. C. (1962). Trichuris Dysentery Syndrome. *West Indies Medical Journal* **11**, 235-9.

RIHET, P., DEEURE, C. E., BURGOIS, A., PRATA, A. & DESSAIN, A. J. (1991) Evidence for an association between human resistance to *Schistosoma mansoni* and high anti-larval IgE levels. *European Journal of Immunology* **21**, 2679-2686.

ROSSIGNOL, J. F. & MAISONNEUVE, H. (1984). Benzamidazoles in the treatment of trichuriasis: A Review. *Annals of Tropical Medicine and Parasitology* **78**, 135-144.

ROUSSEAU, D., LE FICHOUX, Y., STEIN, X., SUFFIA, I., FERRUA, B. & KUBAR, J. (1997). Progression of visceral leishmaniasis due to leishmania infantum in BALB/c mice is markedly slowed by prior infection with *Trichinella spiralis*. *Infection and Immunity* **65,** 4987-4983.

SABIN, E. A., ARAUJO, M. I., CARVALHO, E. M. & PEARCE, E. J. (1996). Imparment of tetanus toxoid-specific Th1-like immune responses in humans infected with *Schistosoma mansoni*. *Journal of Infectious Diseases* **173,** 269-272.

SHER, A. & COFFMAN, R.L. (1992). Regulation of immunity to parasites by T cells and T cell-derived cytokines. *Annual Reviews in Immunology* **10,** 385-409

SIMEON, D. & GRANTHAM-MCGREGOR, S. (1990). Nutritional deficiences and childrens behaviour and mental development. *Nutrition Research Reviews* **3,** 1-24.

SHIRAKAWA T, ENOMOTO, T., SHIMAZU, S. & HOPKIN, J. M. (1997). The inverse association between tuberculin responses and atopic disorder. *Science* **275,** 77-79.

STEWART, G. R., BOUSSINESQ, M., COULSON, T, ELSON, L, NUTMAN, T. & BRADLEY, J. E. (1999). Onchocerciasis modulates the immune response to mycobacterial antigens. *Clinical Experimental* Immunology **117,** 517-523.

TAYLOR, M. D., BETTS, C. J. & ELSE, K. J. (2000). Peripheral cytokine responses to *Trichuris muris* reflect those occurring locally at the site of infection. *Infection and Immunity* **68,** 1815-1819.

URBAN, J. F., MADDEN, K. B., SVETIC, A., CHEEVER, A., TROTTA, P. P., GAUSE, W. C., KATONA, I. D. & FINKELMAN, F (1992). The importance of Th2 cytokines in protective immunity to nematodes. *Immunological Review* **12,** 205-220.

Chapter 9: THE IMMUNOBIOLOGY OF HOOKWORM INFECTION.

D.I. Pritchard[1], R.J. Quinnell[2], P.J. Hotez[3], J.M. Hawdon[3] and A. Brown[1].
[1]Boots Science Building, School of Pharmacy, University of Nottingham, UK; [2]School of Biology, University of Leeds, UK; [3]Department of Microbiology and Tropical Medicine, George Washington University, Washington, D.C., USA.
e-mail: David.Pritchard@nottingham.ac.uk

1. INTRODUCTION

The hookworm of humans (*Necator americanus* and *Ancylostoma duodenale*) are small (9-13 mm by 0.35 - 0.6 mm) in their adult stage. They feed on blood and intestinal wall tissue, producing anti-haemostatic materials (Cappello *et al.* 1993; Furmidge *et al.* 1995; Stanssens *et al.* 1996; Chadderdon & Cappello, 1999; Del Valle *et al.* 1999) and possibly exist under some conditions of immune privilege (Pritchard & Brown, 2001). The infective larvae traverse the skin and the lungs before reaching the gut in the case of *Necator americanus* (Pritchard & Brown, 2001). *Ancylostoma duodenale* infects through the skin, but also orally, and may enter a stage of suspended animation or hypobiosis as a larva prior to resuming its life cycle (Schad *et al.* 1973). Transmammmary infections may also occur for *A. duodenale* infections (Hotez & Pritchard, 1995).

2. MOLECULAR PATHOGENESIS OF HOOKWORM INFECTIONS

Hookworm infection can be a major cause of iron deficiency and anaemia in developing countries, depending on the intensity of infection (Pritchard *et al.* 1991; Hotez & Pritchard, 1995 and see Chapter 3). By using radioactive tracers, it has been possible to estimate the amount of blood lost per day to an individual adult hookworm. These estimates vary with species, with 0.2 ml lost per day per female *Ancylostoma duodenale*

(10-13 mm long, 10000-25000 eggs per day (e.p.d.)), compared to 0.04 ml per day for the smaller and less fecund female *N. americanus* (9-11 mm long, 5000-10000 e.p.d. (Crompton, 2000)). The severity of hookworm anaemia will thus depend on the parasite burden and species of parasite, as well as dietary iron intake and losses due to other causes. Typically, worm burdens above 5000 eggs per gram (e.p.g.) (equivalent to 50-150 worms) are associated with a reduction in haemoglobin concentration, though in pregnant women burdens as low as 1000 e.p.g. may cause anaemia. Effects on iron stores, as measured by serum ferritin levels, are apparent at even lower worm burdens (Pritchard *et al.* 1991). Because hookworm infection often occurs together with other infections, particularly malaria, the relative importance of hookworm in the causation of anaemia can be difficult to quantify. However, recent studies from Kenya & Nepal, using attributable fraction methods, have shown that hookworm may cause 30-50 % of moderate to severe anaemia in pregnant women (Shulman *et al.* 1996; Dreyfuss *et al.* 2000).

In China it is still common to identify heavily infected patients with hookworm anaemia. These patients are frequently cachectic and are suffering from negative nitrogen balance. Of interest, the clinical cases of hookworm anaemia and disease appear largely among the elderly in southern China. High hookworm burdens with greater than 20,000 e.p.gs are found predominantly in remote rural areas of Sichuan, Yunnan and Hainan provinces.

A developing knowledge of the identity of molecules important in blood-feeding (Table 9.1) has raised the possibility of vaccination against pathology, rather than infection. It is not known whether naturally-infected humans mount an effective anti-pathology immune response, although antibody responses to a number of molecules involved in blood-feeding have been observed. However, the potential for such vaccination has been shown in laboratory models. For instance, vaccination of hamsters with soluble extracts of *A. ceylanicum* results in resistance to anaemia, but not to hookworm infection (Bungiro *et al.* 2001).

In contrast, vaccination with neutrophil inhibitory factor (NIF) reduced worm fecundity with no effect on pathology (Ali *et al.* 2001; see below). However, an anti-pathology effect was seen in animals vaccinated with irradiated hookworm (*Necator*) larvae, as assessed by reduced haemorrhage and albumin release in the lungs (Culley *et al.* 2001).

It has been argued that hookworm infection may be beneficial in some cases (Pritchard & Brown, 2001). The idea that there is either an association or protection from asthma resulting from hookworm infection is an old one

Table 9.1. Putative anti-haemostatic molecules from hookworms.

A number of molecules believed to be involved in the anti-haemostatic strategies of *Necator americanus* and *Ancylostoma spp* are detailed. For *Necator* these include inhibitors of blood coagulation factors Xa (Furmidge *et al.* 1995) and XIIIa (Brown *et al.* 2001), including calreticulin (Pritchard *et al.* 1999), which has been shown to inhibit Xa activity *in vitro* (unpublished data). The secreted adult and larval necepsins degrade fibrinogen (Girdwood, 1999) while as yet uncharacterised inhibitors of platelet aggregation (Furmidge *et al.* 1995) and platelet binding to von Willibrand Factor have also been identified. *Ancylostoma* antihaemostatic strategies like *Necator* include the inhibition of Factor Xa (Cappello *et al.* 1993), inhibition of platelet aggregation (Chadderdon & Cappello, 1999) and inhibition of the VIIa/Tissue Factor complex (Stanssens *et al.* 1996).

Life cycle stage	***Necator americanus***	***Ancylostoma***
L_3	Necepsin 2 (fibrinogenolytic) Calreticulin (Factor Xa inhibitor?)	?
L_4	Calreticulin (Factor Xa inhibitor?)	?
Adult	Factor Xa inhibitor Factor XIIIa inhibitor Inhibitor of platelet aggregation Inhibitor of platelet binding to von Willibrand Factor Necepsin 1 (fibrinogenolytic) Calreticulin (Factor Xa inhibitor?)	Factor Xa inhibitor (AcAP5, AcAP6) Factor VIIa/Tissue Factor inhibitor (AcAP2) Inhibition of platelet aggregation by blocking GPIIb/IIIa and GPIa/IIa

that remains controversial although, recently Scrivener *et al.* (2001), demonstrated that hookworm infection reduces the risk of respiratory wheeze in economically developing populations and prevents the symptoms of asthma in atopic subjects in rural environments. Conversely, new evidence from Bentwich (Bentwich *et al.* 2000; Bentwich *et al.* 2000) and colleagues, who examined Ethiopian refugees arriving in Israel, suggest that *Necator* infections may predispose to intercurrent viral infections including HIV (and see Chapter 16). In the early part of this century it was shown among military recruits that *Necator* predisposed to intercurrent measles infection. Possibly because of the host T helper type 2 (Th2) bias that *Necator* introduces into its human host, individuals are less able to mount effective T helper type 1 (Th1) antiviral responses. Further work needs to be done to support the view that HIV and other viruses are opportunistic pathogens of hookworm patients. However, the implication of such findings are enormous and suggest the possibility that hookworm and other geohelminths might partially account for the rapid spread of HIV in the developing nations of Africa and India.

3. HOOKWORMS AND THE IMMUNE SYSTEM

During their migration and establishment in humans, hookworms are at all stages of their life cycles in intimate contact with components of the immune system (Hotez & Pritchard, 1995; Pritchard & Brown, 2001). Each of these compartments is capable of vigorous immune reactivity, evidenced by atopic and contact delayed-type hypersensitivity (DTH) reactions in the skin to allergens (Bos & Kapsenberg, 1993), and lung and gut reactivity to allergic and infectious challenges (Lewis & Griffin, 1995; Culley *et al.* 2001). These are not sites of immune privilege, yet the highly antigenic hookworms survive in sufficient numbers to reproduce and perpetuate their life cycles.

However, it is difficult to gauge the infectious success rate of these parasites. Exposure to infective stages is difficult to quantify, as is the degree of attrition in the tissues during migration. Figures have been ascribed to adult life-span in the gut (Hoagland & Schad, 1978), and *Necator* is reputed to survive as an adult for an average of five years to a maximum of 13 years, with *Ancylostoma* surviving on average 12 months. Based on the relative life expectancies of the hookworms coupled with the observation that *Ancylostoma* is more pathogenic and causes more blood loss than *Necator*; Hoagland and Schad (1978) have suggested that the former is more 'opportunistic' in its host-parasite relationship.

The adult worm thus has a long term haematophagous existence (see Table 9.1 for details of putative anti-haemostatics) in the gut while immune stimulatory and/or immune regulatory larval stages continue to enter the immunological compartments of the host. The infection continues despite the presence of proteins that are presumably cross-reactive immunologically between the different life cycle stages (Table 9.2).

4. IMMUNE-EPIDEMIOLOGY OF HOOKWORM INFECTION

4.1. Age-prevalence and age-intensity profiles

Age-prevalence and age-intensity profiles for hookworm infection typically show a rise in childhood to a peak or plateau in teenage years or adulthood (Anderson, 1986). Thus the highest intensity (mean worm burden), and greatest pathology, of hookworm infection are usually seen in adults. In this respect, hookworm epidemiology is distinctly different from that of other geohelminths, such as *Ascaris* and *Trichuris*, where prevalence and intensity are usually highest in children. For instance, in China's Hainan province (an island in the South China Sea), *Necator* causes disease predominantly in the middle aged and elderly populations, whereas *Ascaris* infection predominates among school age children (Ghandi *et al.* 2001). Here, the *Neactor* age-intensity profile is increasing (monotonic), but in some areas of the world convex (peaked) profiles are seen. Such convex profiles are more often seen for *Ancylostoma* or mixed species infections than for *Necator*. For instance, in Anhui, China's poorest eastern province, *Ancylostoma* infections peak in middle age (Wang *et al.* 1999), and in Paraguay, where mixed infection occurs, 5-14 year old children have the heaviest worm burdens (Labiano-Abello *et al.* 1999).

The intensity of hookworm infection will depend on the balance between two population processes, the rate of acquisition of worms by the host (rate of exposure to infective stages) and the rate of loss (worm mortality rate; Anderson, 1986). Thus the shape of the age-intensity profile will depend on the relationships between exposure and host age, and worm mortality and host age. If neither exposure nor worm mortality vary with host age, a monotonic age-intensity profile is expected. In contrast, convex profiles can be generated if exposure is lower in adults than children, or the

Table 9.2. Antigens of importance present in different life cycle stages of *N. americanus.*

Antigens of importance are expressed at different stages of the life cycle of *N. americanus*. Acetylcholinesterase, (Pritchard *et al.* 1991) possibly involved in the inhibition of gut peristalsis, and necpain (Pritchard *et al.* 2001), believed to play a role in haemoglobin degradation are both expressed in the gut dwelling and haematophagus L_4 and adult stages. Similarly, necepsins 1 and 2 (Girdwood, 1999) also believed to play a role in haemoglobin degradation are expressed in L_4 and adults. In addition necepsin 2 also degrades skin macromolecules during larval skin penetration (Brown *et al.*, 1999) and is expressed in *Necator* larvae. The major allergen calreticulin (Pritchard *et al.* 1999; Kasper *et al.* 2001) is expressed in all 3 life cycle stages

Antigen	***L_3***	***L_4***	***Adult***
Acetylcholinesterase	X	✓	✓
Calreticulin	✓	✓	✓
Necepsin 1	X	✓	✓
Necepsin 2 (Girdwood, 1999)	✓	✓	✓
Necpain (Necator cathepsin B)	X	✓	✓

worm mortality rate is higher in adults than children. The different patterns seen for hookworm versus other geohelminth infections may reflect age-related differences in exposure. Though estimates of exposure to geohelminth infections are very hard to obtain, it is conceivable that exposure to the skin-penetrating infective stages of hookworm will be higher in adults than children, whereas oral exposure to *Ascaris* and *Trichuris* is thought to be highest in children.

What can we conclude about acquired immunity from age-intensity profiles? Acquired immunity is likely to increase parasite death rates in older hosts, who have been exposed to infection for longer, and so may generate convex age-intensity profiles. However, mathematical modelling has shown that, even if acquired immunity is operating, there may be monotonic age-intensity profiles (Woolhouse, 1992). Interpretation of age-intensity profiles is further complicated by the unknown (and perhaps increasing) relationship between exposure and age. Stronger evidence for an effect of acquired immunity has come from the analysis of the relative convexity of many age-intensity profiles from diverse populations. For hookworms, and other helminths, the age at which the peak intensity is seen can be shown to be inversely related to the strength of transmission: where transmission is more intense, the age-intensity profile peaks earlier (i.e. is more convex). This pattern, termed a 'peak shift', is strongly suggestive of a role for acquired immunity in determining the intensity of infection (Woolhouse, 1998).

4.2. Immune responses to hookworm infection

Human hookworm infection, like all helminth infection, results in a strong Th2 immune response, with high levels of eosinophils and antibodies, particularly specific and non-specific IgE. Specific IgG responses have been demonstrated against cuticular proteins (Pritchard *et al.*, 1988), cathepsin B and necepsin 1 (Brown, 2000) and acetylcholinesterase (Brown & Pritchard, 1993) while specific IgE has been recorded against the hookworm allergen calreticulin (Pritchard *et al.* 1999). Recent cellular studies in Papua New Guinea have shown that, as expected, infected individuals produce IL-4 in response to hookworm antigen. However, most people also mount a proliferative and IFN-γ response to infection, suggesting a mixed Th1/Th2 cytokine response (Quinnell, 2001). Some immune responses, such as the IgG4 antibody response to crude parasite antigens, appear to correlate with prevalence and intensity (Palmer *et al.* 1996; Xue *et al.* 2001), and may one day be used as a marker for infection in patients unwilling to provide faecal samples. In contrast, there is evidence from Papua New Guinea that Th2 responses may be protective, as levels of both total and anti-hookworm IgE are negatively correlated with hookworm size and fecundity (Pritchard *et al.* 1995). Similarly, negative correlations have been observed between hookworm burden, particularly in adults, and immune responses, such as IFN-γ responses and anti-hookworm IgG, IgM and IgE antibody levels (Quinnell *et al.* 1995; Quinnell, 2001). Such correlations may indicate either that these are protective immune responses, or that they are down-regulated in heavy infections, or both. Immune-modulation is known to occur, as anti-hookworm proliferative responses rise after chemotherapy. However, these studies raise the interesting possibility that both antibody and Th1 responses may reduce worm burden, whilst Th2 responses reduce worm fecundity. Intriguingly, two studies have shown that individuals who have received BCG vaccination, which may bias towards Th1 responsiveness, have a lower prevalence of hookworm infection than unvaccinated individuals (Barreto *et al.* 2000; Elliott *et al.* 1999). Laboratory studies clearly show the potential for separate anti-fecundity and anti-worm burden immunity; for instance, vaccination of hamsters with *A. ceylanicum* neutrophil inhibitory factor reduces worm fecundity, but not worm burden (Ali *et al.* 2001). One possibility is that protective anti-larval immunity is largely Th1, whilst anti-adult responses are largely Th2. Studies are underway to determine whether antigen-specific antibodies might correlate with resistance.

4.3 Variation in worm burden between individuals - overdispersion and predisposition

In common with other helminth infections, hookworms are highly aggregated, with many hosts having few worms, and only a few hosts having heavy burdens (Anderson & May, 1991). Typically, 20 % of the population have 80 % of the worms. This pattern has some important consequences: in particular, only a proportion of the population will have severe pathology. Variation between people in their exposure to infective stages may be important in generating aggregation, though variation in protective immunity may also be involved. There is also strong evidence from treatment and reinfection studies that certain individuals are predisposed to heavy or light infection (Keymer & Pagel, 1990), suggesting that there are consistent differences through time between people in their exposure or immunity (also see Chapter 1). The relative importance of exposure or immunity in generating predisposition is not known. However, recent studies have suggested some immunological effect, illustrated by predisposition to high or low worm weight, as well as worm burden (Quinnell *et al.* 2001). There is also evidence for genetic control of human hookworm burdens, which suggests that genetic factors, perhaps related to immunity, are involved in predisposition (Williams-Blangero *et al.* 1997) (see Chapter 10).

5. IMMUNE EVASION AND MODULATION BY HOOKWORMS

The immunological relationship between hookworms and humans is complex to say the least. How does a highly antigenic organism (Table 9.3 lists the parasite components and secretory products known to be antigenic during human infection) survive in an immunologically hostile environment? Is there any evidence to suggest that hookworms subvert the immune system? Table 9.4 lists the immune evasion strategies possibly employed by hookworms.

5.1. Immune evasion by larval stages

It is apparent that the sheath (cast cuticle of pre-infective larval stage) of the parasite may afford a degree of early protection, particularly in pre-exposed and immunologically primed individuals, as the L_3 stage may carry the antigenic sheath into the skin during infection (Kumar & Pritchard, 1992). Furthermore, hookworms possess potent collagen-binding proteins, albeit identified using cDNA library and phage display technology and affinity panning onto human collagen, when preferential affinity for host collagen was shown. The secretion of such proteins by larval stages (to be demonstrated) would serve to leave a false antigenic trail behind the migrating parasite, almost like a slug trail, thus diverting precious immunological resources in the skin to a parasite protein bound to host collagen at this important immunological effector site. L_3 extracts have also been shown to suppress mitogen-induced T cell proliferation, albeit in a rodent system, an attribute that would surely assist infective larvae to survive and to re-infect primed hosts. Whatever the strategy employed by larval stages to evade immunity, this can clearly be overcome under some circumstances, as evidenced by the ability to vaccinate with bolus infections of live L_3 larvae (Brown, 2000; Culley *et al.* 2001). This observation has important implications for future vaccination strategies.

Table 9.3. Parasite components and secretory products known to be antigenic during human infection.

Many molecules secreted by both *Necator* and *Ancylostoma spp* are immunogenic during infection. These include for *Necator*, excretory-secretory (ES) products (Carr & Pritchard, 1987), cuticular proteins (Pritchard *et al.* 1988), calreticulin (Pritchard *et al.* 1999), necpain (Pritchard *et al.* 2001), necepsins 1 and 2 (Girdwood, 1999) and acetylcholinesterase (Brown & Pritchard, 1993). In *Ancylostoma* infection immunogenic proteins include ES products, the Asp proteins (Sen *et al.*, 2000) and a major secretory antigen Ac68.

Life cycle stage	***Necator americanus***	***Ancylostoma***
L_3	Cuticular molecules Secretory products Necepsin 2 Calreticulin	Secretory products Asp-1 Asp-2
L_4	Cuticular molecules Secretory products	Secretory products Asp-2?
Adult	Cuticular molecules Secretory products Necepsin 1 Calreticulin Necpain AChE	Secretory products Asp-2 Ac68 (Secretory antigen)

The predominant proteins released by *Ancylostoma* larvae after host stimulation have now been isolated and their genes cloned. It is presumed that these gene products are released by the L_3 upon host entry. The two most abundant molecules are cysteine rich secretory proteins (CRISPs) known as the ASPs (*Ancylostoma* secreted proteins). Asp-1 is a 45 kDa non-glycosylated polypeptide (Hawdon *et al.* 1996), and Asp-2 is a glycosylated 24 kDa protein (Hawdon *et al.* 1999). Both CRISPs have regions of amino acid sequence similar to insect venom proteins; Asp-1 is a heterodimorphic repeat of an Asp-2 like monomer (Hawdon *et al.* 1999). The function of the Asps is still unknown, although both the monomorphic and heterdimorphic forms are conserved among the hookworms including *Necator* (Zhan *et al.* 1999). The third major protein released by *Ancylostoma* L_3 is a 60 kDa zinc metalloprotease known as MTP. The cDNA for MTP was recently cloned and found to belong to the astacin family of metalloproteases (Zhan *et al.* 2001).

Table 9.4. Immune evasion strategies possibly employed by hookworms

Hookworms secrete a number of proteins which are used to evade the immune system and establish a site of immune privilege at the site of attachment. These include casting off the L_3 sheath to divert the immune system (Kumar & Pritchard, 1992), the anti oxidant shield to combat the superoxide anion (O_2^-) produced by immune effector cells (Pritchard & Brown, 2001), an eotaxin degrading metallo-proteinase which reduces the infiltration of eosinophils to the site of attachment and pro-apoptotic factors (kaliseptines/kalicludines) that kill activated T cells. *Ancylostoma* spp secrete a neutrophil inhibitory factor (NIF), which inhibits CD11b/18 dependent leucocyte function (Moyle *et al.* 1994) and may have important therapeutic potential (Rahman *et al.* 2000).

Life cycle stage	***Necator americanus***	***Ancylostoma***
L_3	Nec H1 collagen binding protein Discarded sheath	Not known
L_4	Not known	Not known
Adult	Calreticulin Anti-oxidant shield Eotaxin metalloproteinase Pro-apoptotic factor Kaliseptines/kalicludines? Nec H1 collagen binding protein	Neutrophil inhibitory factor (NIF)

Very little is known about the immune evasion strategies employed by the L_4 stage, either in the lung or upon its early arrival in the gut. Given the key role played by eosinophils in the lung in protecting against larval trans-migration (Culley *et al.* 2001), it will be important to search for the eotaxin metalloproteinase recently discovered in adult stages (Culley *et al.* 2000). The recent survey of EST's (Wellcome Trust Beowulf Initiative) from an L_4 cDNA library is beginning to shed light on the molecular capability of this stage to subvert immunity.

5.2. Potential immune evasion molecules associated with adult stages

A larger number of putative immune evasion molecules have been discovered from adult hookworms, particularly *Necator*, primarily because of the biomass of material available (sparse as opposed to meagre!) for study and the fact that an early expressed sequence tag or EST project was conducted on the adult stage of this species. It is equally possible that some if not all of these activities are expressed at all stages of the life cycle.

5.2.1. Calreticulin (*Necator*).

Calreticulin-like protein was discovered during the screening of an adult *Necator* cDNA library with plasma from infected individuals from Papua New Guinea and a second antibody designed to detect IgE binding (Pritchard *et al.* 1999). The aim was to identify allergens associated with possible immune protection against hookworm infection. Having been duly identified and detected in all stages of the life cycle, a recombinant calreticulin was assessed for its ability to interact with the complement system (Kasper *et al.* 2001), given the association of calreticulin with C1q in systemic lupus erythematosus (Eggleton *et al.* 1997; Kishore *et al.* 1997). Calreticulin is also implicated in cytoplasmic signalling events following association with integrins (Reilly *et al.* 2000). An investigation of possible interactions with the signalling domains of integrins, in particular αIIb subunit, wild type $\alpha 2$, $\alpha 5$, and αv subunits was undertaken. In each case, a close association with these important immunologically active molecules was seen, suggesting an immune modulatory role for calreticulin. However, calreticulin has not yet been conclusively proven to be secreted by hookworms, although immune-reactive material does appear in secretory products. Calreticulin is, however, found in tick saliva (Jaworski *et al.* 1996) and on the outer surface of cell membranes (Gray *et al.* 1995), undermining its reputation as solely chaperone resident in the endoplasmic reticulum (Krause & Michalak, 1997). Given its potential importance to immune evasion, parasite metabolism and its cross-stage expression, calreticulin was recently nominated as a hookworm vaccine candidate (Hotez *et al.* 1999).

5.2.2 Eotaxin metalloproteinase and the anti-oxidant shield (*Necator*).

Of the immunological responses elicited by infection, the Th2 response, with its associated IgE and eosinophilia, appears to be crucial to protection against helminth infection. It would be logical for a successful parasite such as a hookworm to have evolved a capacity to deal with at least some components of this seemingly compartmentalised immune network, particularly as it can be argued that hookworms are manifestly allergenic (Pritchard, 1993; Pritchard & Walsh, 1995; Pritchard *et al.* 1999). This would indeed appear to be the case, in that adult *Necator* at least possesses a metalloproteinase activity capable of specifically cleaving eotaxin 1 (Culley *et al.* 2000). Cleavage of eotaxin by this metalloproteinase prevents the infiltration of radiolabelled eosinophils in hamster skin. Similarly, MEP-1 a gut zinc metallo-proteinase localised to the gut brush border membrane of *A. caninum* (Jones & Hotez, 2001) is also being investigated as a possible enzyme that may cleave eotaxin.

The lack of activity against eotaxin 2 suggests host adaptation to hookworm infection. However, any antibody and complement-primed eosinophils overcoming this defence to engage the parasite with their respiratory burst would have their effect neutralised by the combined secretion of a superoxide dismutase (Brophy *et al.* 1995) and glutathione-S-transferase (Brophy *et al.* 1995), providing the parasite with an anti-oxidant shield (Pritchard & Brown, 2001).

The status of the stand-alone SOD in *Necator* still poses a few questions. It has been argued that helminths are relatively resistant to hydrogen peroxide (Devine, 1995), but hard evidence is lacking. What is likely is that any hydrogen peroxide generated will have a cytotoxic effect on infiltrating leucocytes, and may act as a chemical defence against immunological attack.

5.2.3 T cell toxins (*Necator*).

Hookworm secretions induce apoptosis in activated human T cells and T cell lines (Chow *et al.* 2000). This could clearly be important to parasite survival, given the pivotal pole of the T cell in parasite expulsion. The apparent selectivity of action against activated cells, if operative *in vivo*, could result in sites of localised privilege in the absence of overt toxicity against bystander leukocytes, a mechanism benefiting both host and parasite alike.

The molecules responsible for this effect have yet to be identified but are of low molecular mass. Coincidentally, the peptidic kaliseptines identified in an EST project (Daub *et al.* 2000) have the potential to interfere with T cell function by modulating Kv1.3 channel activity. Although hookworm kaliseptines have yet to be proven to possess such activity, sea anenome kaliseptines certainly do (Schweitz *et al.* 1995), and hookworm secretions modulate human T cell activity in a manner suggestive of Kv1.3 involvement (C. Jagger, *personal communication*). Following engagement of the T cell receptor by mitogen, T cells typically show a dramatic increase in their intracellular calcium level ($[Ca^{2+}]_i$) which often occurs as a series of pronounced oscillations (Berridge *et al.* 1998). As a direct consequence of these oscillations, factors such as NF-AT enter the nucleus and activate specific genes for products (such as IL-2) which amplify the immune response. When human PBMCs are exposed to *N. americanus* excetory-secretory (ES) products following the addition of mitogen, the $[Ca^{2+}]_i$ in activated cells is reduced. The discovery in *N. americanus*, of a family of mRNAs for kaliseptine-like molecules suggests that the reduction in $[Ca^{2+}]_i$ seen in the presence of hookworm ES products, is due to blockade of regulatory K^+ channels by factors present in the *N. americanus* secretions. It is also worth noting that sites of inflammation have potassium ion concentrations recorded at 10 mM in excess of normal. Such concentrations can induce T cell de-polarisation, and potassium efflux through Kv1.3, leading to integrin–mediated cell adhesion and migration. T cell toxins such as kaliseptines could certainly interfere with T cell physiology by blocking Kv1.3.

5.2.4 Neutrophil inhibitory factor (NIF-*Ancylostoma*).

NIF was initially discovered in extracts from the dog hookworm *Ancylostoma caninum* (Moyle *et al.* 1994), and later in *Ancylostoma ceylanicum*, although it is not found in *Necator*. NIF is of molecular mass 41 kDa and has an affinity for the I domain of the integrin CD11b/18, which results in its ability to prevent neutrophil adhesion and activation. Its potency in such assays has led to its application to re-perfusion injury in humans, reinforcing the belief that parasites remain a relatively untapped source of immune modulatory molecules. With regard to its role in the host parasite relationship, NIF would appear to be secreted by *Ancylostoma ceylanicum*, and it's secretions possess anti-neutrophil activity *in vitro* (Ali *et al.* 2001). Furthermore, animals can be vaccinated with NIF against challenge infection; vaccination results in a significant reduction (85.8 %) in worm fecundity by 21 days post challenge infection, indicating the value of the molecule to the genus *Ancylostoma*.

6. VACCINATION

Although they have been available for decades, the benzimidazole anthelminthics have failed to control hookworm in endemic areas. Mebendazole first came into the market in 1972 and albendazole in 1983. One of the major reasons for this failure are the high rates of re-infection that occur following treatment (Quinnell *et al.* 1993; Quinnell *et al.* 2001). A World Health Organisation- sponsored study in Tanazania found that post-treatment re-infection occurs within four to12 months, usually to pre-treatment levels (Albonico *et al.* 1995). Also of concern is the potential for emerging benzimidazole anthelminthic drug resistance. The first reported failure of a benzimidazole to treat hookworm was reported by DeClercq *et al.* in 1997 although sporadic reports of a similar nature are now emerging from China. Because benzimidazole drug resistance can occur following a point mutation in the parasite tubulin allele, there is a worry that rapid resistance might occur in a similar way to the widespread drug resistance that now threatens the sheep and cattle industry.

As an alternative or complementary approach to control, there are some efforts underway to develop recombinant vaccines against hookworm (Sabin Hookworm Vaccine Initiative; http://www.sabin.org). The potential efficacy of anti-hookworm vaccines was first demonstrated in principle in the 1930s using live normal or irradiated infective L_3 of *Ancylostoma* (Hotez *et al.* 1996). Recently, it has been shown that vaccination with irradiated *Necator* larvae confers almost complete immunity to challege infection (Brown, 2000; Culley *et al.* 2001). Immunity is associated with the Th2 phenotype i.e. high levels of IgG1, IgE and IL-5 and a pronounced eosinophilia. In addition, vaccination with irradiated larvae reduced the pathology associated with the passage of larvae through the lungs. The presence of larvae in the lungs also induces the production of the chemokine attractants eotaxin and MIP-1α, although the levels of these chemokines are not enhanced by vaccination with irradiated larvae (Culley *et al.* 2001). Similarly, lung worm reductions of up to 31 % have been observed following vaccination with larval ES products (Girod *et al.* 2001). Based on the success of these L_3 vaccines, efforts are underway to identify the major antigens associated with larval vaccination, possibly including the Asps, MTP (Hotez *et al.* 1999), calreticulin (Kasper *et al.* 2001) and proteinases associated with skin penetration such as necepsin 2. Asp-1 appears to be a particularly attractive candidate in this regard (Ghosh *et al.* 1996; Ghosh & Hotez, 1999; Liu *et al.* 2000). A second approach to vaccination relies on targeting adult worm products that are critical for parasite survival at the site of attachment. Among these might include MEP-1 (Jones & Hotez, 2001), a gut derived antigen that might elicit protective antibodies similar to some of the current tick vaccines (Willadsen & Kemp, 1988). Tables 9.5 and 9.6 list the potential vaccine candidates for both necatoriasis and ancylostomiasis.

7. CONCLUSIONS

The applied immunologist will remain interested in the strategies used by the hookworm parasites to modulate the human immune system. This knowledge will surely be used by the vaccinologist to further the quest for long-lasting protection against hookworm infection where infection intensities warrant intervention. These goals will be furthered by the selective exploitation of information becoming available from hookworm genomics initiatives and the application of functional genomics in the field. Furthermore, the fact that some hookworms have evolved to modulate immunity, to the extent that field studies are now beginning to show solid evidence for protective effects against atopic symptoms (Scrivener *et al.* 2001), raises the possibility that further hookworm products such as NIF may be exploited therapeutically (Rahman *et al.* 2000).

Table 9.5. Vaccine candidates: necatoriasis.

A number of antigens are being considered as possible vaccines against *N. americanus* (Sabin Vaccine Institute - Hookworm Vaccine Initiative; http://www.sabin.org). The table lists the current molecules under investigation and the life cycle stage in which it is expressed.

Vaccine candidate	***Nature of molecule***	***Coverage of life cycle***
Necepsin 1	Fibrinogenolytic aspartyl haemoglobinase	Adult
Necepsin 2	Aspartyl proteinase- (skin penetration)	L_3-Adult
Necpain	Thiol proteinase in gut	L_4-Adult
Calreticulin	Molecular chaperone Immune modulant	L_3-Adult
Eotaxin metalloproteinase	Chemokine proteinase	Adult
Asp-1	Secreted antigen (Function unknown)	L_3
Asp-2	Secreted antigen (Function unknown)	L_3-Adult

Table 9.6. Vaccine candidates: ancylostomiasis.

List of the proteins currently under investigation as possible vaccines against *Ancylostoma* infection and the life cycle stage in which they are expressed (Sabin Vaccine Institute - Hookworm Vaccine Initiative; http://www.sabin.org).

Vaccine candidate	***Nature of molecule.***	***Coverage of life cycle***
MEP-1	Zinc metallo proteinase	L_3-Adult
Asp-1	Secreted antigen (Function unknown)	L_3
Asp-2	Secreted antigen (Function unknown)	L_3-Adult
MTP	Zinc metallo proteinase	L_3
NIF	Neutrophil inhibitor	Adult
Cathepsin B	Thiol proteinase	Adult

REFERENCES

ALBONICO, M., SMITH, P. G., ERCOLE, E., HALL, A., CHWAYA, H. M., ALAWI, K. S. & SAVIOLI, L. (1995). Rate of reinfection with intestinal nematodes after treatment of children with mebendazole or albendazole in a highly endemic area. *Transactions of the Royal Society of Tropical Medicine and Hygiene* **89**, 538-541.

ALI, F., STANSSENS, P., TIMOTHY, L. M., SOULE, H. R., BROWN, A. & PRITCHARD, D. I. (2001). Vaccination with secreted Neutrophil Inhibitory Factor (NIF) reduces the fecundity of the hookworm *Ancylostoma ceylanicum*. *Parasite Immunology* **23**, 237-249.

ANDERSON, R. M. (1986). The population dynamics and epidemiology of intestinal nematode infections. *Transactions of the Royal Society of Tropical Medicine and Hygiene* **80**, 686-696.

ANDERSON, R. M. & MAY, R. M. (1991). *Infectious Diseases of Humans: Dynamics and Control*, Oxford University Press, Oxford.

BARRETO, M. L., RODRIGUES, L. C., SILVA, R. C. R., ASSIS, A. M. O., REIS, M. G., SANTOS, C. A. S. T. & BLANTON, R. E. (2000). Lower hookworm incidence, prevalence, and intensity of infection in children with a Bacillus Calmette-Guerin vaccination scar. *Journal of Infectious Diseases* **182**, 1800-1803.

BENTWICH, Z., KALINKOVICH, A., BORKOW, G., STEIN, M., QIBIN, L. & WEISMAN, Z. (2000). Helminthic infections have a major impact on pathogenesis of AIDS and development of HIV protective vaccines in the developing countries. *AIDS* **14**, P377.

BENTWICH, Z., WOLDAY, D., MARIAM, Z. G., MAAYAN, S.& LANDAY, A. (2000). Treatment of intestinal worms decreases HIV plasma viral load. *AIDS* **14**, P376.

BERRIDGE, M. J., BOOTMAN, M. D. & LIPP, P. (1998) Calcium-a life and death signal. *Nature* **395**, 645-648.

BOS, J. D. & KAPSENBERG, M. L. (1993). The skin immune system: progress in cutaneous biology. *Immunology Today* **14**, 75-78.

BROPHY, P. M., PATTERSON, L. H., BROWN, A. & PRITCHARD, D. I. (1995). Glutathione S-transferase (GST) expression in the human hookworm *Necator americanus*: potential roles for excretory-secretory forms of GST. *Acta Tropica* **59**, 259-263.

BROPHY, P. M., PATTERSON, L. H. & PRITCHARD, D. I. (1995). Secretory nematode SOD-offensive or defensive? *International Journal for Parasitology* **25**, 865-866.

BROWN, A. (2000). *Necator americanus*: characterisation of secreted proteinases and vaccine development. PhD, Nottingham Trent University.

BROWN, A., GIROD, N., BILLETT, E. E. & PRITCHARD, D. I. (1999). *Necator americanus* (human hookworm) aspartyl proteinases and digestion of skin macromolecules during skin penetration. *American Journal of Tropical Medicine and Hygiene* **60**, 840-847.

BROWN, A., GIROD, N. & PRITCHARD, D. I. (2001). The Characterisation and purification of Factor Xa and XIIIa inhibitors from the human hookworm *Necator americanus*. Manuscript in preparation.

BROWN, A. & PRITCHARD, D. I. (1993). The immunogenicity of hookworm (*Necator americanus*) acetylcholinesterase (AChE) in man. *Parasite Immunolgy* **15**, 195-203.

BUNGIRO, R. D. J., GREENE, J., KRUGLOV, E. & CAPPELLO, M. (2001). Mitigation of hookworm disease by immunization with soluble extracts of *Ancylostoma ceylanicum*. *Journal of Infectious Diseases* **183**, 1380-1387.

CAPPELLO, M., CLYNE, L. P., MCPHEDRAN, P. & HOTEZ, P. J. (1993). *Ancylostoma* Factor Xa inhibitor: Partial purification and its identification as a major hookworm derived anticoagulant *in vitro*. *Journal of Infectious Diseases* **167**, 1474-1477.

CARR, A. & PRITCHARD, D. I. (1987). Antigen expression during development of the human hookworm, *Necator americanus* (Nematoda). *Parasite Immunology* **9**, 219-234.

CHADDERDON, R. C. & CAPPELLO, M. (1999). The hookworm platelet inhibitor: Functional blockade of integrins GPIIb/IIIa (alpha(IIb)beta(3)) and GPIa/IIa (alpha(2)beta(1)) inhibits platelet aggregation and adhesion *in vitro*. *Journal of Infectious Diseases* **179**, 1235-1241.

CHOW, S. C., BROWN, A. & PRITCHARD, D. (2000). The human hookworm pathogen *Necator americanus* induces apoptosis in T lymphocytes. *Parasite Immunology* **22**, 21-29.

CROMPTON, D. W. T. (2000). The public health importance of hookworm disease. *Parasitology* **121** (Suppl), S39-S50.

CULLEY, F. J., BROWN, A., CONROY, D. M., SABROE, I., PRITCHARD, D. I. & WILLIAMS, T. J. (2000). Eotaxin is specifically cleaved by hookworm metalloproteases preventing its action *in vitro* and *in vivo*. *Journal of Immunology* **165**, 6447-6453.

CULLEY, F. J., BROWN, A., GIROD, N., PRITCHARD, D. I. & WILLIAMS, T. J. (2001). Innate and cognate mechanisms of pulmonary eosinophillia in helminth infection. *Infection and Immunity*. In Press.

DAUB, J., LOUKAS, A., PRITCHARD, D. I. & BLAXTER, M. (2000). A survey of genes expressed in adults of the human hookworm *Necator americanus*. *Parasitology* **120**, 171-184.

DECLERCQ, D., SACKO, M., BEHNKE, J., GILBERT, F., DORNY, P. & VERCRUYSSE, J. (1997). Failure of mebendazole in treatment of human hookworm infections in the southern region of Mali. *American Journal of Tropical Medicine and Hygiene* **57**, 25-30.

DEL VALLE, A., CHADDERDON, R. C. & CAPPELLO, M. (1999). The hookworm platelet inhibitor blocks fibrinogen binding to the platelet integrin GPIIb/IIIa (alpha(IIb)beta(3)). *Pediatric Research* **45**, 931.

DEVINE, M. (1995). The reponse of nematodes to stress *in vitro*. PhD. University of Nottingham.

DREYFUSS, M. L., STOLTZFUS, R. J., SHRESTHA, J. B., PRADHAN, E. K., LECLERQ, S. C., KHATRY, S. K., SHRESTHA, S. R., KATZ, J., ALBONICO, M. & WEST, K. P. J. (2000). Hookworms, malaria and vitamin A deficiency contribute to anemia and iron deficiency among pregnant women in the plains of Nepal. *Journal of Nutrition* **130**, 2527-2536.

EGGLETON, P., REID, K. B. M., KISHORE, U. & SONTHEIMER, R. D. (1997). Clinical relevance of calreticulin in systemic lupus erythematosus. *Lupus* **6**, 564-571.

ELLIOTT, A. M., NAKIYINGI, J., QUIGLEY, M. A., FRENCH, N., GILKS, C. F. & WHITWORTH, J. A. G. (1999). Inverse association between BCG immunisation and intestinal nematode infestation among HIV-1-positive individuals in Uganda. *Lancet* **354**, 1000-1001.

FURMIDGE, B. A., HORN, L. A. & PRITCHARD, D. I. (1995). The anti-haemostatic strategies of the human hookworm *Necator americanus*. *Parasitology* **112**, 81-87.

GHANDI, N. S., CHEN, J. Z., KHOSHNOOD, K., XIN, F. Y., LI, S. W., LIU, Y. R., ZHAN, B., XUE, H. C., TONG, C. J., WANG, Y., WANG, W. S., HE, D. X., CHEN, C., XIAO, S. H., HAWDON, J. M. & HOTEZ, P. J. (2001). Epidemiology of *Necator americanus* hookworm infections in Xiulongkan Village, Hainan Province, China: High prevalence and intensity among middle-aged and elderly residents. *Journal of Parasitology* **84**. In Press.

GHOSH, K., HAWDON, J. & HOTEZ, P. (1996). Vaccination with alum-precipitated recombinant *Ancylostoma*-secreted protein 1 protects mice against challenge infections with infective hookworm (*Ancylostoma caninum*) larvae. *Journal of Infectious Diseases* **174**, 1380-1383.

GHOSH, K. & HOTEZ, P. (1999). Antibody dependent reductions in mouse hookworm burden after vaccination with *Ancylostoma caninum* secreted protein 1. *The Journal of Infectious Diseases* **180**, 1674-1681.

GIRDWOOD, K. (1999). Aspartic proteinases and aspartic proteinase inhibitors from parasitic nematodes. PhD. University of Wales.

GIROD, N., BROWN, A. P., BILLETT, E. E. & PRITCHARD, D. I. (2001). Successful vaccination against *Necator americanus* and the immunological phenotype of vaccines. Manuscript in preparation.

GRAY, A. J., PARK, P. W., BROEKELMANN, T. J., LAURENT, G. J., STENMARK, K. R. & MECHAM, R. P. (1995). The mitogenic effects of the Bβ chain of fibrinogen are mediated through cell surface calreticulin. *Journal of Biological Chemistry* **270**, 26602-26606.

HAWDON, J. M., JONES, B. F., HOFMANN, D. R. & HOTEZ, P. J. (1996). Cloning and characterisation of *Ancylostoma* secreted protein. *The Journal of Biological Chemistry* **271**, 6672-6678.

HAWDON, J. M., NARASIMHAN, S. & HOTEZ, P. J. (1999). *Ancylostoma* secreted protein 2: cloning and characterization of a second member of a family of nematode secreted proteins from *Ancylostoma caninum*. *Molecular and Biochemical Parasitology* **99**, 149-165.

HOAGLAND, K. E. & SCHAD, G. A. (1978). *Necator americanus* and *Ancylostoma duodenale*: Life history parameters and epidemiological implications of two sympatiric hookworms on humans. *Experimental Parasitology* **44**, 36-39.

HOTEZ, P. J., GHOSH, K., HAWDON, J. M., NARASIMHAN, S., JONES, B., SHUHUA, X., SEN, L., BIN, Z., HAECHOU, X., HAINAN, R., HENG, W. & KOSKI, R. A. (1999). Experimental approaches to the development of a recombinant hookworm vaccine. *Immunological Reviews* **171**, 163-172.

HOTEZ, P. J., HAWDON, J. M., CAPPELLO, M., JONES, B. F., GHOSH, K., VOLVOVITZ, F. & SHUHUA, X. (1996). Molecular approaches to vaccinating against hookworm disease. *Pediatric Research* **40**, 515-521.

HOTEZ, P. J. & PRITCHARD, D. I. (1995). Hookworm Infection. *Scientific American* **272**, 42-48.

JAWORSKI, D. C., HIGGINS, J. A., RADULOVIC, S., VAUGHAN, J. A. & AZAD, A. F. (1996). Presence of calreticulin in vector fleas (Siphonaptera). *Journal of Medical Entomology* **33**, 482-489.

JONES, B. F. & HOTEZ, P. J. (2001). Molecular cloning and characterization of Ac-MEP-1 a developmentally regulated gut luminal metalloendopeptidase from adult *Ancylostoma caninum* hookworm. *Molecular and Biochemical Parasitology*. In press.

KASPER, G., BROWN, A., EBERL, M., VALLAR, L., KIEFFER, N., BERRY, C., GIRDWOOD, K., EGGLETON, P., QUINNELL, R. J. & PRITCHARD, D. I. (2001). A calreticulin-like molecule from the human hookworm *Necator americanus* interacts with C1q and the cytoplasmic signalling domains of some integrins. *Parasite Immunology* **23**, 141-152.

KEYMER, A. E. & PAGEL, M. (1990). Predisposition to helminth infection. In *Hookworm disease: current status and new directions* (ed. Schad, G.A. & Warren, K.S), pp.177-209. Taylor and Francis, London.

KISHORE, U., SONTHEIMER, R. D., SASTRY, K. N., ZAPPI, E. G., HUGHES, G. R. V., KHAMASHTA, M. A., REID, K. B. M. & EGGLETON, P. (1997). The systemic lupus erythematosus (SLE) disease autoantigen-Calreticulin inhibit C1q association with immune complexes. *Clinical and Experimental Immunology* **108**, 181-190.

KRAUSE, K. H. & MICHALAK, M. (1997). Calreticulin. *Cell* **88**, 439-443.

KUMAR, S. & PRITCHARD, D. I. (1992). Skin penetration by ensheathed third stage infective larvae of *Necator americanus*, and the host's immune response to larval antigens. *International Journal for Parasitology* **22**, 573-579.

LABIANO-ABELLO, N., CANESE, J., VELAZQUEZ, M. E., HAWDON, J. M., WILSON, M. L. & HOTEZ, P. J. (1999). Epidemiology of hookworm infection in Itagua, Paraguay: a cross sectional study. *Memorias Do Instituto Oswaldo Cruz* **94**, 583-586.

LEWIS, D. J. M. & GRIFFIN, G. E. (1995). Immunology of the gastrointestinal tract. *Current Opinions in Infectious Diseases* **8**, 368-373.

LIU, S., GHOSH, K., ZHAN, B., SHAN, Q., THOMPSON, M. G., HAWDON, J., XIAO, S. H., KOSKI, R. A. & HOTEZ, P. J. (2000). Hookworm burden reductions in BALB/c mice vaccinated with Ancylostoma secreted protein 1 (ASP-1) from *Ancylostoma duodenale, A. caninum* and *Necator americanus*. *Vaccine* **18**, 1096-1102.

MOYLE, M., FOSTER, D. L., MCGRATH, D. E., BROWN, S. M., LAROCHE, Y., DE MEUTTER, J., STANSSENS, P., BOGOWITZ, C. A., FRIED, V. A., ELY, J. A., SOULE, H. R. & VLASUK, G. P. (1994). A hookworm glycoprotein that inhibits neutrophil function is a ligand of the integrin CD11b/CD18. *Journal of Biological Chemistry* **269**, 1008-1015.

PALMER, D. R., BRADLEY, M. & BUNDY, D. A. (1996). IgG4 responses to antigens of adult *Necator americanus*: potential for use in large-scale epidemiological studies. *Bulletin of the World Health Organization* **74**, 381-386.

PRITCHARD, D. I. (1993). Atopy and helminth parasites. *International Journal for Parasitology* **23**, 167-168.

PRITCHARD, D. I. & BROWN, A. (2001). Is *Necator americanus* approaching a mutualistic symbiotic relationship with humans? *Trends in Parasitology* **17**, 169-172.

PRITCHARD, D. I., BROWN, A., KASPER, G., MCELROY, P., LOUKAS, A., HEWITT, C., BERRY, C., FÜLLKRUG, R. & BECK, E. (1999). A hookworm allergen that strongly resembles calreticulin. *Parasite Immunology* **21**, 439-450.

PRITCHARD, D. I., BROWN, A., NOWELL, M., HEWITT, C., GIRDWOOD, K., BERRY, C., BECK, E. & FÜLLKRUG, R. (2001). The characterisation of a cathepsin B like enzyme from the human hookworm *Necator americanus*, Submitted.

PRITCHARD, D. I., LEGGETT, K. V., ROGAN, M. T., MCKEAN, P. G. & BROWN, A. (1991). *Necator americanus* secretory acetylcholinesterase and its purification from excretory-secretory products by affinity chromatography. *Parasite Immunology* **13**, 187-199.

PRITCHARD, D. I., MCKEAN, P. G. & ROGAN, M. (1988). Cuticle preparations from *Necator americanus* and their immunogenicity in the infected host. *Molecular and Biochemical Parasitology* **28**, 275-283.

PRITCHARD, D. I., QUINNELL, R. J., MOUSTAFA, M., MCKEAN, P. G., SLATER, A. F. G., RAIKO, A., DALE, D. D. S. & KEYMER, A. E. (1991). Hookworm (*Necator americanus*) infection and storage iron depletion. *Transactions of the Royal Society of Tropical Medicine and Hygiene* **85**, 235-238.

PRITCHARD, D. I., QUINNELL, R. J. & WALSH, E. A. (1995). Immunity in humans to *N. americanus*: IgE, parasite weight and fecundity. *Parasite Immunology* **71**, 71-75.

PRITCHARD, D. I. & WALSH, E. A. (1995) The specificity of the human IgE response to *Necator americanus*. *Parasite Immunology* **17**, 605-607.

QUINNELL, R. J. (2001). Proliferative and cytokine responses to human *Necator americanus* infection before and after chemotherapy. Manuscript in Preparation.

QUINNELL, R. J., GRIFFIN, J., NOWELL, M. A., RAIKO, A. & PRITCHARD, D. I. (2001). Predisposition to hookworm infection in Papua New Guinea. *Transactions of the Royal Society of Tropical Medicine and Hygiene* **95**, 139-142.

QUINNELL, R. J., WOOLHOUSE, M. E. J., WALSH, E. A. & PRITCHARD, D. I. (1995). Immunoepidemiology of human nectoriasis: correlations between antibody responses and parasite burdens. *Parasite Immunology* **17**, 313-318.

QUINNELL, R. J., SLATER, A. F. G., TIGHE, P. J., WALSH, E. A., KEYMER, A. E. & PRITCHARD, D. I. (1993). Reinfection with hookworm after chemotherapy in Papua New Guinea. *Parasitology* **106**, 379-385.

RAHMAN, X. N., MINSHALL, R. D., TIRUPPATHI, C. & MALIK, A. B. (2000). Beta2-integrin blockade driven by E-selectin promoter prevents neutrophil sequestration and lung injury in mice. *Cirulation Research* **87**, 254-260.

REILLY, D., MORAN, N., FITZGERALD, D. J., PRITCHARD, D. I., VALLAR, L. & KIEFFER, N. (2000). Calreticulin interaction with the alpha(IIb) cytoplasmic tail of the platelet integrin alpha(IIb)beta(3.). *Blood* **96**, 2681.

SCHAD, G. A., CHOWDBURY, A. B., DEAN, C. G., KOCHAR, V. K., NAWALINSKI, T. A., THOMAS, J. & TONASICA, J. A. (1973). Arrested development in human hookworm infection: An adaptation to a seasonally unfavourable external environment. *Science* **180**, 502-504.

SCHWEITZ, H., BRUHN, T., GUILLEMARE, E., MOINIER, D., LANCELIN, H.-M., BERESS, L. & LAZDUNSKI, M. (1995). Kalicludines and Kaliseptine: two different classes of sea anemone toxins for voltage-sensitive K+ channels. *Journal of Biological Chemistry* **270**, 25121 - 25126.

SCRIVENER, S., YEMANEBERHAN, H., ZEBENIGUS, M., TILAHUN, D., GIRMA, S., ALI, S., MCELROY, P., CUSTOVIC, A., WOODCOCK, A., PRITCHARD, D. I., VENN, A. & BRITTON, J. (2001). Independent effects of intestinal parasite infection and domestic allergen exposure on the risk of wheeze in Ethiopia. *Lancet*. In press.

SEN, L., GHOSH, K., ZHAN, B., QIANG, S., THOMPSON, M. G., HAWDON, J. M., KOSKI, R. A., XIAO, S. H. 0. & HOTEZ, P. J. (2000). Hookworm burden reductions in BALB/c mice vaccinated with recombinant Ancylostoma secreted proteins (ASPs) from *Ancylostoma duodenale, Ancylostoma caninum* and *Necator americanus. Vaccine* **18**, 1096-1102.

SHULMAN, C. E., GRAHAM, W. J., JILO, H., LOWE, B. S., NEW, L., OBIERO, J., SNOW, R. W. & MARSH, K. (1996). Malaria is an important cause of anaemia in primigravidae: Evidence from a district hospital in costal Kenya. *Transactions of the Royal Society of Tropical Medicine and Hygiene* **90**, 535-539.

STANSSENS, P., BERGUM, P. W., GANSEMANS, Y., JESPERS, L., LAROCHE, Y., HUANG, S., MAKI, S., MESSENS, J., LAUWEREYS, M., CAPPELLO, M., HOTEZ, P. J., LASTERS, I. & VLASUK, G. P. (1996). Anticoagulant repertoire of the hookworm *Ancylostoma caninum. Proceedings of the National Academy of Sciences of the United States of America* **93**, 2149-2154.

VERHEUGEN, J. A. H., LE DIEST, F., DEVIGNOT, V. & KORN, H. (1997). Enhancement of calcium signalling and proliferation responses in activated human T lymphocytes. Inhibitory effects of K+ channel block by charybdotoxin depend on the T cell activation state. *Cell Calcium* **21**, 1-17.

WANG, Y., SHEN, G. J., WU, W. T., XIAO, S. H., HOTEZ, P. J., LI, Q. Y., XUE, H. C., X.M., Y., LIU, X. M., ZHAN, B., HAWDON, J. M., CHOU, L., JI HONG, H. C. M. & FENG, Z. (1999). Epidemiology of human ancylostomiasis in Nanlin County (Zhongzhou Village), Anhui Province China. 1. Prevalence, intensity and hookworm species identification. *Southeast Asian Journal of Tropical Medicine and Public Health* . In press.

WILLADSEN, P. & KEMP, D. H. (1988). Vaccination with 'concealed' antigens for tick control. *Parasitology Today* **4**, 196-198.

WILLIAMS-BLANGERO, S., BLANGERO, J. & BRADLEY, M. (1997). Quantitative genetic analysis of susceptibility to hookworm infection in a population from rural Zimbabwe. *Human Biology* **69**, 201-208.

WOOLHOUSE, M. E. J. (1992). A theroretical framework for the immunoepidemiology of helminth infection. *Parasite Immunology* **14**, 563-578.

WOOLHOUSE, M. E. J. (1998). Patterns in parasite epidemiology: the peak shift. *Parasitology Today* **14**, 428-434.

XUE, H. C., WANG, Y., XIAO, S. H., LIU, S., WANG, Y., SHEN, G. J., WU, W. T., ZHAN, B., DRAKE, L., FENG, Z. & HOTEZ, P. J. (2001). Epidemiology of human ancylostomiasis among rural villagers in Nanlin County (Zhongzhou Village), Anhui Province, China: II. Seroepidemiological studies of the age relationships of serum antibody levels and infection status. *Southeast Asian Journal of Tropical Medicine and Public Health.* In press.

ZHAN, B., HAWDON, J., SHAN, Q., REN, H. N., QIANG, H. Q., HU, W., XIAO, S. H., LI, T. H., GONG, X., FENG, Z. & HOTEZ, P. (1999). Ancylostoma secreted protein 1 (ASP-1) homologues in human hookworms. *Molecular and Biochemical Parasitology* **98**, 143-149.

ZHAN, L. L., ZHANG, B. H., TAO, H., XIAO, S. H., HOTEZ, P., ZHAN, B., LI, Y. Z., LI, Y., XUE, H. C., HAWDON, J., YU, H., WANG, H. & FENG, Z. (2001). Epidemiology of human geohelminth infections (ascariasis, trichuriasis, necatoriasis) in Lushui and Puer Counties, Yunnan Province, China. *Southeast Asian Journal of Tropical Medicine and Public Health.* In press.

Chapter 10
HUMAN HOST SUSCEPTIBILITY TO INTESTINAL WORM INFECTIONS

Sarah Williams-Blangero and John Blangero
Dept of Genetics, Southwest Foundation for Biomedical Research, San Antonio, Texas, USA
e-mail: sarah@darwin.sfbr.org

1. INTRODUCTION

The soil-transmitted intestinal helminths (hookworm, roundworm, and whipworm) are major international health concerns, affecting over a quarter of the world's population. Epidemiological studies have shown that susceptibility to these parasites generally aggregates within families. This evidence, in combination with the many empirical observations that worm burden is overdispersed (*i.e.*, a small proportion of individuals generally harbors a large percentage of a population's total worm burden), and the fact that certain individuals have a tendency to repeatedly develop high worm burdens after anthelminthic therapy, suggests that genetic factors may play an important role in determining risk for helminthic infections.

Relatively few genetic studies of susceptibility to infectious diseases have been conducted in human populations. However, recent developments in statistical and molecular genetics have created an exciting research environment where it is now possible to explore in detail the genetic and environmental factors influencing susceptibility to a broad range of infectious diseases. These developments have opened up a great range of opportunities in infectious disease research (Abel & Dessein, 1997; Dessein *et al.* 2001; Hill, 1996, 1998).

Recent studies have found evidence of significant genetic effects on susceptibility to many infectious diseases, including schistosomiasis (Abel *et al.* 1991; Marquet *et al.* 1996, 1999), leprosy (Abel *et al.* 1995), malaria (Abel *et al.* 1992; Garcia *et al.* 1998; Rihet *et al.* 1998), hookworm infection (Williams-Blangero, Blangero & Bradley, 1997a), roundworm infection (Williams-Blangero *et al.* 1999), and *Trypanosoma cruzi* infection (Williams-Blangero *et al.* 1997b).

The use of large scale genomic screens is revolutionizing the study of susceptibility to disease, providing a mechanism to localize (*e.g.*, Marquet *et al.* 1996), and ultimately identify the specific genes involved in determining susceptibility to infectious diseases.

2. OVERDISPERSION OF WORM BURDENS IN HUMAN POPULATIONS: EVIDENCE OF PREDISPOSITION

The three major helminthic infections covered in this volume all exhibit a characteristic pattern of overdispersion in which these parasites tend to be aggregated in a relatively small proportion of the population (Anderson & May, 1985; Anderson & Medley, 1985) (see Chapter 1). For example, in a study of hookworm, whipworm, and roundworm in an Iranian population, Croll and Ghadirian (1981) determined that 1-3% of individuals in the population carried between 11% and 84% of the worms. Other reports have suggested that more than 70% of the parasites are frequently found in less than 10% of available hosts (Anderson, 1982; Anderson & May, 1982, 1985; Anderson & Medley, 1985). This aggregation of infections in a small fraction of the population has been found repeatedly in studies of hookworm (Schad & Anderson, 1985; Bradley *et al.* 1992), roundworm (Elkins *et al.* 1986; Thein-Hlaing, 1985; Thein-Hlaing *et al.* 1987; Forrester *et al.* 1988), and whipworm (Bundy *et al.* 1987; Forrester *et al.* 1988).

Many investigators have interpreted this overdispersion of parasites to reflect predisposition of certain individuals to infection. Significant correlations between pre- and post-treatment parasite loads suggest the involvement of innate host factors in determining this pattern (McCallum, 1990). For hookworm, roundworm, and whipworm, there is substantial evidence that individuals showing high parasite loads prior to treatment demonstrate the highest loads after a period of reinfection (Haswell-Elkins *et al,* 1987; Schad & Anderson, 1985; Bundy, 1986; Bundy & Medley, 1992; Forrester *et al.* 1990).

The possibility that this predisposition is a function of genetic susceptibility to parasitic infection has been raised by a number of authors (Schad & Anderson, 1985; Anderson & Medley, 1985). While there have been few formal genetic studies of human susceptibility to geohelminthic infections, the presence of significant household or family effects on patterns of geohelminthic infection has been identified in studies of ascariasis (Chai,

Seo & Lee, 1983; Williams, Burke & Hendley, 1974; Forrester *et al.* 1988) and trichuriasis (Forrester *et al.* 1988).

3. GENETIC STUDIES OF SUSCEPTIBILITY TO HELMINTHIC INFECTION IN HUMANS

Studies of the genetic components of susceptibility to and response to helminthic infection in man are limited. A familial or household patterning to helminth loads has been noted frequently (*e.g.*, Forrester *et al.* 1988, 1990; Chan, Bundy & Kan, 1994; Conway *et al.* 1995), but specific investigations of the role of genetic factors in generating such patterns have been limited. The major deficiency of epidemiological examinations of these familial or household aggregation patterns has been the application of non-specific statistical methods and inadequate sampling designs for separating out genetic and shared environmental influences on observed patterns.

Several association studies have suggested that genetic factors may influence susceptibility to helminthic infections in humans. For example, an analysis of 48-hour roundworm loads determined in Nigerian children between the ages of 5 and 16 years suggested a role for the MHC in determining resistance to infection (Holland *et al.* 1992). Recently, an association between polymorphisms in the э-adrenoreceptor gene and *Ascaris* egg loads was found in a group of 126 Venezuelan children (Ramsey *et al.* 1999). One of these polymorphisms appears to account for 25% of the observed variation in *Ascaris* egg counts. If this finding is true, then the chromosome 5q region where the *SM1* gene for schistosomiasis (Marquet *et al.* 1996, 1999) was found may also have a locus that influences *Ascaris* infection.

4. GENETIC EPIDEMIOLOGICAL STUDIES OF HELMINTHIC INFECTION

The field of genetic epidemiology is a rapidly expanding area of genetic research which utilizes statistical tools to quantify and localize genetic effects on complex traits. These tools are ideally suited to refining our knowledge of the genetic factors involved in determining epidemiological patterns of helminthic infections in human populations (Williams-Blangero *et al.* 1996a).

5. UTILIZING AVAILABLE EPIDEMIOLOGICAL DATA FOR GENETIC EPIDEMIOLOGICAL STUDIES

Existing epidemiological studies of helminthic infections represent a rich potential source of data for genetic epidemiological studies. Because epidemiological studies frequently are household based, genetic information is embedded in the data gathered. Households frequently consist of nuclear or extended families whose within-household relationships can be fairly easily reconstructed from existing information. The difficulty with utilizing existing data lies in the limited ability to reconstruct relationships between households. The resulting overlap between household membership and family membership makes it impossible to completely differentiate between household and genetic effects. However, genetic epidemiological studies of existing databases can provide strong clues as to whether or not genetic influences are present and whether or not a full scale genetic study is worth pursuing.

For example, utilizing available epidemiological data from a rural population in Zimbabwe, we were able to reconstruct pedigrees adequate for quantitative genetic analysis of the information on hookworm generated for this population. Quantitative genetic analysis of the existing data on hookworm burden demonstrated the presence of significant genetic effects on this helminthic infection (Williams-Blangero *et al.* 1997a). Quantitative measures of hookworm eggs per gram of faeces as determined by the Kato thick smear technique were available for 279 individuals who could be assigned to 62 pedigrees and 10 independent individuals. Utilizing a variance decomposition approach, we demonstrated the heritability of hookworm load to be 0.37 ∀ 0.06 ($p < 0.0001$) in this population (Williams-Blangero *et al.* 1997a). This significant heritability indicated that approximately 37% of the variation in hookworm eggs per gram of faeces was attributable to genetic factors in this population.

A similar analysis was performed utilizing data on whipworm burden as assessed by egg counts determined for a population in Jiangxi, China (Williams-Blangero *et al.* 1996a). Information was available for 788 individuals. Existing demographic and household membership information, allowed assignment of these individuals to a total of 205 pedigrees suitable for genetic analysis. In the Jiangxi data set, we estimated the heritability of *Trichuris* egg counts to be 0.287 ∀ 0.083. Shared household environment also had a relatively modest effect.

While these two small studies suggest that extant epidemiological data may be of use for making general inferences about the potential for genetic studies of human host susceptibility to helminthic infection and worm burden, they also suffer from the defect of not being optimally designed to detect such genetic effects, and to discriminate between genetic and shared environmental effects. In general, studies of extended pedigrees are more powerful for detecting the effects of genes and for localizing them to chromosomal locations (Williams-Blangero *et al.* 1999), than are the nuclear family based studies that are usually captured by the traditional focal household designs of epidemiology. Therefore, a more powerful genetic study would utilize very large pedigrees that encompass a large number of separate households. This situation enables discrimination between the effects of genes and those of environment. Given that helminthic infections are primarily a problem in underdeveloped and developing nations, there are often genetically isolated populations with large extended pedigrees available that may facilitate powerful study designs for finding genes influencing human host susceptibility to infection.

6. A GENETIC EPIDEMIOLOGICAL STUDY OF HELMINTHIC INFECTIONS: THE JIRI HELMINTH PROJECT

Because of the lack of detailed family-based studies examining the genetic basis of human helminthic infections, we established the longitudinal Jiri Helminth Project in 1995 in Jiri, a rural area of eastern Nepal. The first major accomplishment of this project was the creation of a field site and recruitment of a staff, both of which have excellent capabilities for assessing helminthic burden on a large scale. Jiri is an area of approximately 230 square kilometers, located 190 kilometers east of the capital of Nepal, Kathmandu. The region is named for the focal population of the study, a Tibeto-Burman speaking ethnic group called the Jirels. Ethnohistorical accounts and population genetic studies support the folk belief that the Jirels represent a hybrid population that was derived from Sherpas and Sunwars approximately 10-11 generations ago (Blangero, 1987). Population genetic studies have shown that since the founding event, there has been very little gene flow (less than 1% per generation) into the population from either of the parental populations or other groups in the region. However, inbreeding within seven generations of relationship is actively avoided. In 1985, the

Jirel population comprised approximately 3500 individuals all located in the Jiri region. The current Jirel census size for the seven villages sampled in the Jiri Helminth Project is approximately 3000 individuals.

6.1 Sampling Design

We enrolled subjects over an initial two-year period. The original target sample size was 1000 individuals. However, due to the success of community outreach, we sampled 1,261 individuals in the first year. Each individual was examined twice over the two year period, with an approximately 1 year interval between the initial and follow-up exam. The final sample consisted of 659 females and 602 males. The mean age at examination was 25.4 years with a standard deviation of 18.9 years, and a range from 3 years to 85 years of age. The relevant aspects of the sampling protocol included: (1) two consecutive days of faecal samples for quantitation of egg counts, (2) blood draw for DNA extraction, hematological, and plasma marker analyses, and (3) following ingestion of the anthelminthic drug albendazole (400mg), all stools were collected for a period of 96 hours for direct worm counts.

Of the 1,261 individuals examined, 1,261 provided faecal samples for egg counts, 1,205 provide blood samples, and 1,007 provided 96 hour stool samples. In the second year of the study, 1,002 of these individuals were re-examined and provided two days of small fecal samples for egg quantitation and additional blood samples. A total of 965 individuals provided 96 hour stool samples during their second examination. Of these, 910 had also provided complete 96 hour stool samples in their first examinations. Clearance of worms following albendazole treatment was verified through resampling of a proportion of positive individuals for egg count quantitation only.

6.2 Pedigree Structure

Over the past 15 years, we have collected extensive pedigree information on Jirel family relationships, enabling placement of all of the sampled individuals into a single pedigree. However, every individual in the pedigree was not necessarily related to every other individual in the pedigree. For example, the mother of your children may not be related to the mother of your brother's children, but both women will belong to the same pedigree by virtue of both being related to the grandchildren of your parents. Pedigree relationships in the Jirel population have been verified many times and numerous consistency checks have been performed. Of the 1,261 individuals sampled, 257 are founders (*i.e.*, individuals whose parents are unknown or are not needed to determine additional pedigree links) and the remaining 1,004 individuals are members of 521 sibships ranging in size from 1 to 7 sampled individuals. The mean sibship size is 2.12 individuals.

The Jirel pedigree is remarkably complete and complex. We have determined all of the observed pairwise biological relationships between sampled individuals. There are a total of 2,440 pairs of first degree relatives, of which 1,075 are sib-pairs. Similarly, there are 2,406 second degree relationships, and 3,655 third degree relationships. Overall, there are more than 26,000 pairs of relatives that will provide information for the localization of genes influencing susceptibility to helminthic infection. The Jirel pedigree represents one of the largest and most complete samples of relatives ever collected for a genetic study.

6.3 Household Structure

Because of the likelihood of shared environmental effects due to differential exposure, we obtained data on household composition and residence patterns. The 1,261 sampled individuals resided in a total of 250 households. Household sizes ranged between 1 and 15 with an average of approximately five resident individuals sampled per household. All households were mapped using satellite based global positioning technology. This information allows us to consider spatial correlation in exposure. Given that the large Jirel pedigree is distributed across many relatively large households, we have considerable power to detect the effects of shared environmental variables influencing both susceptibility and disease burden.

6.4 Prevalence of Helminthic Infection in the Jirels

Geohelminthic infections are endemic in the Jiri region, and the Jirel ethnic group exhibits the highest rates of infection among local inhabitants (Williams-Blangero *et al.* 1993). During the first year of the Jiri Helminth Project, we observed a total population prevalence of *Ascaris* infection of

27.2% (Williams-Blangero *et al.* 1999). The prevalence of hookworm infection was 55.4% while that for *Trichuris* infection was 14.4%. Of the total population, 64.7% were infected with at least one of these helminths. Multiple infections were common with 20.1% of individuals harboring more than one type of worm. One year after treatment with albendazole, these prevalences were reduced for hookworm and whipworm infections (33.4% and 7%, respectively), but remained relatively unchanged for *Ascaris* (24.2%).

6.5 Genetic Analysis of Roundworm Burden

Table 10.1 presents the results of our genetic analyses of susceptibility to roundworm infection in the Jirel population which have been previously reported (Williams-Blangero *et al.* 1999). Three measures of worm burden were analysed: eggs per gram of faeces, direct worm count, and worm biomass (*i.e.*, weight). For all traits there is unequivocal evidence for a strong genetic component (heritability (h^2)), accounting for between 30% and 48% of the variation in worm burden (Williams-Blangero *et al.* 1999). There is also substantial evidence for shared environmental factors influencing worm burden. These shared environmental effects (c^2) account for 3 to 22% of the total phenotypic variance (Williams-Blangero *et al.* 1999). The relatively small shared environmental effect can also be seen in the correlations between spouses who are unrelated but living in the same household environment. For all roundworm burden traits, we have found this correlation to be very low (Δ 0.08).

Table 10.1. Results of quantitative genetic analyses of *Ascaris* burden in the Jirel population of eastern Nepal

Trait	h^2	*P*	c^2	*P*
Eggs count (Year 1)	0.291	<0.0001	0.061	0.0016
Eggs count (Year 2)	0.399	<0.0001	0.113	<0.0001
Worm count (Year 1)	0.359	<0.0001	0.030	0.0923
Worm count (Year 2)	0.412	<0.0001	0.223	<0.0001
Worm weight (Year 1)	0.335	<0.0001	0.034	0.0584
Worm weight (Year 2)	0.477	<0.0001	0.125	<0.0001

There is remarkable consistency between the results for egg counts and worm counts within each year. Within a given year, the assessment of worm burden before albendazole treatment (egg counts) reveals the same genetic pattern as worm burden assessed post-albendazole treatment (worm counts) (Williams-Blangero *et al.* 1999). Our egg count measure of intensity of infection is not influenced by the albendazole treatment, while the worm counts represent the success of such treatment.

The changes in the relative variance component estimates from year to year also provide important information. The data determined in the second year represent infection after one year of exposure subsequent to anthelminthic treatment. The heritability estimates are consistently higher for the second year data as compared to those evaluated for the first year data. This is also true for the relative importance of common household effects. This improved resolution of genetic signal (*i.e.*, increase in heritability) reflects a decrease in environmental variability that may be attributable to eliminating variation in the length of the exposure period in the second year data. The data from the second year reflect endpoints uniformly assessed one year after anthelminthic treatment.

The evidence consistently indicates that there are significant genetic influences on susceptibility to *Ascaris*, *Trichuris*, and hookworm infections in humans. It is likely that at least 30% of the total variation in worm burden measures observed in human populations is due to innate genetic factors relating to resistance.

7. INTERACTIONS BETWEEN HOST AND PARASITE GENOMES IN DETERMINING VARIATION IN PARASITE LOADS

Co-evolutionary relationships between hosts and parasites result in interactions between the genetic structures of host populations and parasite populations. The potential for interaction between the genomes of the human host and helminthic parasite is enormous. To maximize their reproductive success, parasites generate a diverse set of defenses against the host immune system. Avoidance of the host immune response may be effected through several mechanisms including antigenic variation, diversionary shedding of immunogenic surface proteins, and the production of specific enzymes to reduce host defenses (Riffkin *et al.* 1996). Some parasites even employ cytokines of the host as growth factors. Importantly, helminths may directly suppress host immunofunction by suppressing specific subsets of cells which then alter the host's Th1/Th2 cytokine profile in a manner that helps to promote the parasites survival (Riffkin *et al.* 1996). Many of these evasionary tools will be influenced by genetic variation.

There is evidence that host-parasite genetic interactions influence disease outcome in both animal models and in humans. Genetic variation in *T. muris* influences the outcome of infection, and response to *T. muris* infection varies in mice with different genetic characteristics (Grencis & Entwistle, 1997; Bellaby *et al.* 1995) (see Chapters 1, 7 and 12). In humans, HLA type has been found to affect response to infection with *Plasmodium falciparum* and the strain of *P. falciparum* has been determined to be nonrandomly distributed among HLA types (Hill *et al.* 1997; Gilbert *et al.* 1998). Thus, there appears to be a complex interaction between host and parasite genetic factors in determining malaria outcomes (Gilbert *et al.* 1998). As Hill *et al.* (1997) have noted, knowledge of such interactions may lead to improved approaches in vaccine development.

A search of the literature revealed that no evaluations of host-parasite genetic interaction effects on human nematode infections have been conducted to date. The hypothesis that genetic variation present in the parasite interacts with genetic variation present in the human host to jointly determine the observed variation in quantitative measures of helminthic worm burden remains to be tested.

7.1 Genotype-by-Environment Interaction in *Ascaris* Worm Burden

Using information on the distances between households enrolled in the Jiri Helminth Project obtained using GPS measurements, we extended our statistical genetic models to examine the role of spatial variation in factors influencing worm burden. We determined that a model allowing an exponential decay in the correlation in worm burden phenotypes among individual hosts as a function of the distance between their dwellings (and hence between the areas in which they spend most of their time) best fit our data. When we allow for this type of spatial autocorrelation in addition to host genetic factors, we obtain variance component models that represent highly significant improvements over those that do not allow for spatial variation. This is true for all of the worm burden phenotypes. When we analyze a composite worm burden phenotype based on averaging the z-scores of each of the original phenotypes, we estimate that approximately 39% of the variation in worm burden is due to host genetic factors and 27%

of the variation is due to spatial variation ($p < 0.00001$). Furthermore, there is strong evidence for an interaction between the host genetic and spatial components so that an additional 27% of the variation can be explained by this interaction. An obvious potential source for the observed spatial variation is genetic variation in the parasite itself.

The observed exponential decay in correlation as a function of spatial distance is consistent with an isolation-by-distance population structure model for *Ascaris*. Such an observation is also consistent with the known information on *Ascaris* population structure (Anderson *et al.* 1995; Anderson & Jaenike, 1997). Our model predicts that approximately half of the genetic kinship among parasites is lost at a distance of one-third of a kilometer, a prediction that is reasonable for a macroparasite such as *Ascaris*. Additional support for parasite genetic variation being the source of the observed spatial variation is provided by the presence of the interaction with human host genetic factors. Although geographic variation in egg density in soil could also lead to the observed spatial patterning, there is no obvious mechanism which would lead to an interaction between density and host genetic factors. Alternatively, interaction between host and parasite genomes is both biologically plausible and likely. However, it will require future efforts directed towards evaluation of polymorphic genetic markers in *Ascaris* to unequivocally determine if the observations on spatial patterning are due to parasite genomic variation.

8. FINDING THE SPECIFIC GENES WHICH INFLUENCE SUSCEPTIBILITY TO HELMINTHIC INFECTIONS

During the past few years, enormous advances have been made in the techniques for finding genetic loci influencing disease-related traits. The recent advent of linkage-based genomic scanning methodologies has greatly increased our ability to find and characterize specific loci influencing complex diseases (Lander & Schork, 1994; Blangero, 1995). The genomic scan approach involves placing random markers every 10cM throughout the genome. Such complete coverage of the genome makes it possible to detect all relevant genes influencing the phenotypes of interest. The approach maximizes the chance of successfully detecting genetic effects if they exist.

Despite the fact that genome scanning approaches have only been implemented within the last five years, already there have been significant results in a number of diseases. For example, genes have been mapped for non-insulin dependent diabetes (Hanis *et al.* 1996; Hanson *et al.* 1998; Duggirala *et al.* 1999), obesity (Comuzzie *et al.* 1997; Duggirala *et al*, 1996; Hanson *et al.* 1998), and alcoholism (Reich *et al.* 1998; Begleiter *et al.* 1998). While applications of genomic scanning to infectious disease susceptibility are few, the genomic approach is likely to lead to new insights as evidence by the finding of quantitative trait loci influencing susceptibility to schistosomiasis (Marquet *et al.* 1996).

In preliminary analyses of genome scan data which included 400 markers per individual from 425 members of the Jirel population, we localized two genes having significant effects on susceptibility to *Ascaris* infection as assessed by egg counts (Williams-Blangero *et al.* 2000).

9. FUTURE ADVANCES IN MOLECULAR GENETICS

New molecular advances will soon be of considerable aid for finding the functional mutations in the positional candidate loci identified via linkage-based genome scans. For example, rapid methods for the detection of single nucleotide polymorphisms (SNPs) will greatly enhance capabilities to fine map disease susceptibility loci that are initially found using STR-based genomic scans. Advances in automated sequencing will also speed up both the isolation of these positional candidate genes and the search for mutations that may be the determinants of human host variation in risk of parasitic infection.

ACKNOWLEDGEMENTS

This research was supported by NIH grants AI37901 and AI44406 to S. Williams-Blangero.

REFERENCES

ABEL, L., DEMENAIS, F., PRATA, A., SOUZA, A.E., & DESSEIN, A. (1991). Evidence for the segregation of a major gene in human susceptibility/resistance to infection by *Schistosoma mansoni*. *American Journal of Human Genetics* **48**, 959-970.

ABEL, L., COT, M., MULDER, L., CARNEVALE, P., & FEINGOLD, J. (1992). Segregation analysis detects a major gene controlling blood infection levels in human malaria. *American Journal of Human Genetics* **50**, 1308-1317.

ABEL, L. & DESSEIN, A.J. (1997). The impact of host genetics on susceptibility to human infectious diseases. *Current Opinion in Immunology* **9**, 509-516.

ABEL, L., VU, D.L., OBERTI, J., NGYEN V.T., VAN, V.C., GUILLOD-BATAILLE, M., SCHURR, E., & LAGRANGE, P.H. (1995). Complex segregation analysis of leprosy in southern Vietnam. *Genetic Epidemiology* **12**, 63-82.

ANDERSON, R.M. (1982). The population dynamics and control of hookworm and roundworm infections. In *Population Dynamics of Infectious Diseases* (ed. Anderson, R.M.), pp. 67-108. Chapman and Hall, London, New York.

ANDERSON, R.M. & MAY, R.M. (1982). Population dynamics of human helminth infections: Control by chemotherapy. *Nature* **297**, 557-563..

ANDERSON, R.M. & MAY, R.M. (1985). Helminth infections of humans: Mathematical models, population dynamics, and control. *Advances in Parasitology* **24**, 1-10.

ANDERSON, R.M. & MEDLEY, G.F. (1985). Community control of helminth infections of man by mass and selective chemotherapy. *Parasitology* **90**, 629-660.

ANDERSON, T.J.C. & JAENIKE, J. (1997). Host specificity, evolutionary relationships, and macrogeographic differentiation among *Ascaris* populations from humans and pigs. *Parasitology* **115**, 325-342.

ANDERSON, T.J.C., ROMERO-ABAL, M.E., & JAENIKE, J. (1995). Mitochondrial DNA and *Ascaris* microepidemiology: The composition of parasite populations from individual hosts, families, and villages. *Parasitology* **110**, 221-229.

BLANGERO, J. (1995). Genetic analysis of a common oligogenic trait with quantitative correlates: Summary of GAW9 results. *Genetic Epidemiology* **12**, 689-706.

BLANGERO, J. (1987). Population Genetic Approaches to Phenotypic Microevolution in the Jirels of Nepal. Ph.D. Dissertation, Case Western Reserve University.

BRADLEY, M., CHANDIWAANA, S.K., BUNDY, D.A.P., & MEDLEY, G.F. (1992). The epidemiology and population biology of *Necatur americanus* infection in a rural community in Zimbabwe. *Transactions of the Royal Society of Tropical Medicine and Hygiene* **86**, 73-76.

BUNDY, D.A.P. (1986). Epidemiological aspects of *Trichuris* and trichuriasis in Caribbean communities. *Transactions of the Royal Society of Tropical Medicine and Hygiene* **80**, 706-718.

BUNDY, D.A.P., COOPER, E.S., THOMPSON, D.E., DIDIER, J.M., ANDERSON, R.M., & SIMMONS, I. (1987). Predisposition to *Trichuris trichiura* infections in humans. *Epidemiology and Infection* **98**, 65-71.

BUNDY, D.A.P. & MEDLEY, G.F. (1992). Immuno-epidemiology of human geohelminthiasis: Ecological and immunological determinants of worm burden. *Parasitology* **104** (Suppl.) S105-S109.

CHAI, J.-Y., SEO, B.-S., & LEE, S.-H. (1983). Epidemiological studies on *Ascaris lumbricoides* reinfection in rural communities in Korea. II. Age-specific reinfection rates and familial aggregation of the reinfected cases. *Korean Journal of Parasitology* **21**, 142-149.

CHAN, L., BUNDY, D.A., & KAN, S.P. (1994). Genetic relatedness as a determinant of Predisposition to *Ascaris lumbricoides* and *Trichuris trichiura* infection. *Parasitology* **108**, 77-80.

CONWAY, D.J., HALL, A., ANWAR, K.S., RAHMAN, M.I., & BUNDY, D.A. (1995). Household aggregation of *Strongyloides stercoralis* infection in Bangladesh. *Transactions of the Royal Society of Tropical Medicine and Hygiene* **89**, 258-261.

CROLL, N. & GHADIRIAN, E. (1981). Wormy persons: Contributions to the nature and patters of overdispersion with *Ascaris lumbricoides, Ancylostoma duodenale, Necator americanus*, and *Trichuris trichiura. Tropical and Geographic Medicine* **33**. 241-248.

DESSEIN A.J., CHEVILLARD, C., MARQUET, S., HENRI, S., HILLAIRE, D., & DESSEIN, H. (2001). Genetics of parasitic infections. *Drug Metabolism and Disposition: The Biological Fate of Chemicals* **29**, 484-488.

ELKINS, D., HASWELL-ELKINS, M.R., & ANDERSON, R.M. (1986). The epidemiology and control of intestinal helminths in the Pulicat Lake region of southern India. I. Study design and pre- and post-treatment observations on *Ascaris lumbricoides* infection. *Transactions of the Royal Society of Tropical Medicine and Hygiene* **80**, 774-792.

FORRESTER, J.E., SCOTT, M.E., BUNDY, D.A.P., and GOLDEN, M.H.N. (1988). Clustering of *Ascaris lumbricoides* and *Trichuris trichiura* infections within households. *Transactions of the Royal Society of Tropical Medicine and Hygiene* **82**, 282-288.

FORRESTER, J.E., SCOTT, M.E., BUNDY, D.A.P., and GOLDEN, M.H.N. (1990). Predisposition of individuals and families in Mexico to heavy infection with *Ascaris lumbricoides* and *Trichuris trichiura. Transactions of the Royal Society of Tropical Medicine and Hygiene* **84**, 272-276.

GARCIA, A., MARQUET, S., BUCHETON, B., HILLAIRE, D., COT, M., FIEVET, N., DESSEIN, A.J., & ABEL, L. (1998). Linkage analysis of blood *Plasmodium falciparum* levels: Interest in the 5q31-q33 chromosome region. *American Journal of Tropical Medicine and Hygiene* **58**, 705-709.

GILBERT, S.C., PLEBANSKI, M., GUPTA, S., MORRIS, J., COX, M., AIDOO, M., KWIATKOWSKI, D., GREENWOOD, B.M., WHITTLE, H.C., & HILL, A.V. (1998). Association of malaria parasite population structure, HLA, and immunological antagonism. *Science* **279**, 1173-1177.

GRENCIS, R.K. & ENTWISTLE, G.M. (1997). Production of interferon-gamma homologue by an intestinal nematods: Functionally significant or interesting artefact? *Parasitology* **111**, 353-357.

HANIS, C.L., BOERWINKLE, E., CHAKRABORTY, R., ELLSWORTH, D.L., CONCANNON, P., STIRLING, B., MORRISON, V.A., WAPELHORST, B., SPIELMAN, R.S., GOGOLIN-EWENS, K.J., SHEPARD, J.M., WILLIAMS, S.R., RISCH, N., HINDS, D., IWASAKI, N., OGATA, M., OMORI, Y., PETZGOLD, C., RIETZCH, H., SCHRODER, H.E., SCHULZE, J., COX, N.J., MENZEL, S., BORIRAJ, V.V., CHEN, X., & BELL, G. (1996). A genome-wide search for human non-insulin-dependent (type 2) diabetes genes reveals a major susceptibility locus on chromosome 2. *Nature Genetics* **13**, 161-166.

HANSON, R.L., EHM, M.G., PETTITT, D.J., PROCHAZKA, M., THOMPSON, D.B., TIMBERLAKE, D., FOROUD, T., KOBES, S., BAIER, L., BURNS, D.K., ALMASY, L., BLANGERO, J., GARVEY, W.T., BENNETT, P.H., & KNOWLER, W.C. (1998). An autosomal genomic scan for loci linked to Type 2 diabetes mellitus and body mass index in Pima Indians: An obesity-diabetes locus at 11q23-25. *American Journal of Human Genetics* **63,** 1130-1138.

HASWELL-ELKINS M., ELKINS, D., & ANDERSON, R.M. (1987). Evidence for predisposition in humans to infection with *Ascaris*, hookworm, *Enterobius*, and *Trichuris* in a South Indian fishing community. *Parasitology* **95**, 323-337.

HILL, A.V. (1996). Genetics of infectious disease resistance. *Current Opinion in Genetics and Development* **6**, 348-353.

HILL, A.V. (1998). The immunogenetics of human infectious diseases. *Annual Review of Immunology* **16**, 593-617.

HILL, A.V., JEPSON, A., PLEBANSKI, M., & GILBERT, S.C. (1997). Genetic analysis of host-parasite coevolution in human malaria. *Philosphical Transactions of the Royal Society of London, B, Biological Sciences* **352**,1317-1325.

HOLLAND, C.V., CROMPTON, D.W., ASAOLU, S.O., CRICHTON, W.B., TORIMIRO, S.E., & WALTERS, D.E. (1992). A possible genetic factor influencing protection from infection with *Ascaris lumbricoides* in Nigerian children. *Journal of Parasitology* **78,** 915-916.

LANDER, E.S. & SCHORK, N.J. (1994). Genetic dissection of complex traits. *Science* **265,** 2037-2048.

MARQUET, S., ABEL, L., HILLAIRE, D., DESSEIN, H. (1999). Full results of a genome wide scan which localises a locus controlling the intensity of infection by *Schistosoma mansomi* on chromosome 5q31-q33. *European Journal of Human Genetics* **7, 88-97**.

MARQUET, S., ABEL, L., HILLAIRE, D., DESSEIN, H., KALIL, J., FEINGOLD, J., WEISSENBACH, J., & DESSEIN, A.J. (1996). Genetic localization of a locus controlling the intensity of infection by *Schistosoma mansoni* on chromosome 5q31-q33. *Nature Genetics* **14**, 181-184.

MCCALLUM, H.I. (1990). Covariance in parasite burdens: The effect of predisposition to infection. *Parasitology* **100,** 153-159.

RAMSEY, C.E., HAYDEN, C.M., TILLER, K.J., BURTON, P.R., HAGEL, I., PALENQUE, M., LYNCH, N.R., GOLDBLATT, J., LESOUEF, P.N. (1999). Association of polymorphisms in the ∃-adrenoreceptor with higher levels of parasitic infection. *Human Genetics* **104**, 269-274.

REICH T., EDENBERG, H.J., GOATE, A., WILLIAMS, J.T., RICE, J.P., VAN EERDEWEGH, P., FOROUD, T., HESSLEBROCK, V., SCHUCKIT, M.A., BUCHOLZ, K., PORJESZ, B., LI, T.K., CONNEALLY, P.M., NURNBERGER, J.I., TISCHFIELD, J.A., CROWE, R.R., CLONINGER, C.R., WU, W., SHEARS, S., CARR, K., CROSE, C., WILLIG, C., & BEGLEITER, H. (1998). Genome-wide search for genes affecting the risk of alcohol dependence. *American Journal of Human Genetics* **81**, 207-215.

RIFFKIN, M., SEOW, H.F., JACKSON, D., BROWN, L., & WOOD, P. (1996). Defence against the immune barrage: Helminth survival strategies. *Immunology and Cell Biology* **74**, 564-574.

RIHET, P., TRAORE, Y., ABEL, L., AUCAN, C., TRAORE-LEROUX, T., & FUMOUX, F. (1998). Malaria in humans: *Plasmodium falciparum* blood infection levels are linked to chromosome 5q31-q33. *American Journal of Human Genetics* **63**, 498-505.

SCHAD, G.A. & ANDERSON, R.M. (1985). Predisposition to hookworm infection in humans. *Science* 228:1537-1540.

SMITH, T., BHATIA, K., BARNISH, G., & ASHFORD, R.W. (1991) Host genetic factors do not account for variation in parasite loads in *Strongyloides fuelleborni kellyi*. *Annals of Tropical Medicine and Parasitology* **5**, 533-537.

THEIN-HLAING (1985). *Ascaris lumbricoides* infection in Burma. In *Ascariasis and its public health significance.* (ed. Cromptom, D.W.T., Nesheim, M.C. & Pawlowski, Z.S), pp. 83-112. Taylor & Francis. London.

THEIN-HLAING, THAN-SAW, & MYINT-LIN (1987). Reinfection of people with *Ascaris lumbricoides* following single, 6-month, and 12-month interval mass chemotherapy in Okpo village, rural Burma. *Transactions of the Royal Society of Tropical Medicine and Hygiene*. **81**, 140-146.

WILLIAMS, D., BURKE, G., & HENDLEY, J.O. (1974). Ascariasis: A family disease. *Journal of Pediatrics* **84**, 853-854.

WILLIAMS-BLANGERO, S., BLANGERO, J., & BRADLEY, M. (1997a). Quantitative genetic analysis of susceptibility to hookworm infection in a population from rural Zimbabwe. *Human Biology* **69**, 201-208.

WILLIAMS-BLANGERO, S., BLANGERO, J., ROBINSON, E.S., ADHIKARI, B.N., UPRETI, R.P., & PYAKUREL, S. (1993). Helminthic infections in Jiri, Nepal: Analysis of age and ethnic group effects. *Journal of the Institute of Medicine (Nepal)* **15**, 210-216.

WILLIAMS-BLANGERO, S., BLANGERO, J., & SUBEDI, J. (1996a). A role for genetic epidemiology in the development of international health care programs for soil transmitted helminthiases. In *Culture, Society, and Illness: Transcultural Perspectives*. (ed Subedi, J & Gallagher, E.), pp. 302-315. Prentice Hall, New York.

WILLIAMS-BLANGERO, S., BLANGERO, J., WIEST, P.M., OLDS, G.R., ZHONG, S., WU, G., & MCGARVEY ST (1996b). Genetic analysis of *Trichuris trichiura* infection intensity in Jiangxi, China. *American Journal of Tropical Medicine and Hygiene* **55**, S154.

WILLIAMS-BLANGERO, S., SUBEDI, J., UPADHAYAY, R.P., MANRAL, D.B., RAI, D.R., JHA, B., ROBINSON, E.S., & BLANGERO, J. (1999). Genetic analysis of susceptibility to infection with *Ascaris lumbricoides*. *American Journal of Tropical Medicine and Hygiene* **60**, 921-926.

WILLIAMS-BLANGERO, S., VANDEBERG, J.L., BLANGERO, J., & TEIXEIRA, A.R. (1997b). Genetic epidemiology of seropositivity for *Trypanosoma cruzi* infection in rural Goiás, Brazil. *American Journal of Tropical Medicine and Hygiene* **57**, 538-543.

WILLIAMS-BLANGERO, S., VANDEBERG, J.L., UPADHAYAY, R.P., RAI, D.R., SUBEDI, J., JHA, B., & BLANGERO, J. (2000). A genome scan for susceptibility to *Ascaris* infection in a Nepalese population. *American Journal of Tropical Medicine and Hygiene* **62**, S163.

Chapter 11
POPULATION GENETICS OF INTESTINAL NEMATODES
The Use of Genetic Markers in Inferring Population Movement

Helen Roberts
Laboratory of Evolutionary Genetics, Department of Biology, University College London, UK.
e-mail: h.roberts@ucl.ac.uk

1. INTRODUCTION

Surprisingly little is known about the genetics of intestinal nematodes despite the genome of *Caenorhabditis elegans* being the first multicellular organism to be sequenced. This chapter will deal with why we should be concentrating on genetics of parasitic gastrointestinal nematodes and how we can use available data to further our understanding of these important organisms.

Two important questions to answer in terms of nematode population dynamics, that we may be able to use population genetics for are: how are worms transmitted, and what is the likelihood of drug resistance arising? Drug resistance will also be mentioned in terms of genetic markers and models of gene flow.

2. THE PROBLEMS

For many years, parasites were taken to be genetically homogenous, with little or no variation within populations. But, as is illustrated in other chapters of this book, there are many interesting aspects of nematode infections which belie this idea. The nature of the infection pattern of the gastrointestinal nematodes within a community showing overdispersal is ubiquitous, and yet there are still no complete explanations for this phenomenon (see Chapter 1). The majority of work has focussed on the role played by the host and environment, but parasite strain variation between and within populations may explain some of this variability, and until now there have been no markers with sufficient resolution to examine this in detail. In the last five years or so, molecular markers have become available for some parasites which have changed this situation. The same markers can be used to look for geographic variation, which. in turn, can be used to assess population movement and migration rates, which will become increasingly important if drug resistance were to become the problem it is in the veterinary situation.

It is unlikely that overdispersal can be accounted for solely by parasite variation; the route of infection alone would count against it in that there are numerous infective stages contaminating the environment, yet overdispersal still occurs. However the contribution of parasite variation may be significant, and until that can be assessed accurately, we will be unable to estimate its impact. The types of infections within a host may also prove important. Do large infections represent a large proportion of one strain, or many different ones? Similarly, after treatment and upon reinfection, when many predisposed people regain similarly high worm burdens, do these consist of one strain or several? The question of transmission foci is also important for treatment regimes. Is the focus of infection the school or the house? And if it is the school, do adults pick up infections from their children or is there a second transmission cycle? Do transmission cycles vary depending upon intensity of infection?

Geographical variation of parasite distribution is considerable, and data are becoming available with the use of Geographic Information System (GIS) and Remote Sensing (RS), showing the global patterns of variation. Some of this patterning is due to environmental factors, such as vegetation, rainfall and annual temperature. But there may be some patterns that result from parasite variation. It may be a question of scale: parasite variation may account for micro-variation, while environmental factors may account for macro-variation.

It is important to remember that parasitic nematodes have high host fidelity: host and parasite are co-evolving, and because the generation time of nematodes is much less than that of humans, it is likely that parasite genetic variation plays an important role in adaptation and survival in the host.

Although many of these questions remain to be answered, this chapter will hopefully show how we have advanced towards the answers, and where future work will lie.

3. POPULATION GENETICS

Population genetics (see below for glossary of terms used here) can be defined as the study of the genetic basis of naturally occurring variation, with the aim of describing and understanding the evolutionary forces that create and maintain variation within a species and which lead to differences between species.

Genetic variation can be quantified in several different ways, the major ones being: polymorphism (proportion of loci at which different alleles can be detected), frequency of different alleles at a given locus, and heterozygosity (proportion of individuals where two alleles can be detected). These data are key to models used to understand parameters of mutation, selection and population size. All these factors become important when looking at parasitic populations, and all are related to treatment regimes. For example, it is important to know mutation rates in case resistance does occur; selection will take place under drug pressure, and may lead to mutations and increase in fitness. Knowing the effective population size will indicate whether localised selective processes will occur, and resistance genes spread. If a population is small, for instance, there will be relatively little population movement between groups.

3.1 Geographical structure

Defined as the non-random mating of individuals with respect to location, geographical structure has received attention for two reasons. Geographical separation is an inescapable fact of biology, and differentiation between populations at a local level may represent the first steps in speciation. *F* statistics are the most common way of summarising structure with genetic variability. Variability is partitioned according to differences in heterozygosity into components of within- and between-population variation. The most cited statistic is the proportion of total heterozygosity (H_T) that is explained by within population heterozygosity (H_S). Other *F* statistics give a measure of inbreeding (F_{IS}) or the proportion of variation explained by levels of population classification (sample site <region <country <continent).

It is very common to see F_{ST} values of 0.9 or more for within population variation. This can be observed on large scales in human populations, such as between Caucasian, African and Chinese populations, as well as finer scale such as between villages of Yanomama Native Americans.

Domestic mice and the fruit fly (*Drosophila melanogaster*) show similar levels. However the Jumping Rodent (*Dipodomys ordii*) has F_{ST} of 0.5 meaning only 50% of variation occurs within populations and therefore suggesting strong racial differentiation, and possibly reproductively isolated species.

An overview of population differentiation (as a preamble to doing phylogenetic analysis) is to use a multifactorial statistical test such as principal component analysis, using the program POPSTR, which can separate or group individuals. This provides a three dimensional view, depicting clustering of genetically similar individuals. However such data do not have strong statistical support and should be used circumspectly. In preference phenograms should be used, which can give more rigorous (bootstrap) support.

There are also some inherent problems with using *F* statistics as a measure of population differentiation. Firstly, the geographic delineations are made arbitrarily which can bias the data, secondly there is large sampling variance, and, thirdly when only one summary statistic is examined, statistical support is not as strong, and much information is lost.

Examining how structure affects the pattern of genetic variation patterns is only one aspect of population genetics. Another is Linkage Disequilibrium (LD), occurring when particular alleles of two genes on the same chromosome occur together more frequently than expected by chance. LD is generated by a process of mutation and selection in a population, and is broken down by recombination. Structure can affect LD by the fact that rapid coalescence within a population generates a high frequency of alleles that are in complete association with each other, and if two populations are examined that are in complete LD, but they are treated as one population, LD may be detected. Different human populations have a mixture of varying amounts, whereby migration patterns allowed interbreeding of previously separated populations, and therefore human commensals and parasites are quite likely to exhibit something similar. This leads to differences in allele frequency generating apparent LD between unlinked loci. So, LD may be apparent when two populations have recently mixed, or a particular pair of alleles confers a selective advantage. It becomes important for disease mapping in humans, but the researcher must also be aware of the problem in parasites. To make sure that markers are not incorrectly identified, in terms of whether being responsible for host susceptibility and drug resistance, LD should be assessed in populations, prior to making inferences from these results.

Different genotypes may be favoured in different places, as determined by the environment. Which leads to the question: Can environmental heterogeneity maintain polymorphism, or has it even been responsible for speciation? In the case of the hookworms, they are remarkably similar morphologically, and have overlapping ranges in some areas. This will be more easily examined with the GIS/RS data as it becomes available. If two species inhabit overlapping habitats, usually that with the highest mean fitness will spread to fixation. If the two habitats do not overlap, then both species will remain discrete. This will depend on host offspring dispersal, and whether there are any hybrid species present. Clines may exist, which represent relatively smooth gradients in allele frequency across geographic area of environmental heterogeneity.

3.2 Genetic Markers

3.2.1 MtDNA

Analysis of mitochondrial DNA (mtDNA) is frequently used to answer questions of population genetics and molecular systematics, although relatively little is known about the evolution of parasite mtDNA, such as nucleotide substitution and rate variation. Nematode mtDNA is highly A+T rich and it appears to be more prone to gene rearrangement and recombination. Blouin *et al.* (1998), provide extensive data concerning the mtDNA of several species of nematode which gives information about whether or not mtDNA will prove to be a useful marker. Using Trichostrongylid nematodes as models, substitution matrices have shown that there is bias towards substitution of C or G to A or T. Because of this mutational bias, it is better to ignore third position sites in molecular comparisons, particularly between species. Phylogeny reconstruction should be done, bearing this in mind, as some phyla will separate (artifactually) based on nucleotide composition more than history. Therefore, and because of the high bias of substitution, this means that mtDNA is a very suitable marker for population genetics, but not useful for phylogenetic analysis. It can be used to look at relationships between closely related species and within species and it has been used to suggest the presence of cryptic species as well, but it should not be used to infer relationships between species with deeper phylogenetic branches. *Trichuris* species can reliably be differentiated by their ITS2 sequences and PCR-linked restriction-fragment-length polymorphism (RFLP) (Oliveiros *et al.* 2000), but it is not possible to identify different members of the same species by these means.

3.2.2 Microsatellites

Microsatellites are very short DNA motifs, typically 2-6 base pairs long, occurring in tandem repeats throughout the genome. Their replication is highly unstable, such that the number of repeats changes at a higher rate than single point mutations (due to polymerase slippage). In humans the average rate for point mutations is about 10^{-8} per generation and for microsatellites 10^{-5} per generation. The advantage DNA variation has over allozyme variation is that different types of mutation have different levels of polymorphism. Generally speaking, less constrained sites (non-coding regions) have higher mutation rates, and even within coding regions, mutations that leave the amino acid unaltered are considered to have higher mutation rates. There is also considerable lack of concordance between allozyme variation and DNA variation: the latter showing a greater range.

Microsatellites have now been developed for *Ascaris* and *Trichuris* and are available as sequences in GenBank, and as part of the Web page at www.ucl.ac.uk/biology/goldstein. There are at present fifteen to twenty available, but this number will increase as more are sequenced. This has improved our ability to examine some of the underlying questions that have been mentioned. The microsatellites are all dinucleotide and vary from simple to compound repeats. They were sequenced from a microsatellite-enriched library and each has so far proved to be species specific. It is unknown which chromosomes they map to, and because of the lack of sequence similarity to other nematodes, it has not been possible to use the *C. elegans* database to look for homologues. However the high degree of polymorphism and high variation in repeat number for the alleles for many of the microsatellites allows population structure to be deduced.

The Tandem Repeat Finder Program (see Table 11.1) can be used to find microsatellites in existing genetic data, as it becomes available. Unfortunately, a lot of genetic data emerging for some parasites (for example *Trichuris muris*) is from EST databases, and although useful in their own way, microsatellites from coding regions are less likely to be polymorphic.

3.2.3 Single Nucleotide Polymorphisms (SNPs)

As the genetic database of the intestinal nematodes becomes more extensive, it will be possible to identify SNPs as they have been for other organisms. Combining SNP and microsatellite data will give not only more detail about individual parasites, but also information about the genome as a whole and assess regions of the genome that are important in drug resistance, or inducing pathogenesis.

Table 11.1. Web Sites and Statistical Programmes of interest

Site	Web Address
National Centre for Biological Information, including site for GenBank and *C.elegans* sequencing project	http://www.ncbi.nlm.nih.gov
PHYLIP – for assembling data for trees	http://evolution.genetics.washington.edu/ phylip.html
TREEVIEW – for drawing trees	http://taxonomy.zoology.gla.ac.uk/rod/ treeview.html
Arlequin – for analysis of population genetics	http://lgb.unige.ch/arlequin/
Microsatellite web base for *Ascaris* and *Trichuris*	http://www.ucl.ac.uk/biology/goldstein
Tandem Repeat Finder program	http://c3.biomath.mssm.edu/trf.html
Human STR Database	http://www.cstl.nist.gov/biotech/strbase/index. htm
POPSTR	Henry Harpending; hxh5@psu.edu
Structure	J.Pritchard: j.pritchard@stats.ox.ac.uk

SNPs of *C .elegans* have been found using four strains isolated from natural wild populations, which were sequenced by the shotgun approach, and checked for SNPs (Koch *et al.* 2000). The majority of polymorphisms encountered do not alter the amino acid encoded, as they are found in the introns or the third base of a codon. Higher levels were found on autosomal arms than around chromosomal centres. Among 24 isolates, most SNPs are shared between strains, and patterns agree with classification of races. At present this type of analysis is not available for the gastrointestinal nematodes, but with collaboration and concerted efforts to sequence the genomes, it is a very real possibility for the future.

4. PARASITE STUDIES

There have been few studies made on human parasites. The most comprehensive so far has been that of Anderson, Romero-Abal & Jaenike, (1993 *et seq.*) who examined *Ascaris* populations mainly in Guatemala, using allozyme data, ribosomal DNA RFLP data and mitochondrial DNA sequence analysis. He and co-workers showed that there was strong differentiation between *Ascaris* from humans and *Ascaris* from pigs indicating low migration rate between human and pig. However in non-endemic areas, when humans have presented with *Ascaris* infection, the worms had ribosomal DNA (rDNA) pattern, found most commonly in populations from pigs, suggesting there is cross infection. This work has been very important in showing that there is not only cross-infection occurring in some areas, but also that it will be possible to examine in more detail the ancestry behind the populations, and estimate whether speciation has taken place in areas of high endemicity (such as Guatemala) whilst in areas of low transmission, the human and pig populations are effectively still both showing ancestral DNA markers.

The use of multiple markers, such as microsatellites, can be used to look at time to ancestral lineages and give estimates of time to the most recent common ancestor (TMRCA). These, alongside coalescence times, estimated from tree branches, will show how much populations differ from one another and how much movement there has been between pig and human populations. Anderson estimated that the main split between the mitochondrial haplotypes he observed, would have been around 600,000 years before present (b.p.) which is a considerable time prior to domestication of the pig. If this time had come after domestication, it would suggest there was more likelihood of movement of parasites between hosts. This is of great importance in light of treatment regimes, as treating domestic pig populations would also become necessary. Drug resistance arising from frequent treatment in pigs would also spread to human populations very quickly.

Zhu *et al.* (1999) sequenced the internal transcribed spacer (ITS-1) region of nuclear ribosomal DNA of *Ascaris* from humans and pigs, and showed there are six nucleotide differences between human and pig-derived worms, two of which were dinucleotide deletions, and the other two were SNPs. They sequenced just seven worms from each host species, so sequencing a higher number will reveal more SNPs and may be able to shed light on their evolutionary significance. To date, one of the issues concerning

A. suum and *A. lumbricoides* is whether they represent two separate species. This is unlikely to be resolved using these types of markers, without cross-mating experiments being carried out. These experiments could be done using a pig model system developed in Denmark (Jungerson *et al.* 1996; see Chapter 7).

In work being carried out here (Tables 11.2 and 11.3), we have been using microsatellites from *Ascaris* and *Trichuris* to look at geographical variation between sites in South America, Central America and Asia. F_{ST} values show, as expected, that the majority of variation is within populations, rather than between populations or between groups (defined as continents).

Table 11.2. Preliminary data from the studies with *Trichuris*: Population pairwise Fst's

	1	2	3	4	5	6
2	-0.021					
3	0.010	0.007				
4	0.026	0.031	0.023			
5	0.038	0.059	0.038	0.009		
6	0.043	0.065	0.067	0.068	0.092	
7	0.055	0.054	0.051	0.054	0.091	0.067

1: China, 2: Viet Nam, 3: Thailand, 4: St Lucia, 5: Jamaica, 6: Colombia, 7:Guatemala

Table 11.3. Preliminary data from the studies with *Ascaris*: Population pairwise Fst's

	1	2	3	4	5	6
2	0.030					
3	0.031	0.020				
4	0.045	0.063	0.052			
5	0.053	0.0	0.053	0.079		
6	0.121	0.123	0.098	0.156	0.108	
7	0.082	0.057	0.050	0.057	0.060	0.106

1: Cameroon, 2: VietNam, 3: Thailand, 4: Bangladesh, 5: StLucia, 6: Colombia, 7: Guatemala

It becomes apparent from these data that not only is it possible to distinguish between geographical populations, but it is also possible to infer geographic movement, as the lower the F_{ST} value, the closer a population is deemed to be related. Therefore, the South American and Central American populations are closely related within their group, but not to each other, in accordance with the differences between the human populations inhabiting these areas. Another interesting fact is that the overall F statistics show that 93% and 95% of the total variation is found within populations rather than between, which is in line with values for human populations. This is in agreement with data for other parasites, such as *Haemonchus contortus*, that have a large effective population size and where there has been considerable population movement due to movement of farm animals (Blouin *et al.* 1995).

4.1 Fitness effect on overall variability

The intestinal nematodes are under constant selection pressure because of the frequency of treatment. Under some control programmes, treatment with albendazole may occur every six to twelve months. Although there are no definite reports of drug resistance yet, reports of reduced efficacy of this drug in some areas are becoming frequent. Drug treatment increases selection pressure, and it is unknown how much this will affect genetic variability. It should be expected that this will decrease substantially, but more importantly, if new mutations arise, which cause resistance, their spread through the population could be rapid (local selective sweep), depending on the mutation rate and the rate of migration. Examining populations of worms following successive drug treatments over several months or years will give us an estimate of this.

A resistant and susceptible line of *Teladorsagia cirumcincta* were examined for relative fitness in terms of egg production, development of larvae, infectivity of larvae, and survival of larvae to adults produced in an infection, and it was found that there was no reduction in fitness associated with resistance to benzimidazole drugs (Elard, Sauve & Humbert, 1998). This would suggest that the reduction of genetic variability under selection does not reduce fitness, therefore, if drug resistance were to arise, it would spread rapidly through the population.

4.2 Transmission and the use of structure programs

Previous studies have found difficulty looking at transmission of nematodes from one host to another mainly because the resolution of the markers had not been high enough. Allozymes and RFLP are unable to show fine detail between individuals. mtDNA is a far more sensitive marker when used for within species variation, but we have found that microsatellites are also useful markers, with the advantage of being easy to type.

For example, below is a preliminary phylogenetic tree to show the differences between individual *Ascaris* from people within five houses in a community in Vietnam (Figure 11.1). Each individual worm can be identified, and using this approach we may be able to focus on the transmission of infection. Each number refers to a person. Five worms have been identified from each person, and households contain three people.

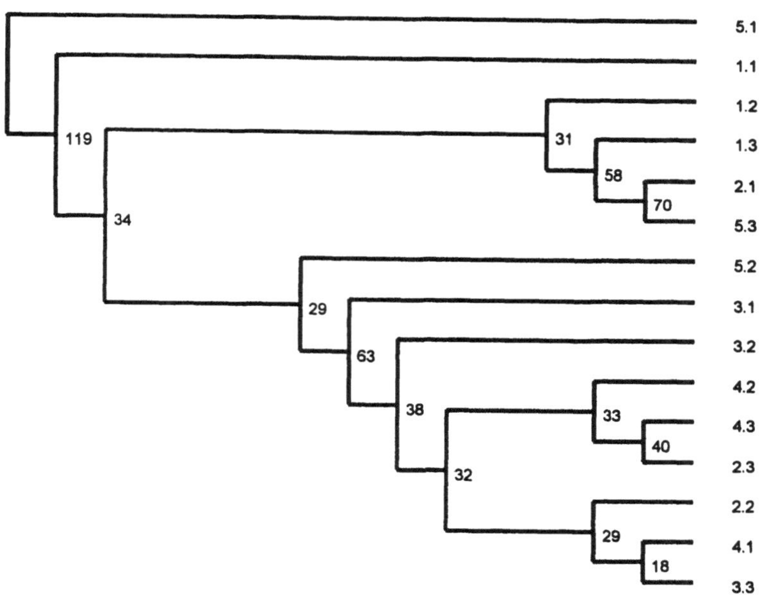

Figure 11.1. Neighbour Joining tree based on Proportion of Shared Alleles for *Ascaris* from individual hosts of the same village using five loci. Each individual lived in one of five houses (first number); three individuals per house (second number) and each with five *A. lumbricoides*. Bootstrap values are indicated, but are so low as to be irrelevant. This indicates that not only does the tree itself have no statistical support, but also that more markers will be required to reverse this. Bootstrap values ARE statistically significant for trees of individuals from a wider geographic area. This illustrates the danger of using too few or non-neutral markers in population genetics.

One problem that arises from this particular study is that the age of the infection is unknown. However, using reinfection data, it will be possible to look at infections that occurred within the last few months. At the same time, it is necessary to collect detailed genealogical data about the hosts involved in the study. Another approach is to use a non-hierarchical clustering simulation program, such as Structure, which can assign each worm to groups, and will allow the researcher to look in far greater detail at transmission. This, however, requires several markers (at least ten microsatellites) before the resolution becomes fine enough.

This brings us to the question of how many markers should be used for studies. It is obvious that, although between five and ten markers are useful for showing individual differences, for many of the statistical programs mentioned in this chapter, the more the better. Ten markers give high bootstrap values for geographical isolates, but not for small-scale structure. Therefore the number required will depend upon the type of study.

5. CONCLUSIONS

Population genetics of the gastrointestinal nematodes is a growing subject that is able to draw on the work being done on other organisms and which will benefit from collaborative sequencing projects. It is becoming increasingly clear that there is considerable information to be gained in this field that will help elucidate some of the problems associated with these infections.

The researcher now has available new methods for screening population genetic structure, using microsatellites and SNPs. New statistical approaches will allow examination of population admixture and genetic variability under selection pressure. With these methods we may be able to answer some of the questions about parasite variation, parasite advantage in the host, the evolution of parasitism and factors affecting transmission.

GLOSSARY

Microsatellites: Repeats of two to six nucleotides arranged tandemly throughout the genome. They have a higher mutation rate than for point mutations, and can increase or decrease in repeat number due to a step wise mutation process. Inserttions or deletions can also arise which eventually lead to the breakdown of the repeat.

Localised Selective Sweep: A new advantageous mutation entering a population that spreads rapidly through the population until fixation and reduces variation at neutral linked markers.

Linkage Disequilibrium: Genetic markers non-randomly associated with each other in a population. Breaks down rapidly with time when unlinked markers are examined.

Coalescence: The process by which two genes may have come from a common ancestor allowing the researcher to look at the ancestral history of genes.

Admixture: Two previously separated populations mix and interbreed. In case-control studies of association, population subdivision or recent admixture of populations can lead to spurious associations between a phenotype and unlinked candidate loci.

REFERENCES

ANDERSON, T.J.C., ROMERO-ABAL, M.E. & JAENIKE, J. (1993). Genetic structure and epidemiology of *Ascaris* populations: Patterns of host affiliation in Guatemala. *Parasitology* **107**,319-334.

ANDERSON, T.J.C. & JAENIKE, J. (1997). Host specificity, evolutionary relationships and macrogeographic differentiation among *Ascaris* populations from humans and pigs. *Parasitology* **115**, 325-342.

ANDERSON, T.J.C., ROMERO-ABAL, M.E. & JAENIKE, J. (1995). Mitochondrial DNA and *Ascaris* microepidemiology: the composition of parasite populations from individual hosts, families and villages. *Parasitology* **110**, 221-229.

ANDERSON T.J.C. (1995). *Ascaris* infections in humans from North America: Molecular evidence for cross-infection. *Parasitology* **110**, 215-219.

BLOUIN, M.S., YOWELL, C.A., COURTNEY, C.H. & DAME, J.B. (1995). Host movement and the genetic structure of populations of parasitic nematodes, *Genetics* **141,** 1007-1014.

BLOUIN, M.S., YOWELL, C.A., COURTNEY, C.H. & DAME, J.B. (1998). Substitution bias, rapid saturation and the use of mtDNA for nematode systematics. *Molecular and Biological Evolution* **15,** 1719-1727.

ELARD, L., SAUVE, C. & HUMBERT, J.F. (1998). Fitness of benzimidazole-resistant and – susceptible worms of *Teladorsagia cirumcincta*, a nematode parasite of small ruminants. *Parasitology* **117**, 571-8

JUNGERSON, G., ERIKSEN, L., NIELSEN, C.G., ROEPSTORFF, A. & NANSEN, P. (1996). Experimental transfer of *Ascaris suum* from donor pigs to helminth naïve pigs. *Journal of Parasitology* **82,** 752-756.

KOCH, R., VAN LUENEN, H.G.A.M., VAN DER HORST, M., THIJSSEN, K.L. & PLASTERK, R.H.A. (2000). Single nucleotide polymorphisms in wild isolates of *Caenorhabditis elegans*. *Genome Research* **10,** 1690-1696.

OLIVEROS, R., CUTILLAS, C., DE ROJAS, M. & ARIAS, P. (2000). Characterisation of four species of *Trichuris* (Nematoda: Enoplida) by their second internal transcribed spacer ribosomal DNA sequence. *Parasitology Research* **86,** 1008-13.

ZHU, X., CHILTON, N.B., JACOBS, D.E., BOES, J. & GASSER, R.B. (1999). Characterisation of *Ascaris* from human and pig hosts by nuclear ribosomal DNA sequences. *International Journal for Parasitology* **29**, 469-478.

Chapter 12
PARASITE STRAIN DIVERSITY AND HOST IMMUNE RESPONSES

Derek Wakelin and Janette E. Bradley
School of Life and Environmental Sciences, University of Nottingham, Nottingham, NG7 2RD, UK
e-mail: D.Wakelin@nottingham.ac.uk

1. INTRODUCTION

Although of greatest importance in tropical and subtropical countries, the major geohelminths of humans (*Ascaris*, Hookworms, *Trichuris*) have a wide distribution that is limited only by environmental conditions and by socio-economic factors. Other intestinal nematodes, eg *Trichinella*, have a near global distribution. Such patterns of distribution, and the nature of the host-parasite relationships established by these nematodes, imply that there must be considerable genotypic and phenotypic diversity within each species. Such diversity may reflect genetic drift, founder population effects, or differences in response to local selection pressures. Knowledge of this diversity is, with some notable exceptions, rather limited, and its correlation with aspects of the immune response, despite an obvious practical and theoretical importance, has received little attention. In this chapter we will describe evidence for genetic variation within these worm populations, consider the evidence for genotypic and phenotypic changes in response to selection, discuss the nature of immune responses to intestinal nematodes and the ways in which genotypic change may affect these, and then describe two case studies in which parasite variation has been related to host immunity.

2. GENETIC VARIATION

Genetic variation in human geohelminths has been most extensively studied in *Ascaris*. Studies carried out by Anderson and co-workers using analyses of mitochondrial DNA have demonstrated considerable intra-population diversity (Anderson, Blouin & Beech, 1998). For example, 42 distinct mitochondrial genotypes were identified in 265 worms taken from humans and pigs in two Guatemalan villages (Anderson *et al.* 1995). Parasites carrying the same genotype were found more frequently than predicted by chance within single individuals. This suggests the operation either of environmental influences (e.g. contact with clusters of genetically related eggs – the most likely explanation) or the operation of a genetic influence, such that eggs of certain genotypes are more likely to mature into adult worms in hosts of a particular genotype. A recent study with *Necator* (Hawdon *et al.*, 2001), using sequences of the mitochondrial cytochrome oxidase 1 gene, found 25 unique haplotypes in some 120 hosts from four villages. However, there was no evidence for non-random distribution of genotypes among hosts within villages.

Genetic variation has been extensively studied in *Trichinella*. Current views are that this genus contains 10 closely related genotypes, of which 7 have been given species status. The extensive variation in this species complex has been investigated by PCR using random oligonucleotide primers and a variety of specific primers from internal transcribed spacers, mitochondrial and ribosomal DNA as well as sequences from known antigens; restriction fragment length polymorphism (RFLP), polymerase chain reaction-restriction fragment length polymorphism (CFLP) and microsatellite markers have also been used. These approaches have not only revealed variations that can be associated with defined genotypes, but have also shown considerable intra-genotype variation (e.g. Nagano *et al.*1999; Wu *et al.* 1999; La Rosa & Pozio, 2000; La Rosa *et al.* 2001). Extensive studies on genetic variation have also been carried out in the trichostrongyle parasites of domestic stock (Gasser & Newton, 2000).

Collectively these data show that genetic polymorphism is a characteristic likely to apply to all intestinal nematodes. Two crucial questions are therefore a) whether the frequency of such polymorphisms can be influenced by selection, and b) whether these polymorphisms might influence the expression of molecules that play a role in the host-parasite relationship.

3. RESPONSE TO SELECTION

A number of older studies examined the possibility that the host range of intestinal nematodes could be altered by passage through abnormal hosts – i.e. by imposing a selection pressure for the ability to survive and reproduce. For example, Haley and colleagues (Haley, 1966; Solomon & Haley, 1966) successfully adapted the rat parasite *Nippostrongylus brasiliensis* to mice and to hamsters by repeated serial passage. The ability of *N brasiliensis* to survive and reproduce is strongly influenced by host immunity even in the natural host and it is possible that adaptation to the mouse involved changes in the worms that altered both physiological and immunological parameters. The human hookworm *Necator* has similarly been adapted to hamsters (Sen & Seth, 1967; Behnke, Paul & Rajasekariah, 1986), but, although *Strongyloides ratti* from rats can infect mice, attempts to achieve enhanced adaptation by repeated passage were unsuccessful (Gemmill, Viney & Read, 2000).

Phenotypic changes in intestinal nematodes in response to host immunity were first reported by Ogilvie and colleagues (Edwards, Burt & Ogilvie, 1971) in studies with *N. brasiliensis.* As immunity developed during a primary infection, worms showed changes in their acetylcholinesterase isoenzyme profile. Worms developing in immune rats also showed changes ('adaptation') that made them less immunogenic in naïve animals (Ogilvie 1972). Similar changes were also seen in worms established as the result of trickle rather than single pulse infection, such worms showing an enhanced ability to survive in the immune host (Jenkins & Phillipson, 1972). Dobson and co-workers carried out a series of experiments in which *Heligmosomoides polygyrus* was passaged through mice with genetically determined differences in resistance or which had acquired resistance to infection through prior infection. Their data showed that this resulted in a number of heritable phenotypic changes. Mice passaged through outbred Q strain mice for 10 generations showed enhanced infectivity for this strain but not for C3H mice, which were more susceptible (Dobson & Owen, 1977). Worms passaged through immune Q mice also showed enhanced infectivity in immune as compared with naïve mice, although the precise outcome was influenced by the mouse strains that had been used (Dobson & Tang 1991). In none of these cases was the phenotypic adaptation observed linked to underlying genotypic change. The most important examples where this has been done concern the development of anthelmintic resistance, particularly in intestinal trichostrongyles of sheep and goats (Sangster, 1999). Resistance to the benzimidazoles in parasites such as *Haemonchus contortus* and *Trichostrongylus colubriformis* is due to a point mutation resulting in the replacement of phenylalanine by tyrosine at position 200 in the β-tubulin isotype 1 gene (Grant & Mascord, 1996). Resistance develops rapidly under the selection pressure imposed by anthelmintic treatment, and there is little reversion even if treatment is withdrawn. This implies that the mutant gene carries no significant fitness costs, although evidence for and against this is not clear-cut. For example, well-controlled experimental studies in sheep using strains of similar origin failed to show significant differences in immunogenicity and pathogenicity between drug-resistant and drug-susceptible strains or in their faecal egg output (Barrett, Jackson & Huntley, 1998). Maingi, Scott & Prichard (1990), however, by comparing moderately and highly resistant strains, showed that parasitological and pathological parameters correlated with increasing resistance. One study with *T. colubriformis* in rabbits, comparing resistant and susceptible strains (Mallet & Hoste, 1995), did report a greater mucosal inflammatory response against the former, and this was associated with reduced parasite fecundity. The strains used were, however, of different geographical origin and this may have been a contributory factor.

4. IMMUNITY TO INTESTINAL HELMINTHS

Work with intestinal nematodes in laboratory rodents has shown clearly that hosts respond strongly to infection and that these responses can lead to significant protection, leading to an accelerated loss of worms from the intestine. A continuing difficulty is identification of the mechanisms through which protection is achieved. There is a general consensus that immunity is T cell dependent, and that T helper 2 (Th2) cells are critically important in initiating and regulating protective responses. However, although the intestine mounts characteristic responses to infection with any parasite, it is likely that those that are effective in generating an effective resistance will differ between different species of nematode. Infections generate a wide range of cellular, serological and inflammatory responses, but few of these have been shown to be causally correlated with protection. It is possible, therefore, that much genotypic variation, even though reflected in phenotypic differences in parasite immunogenicity, will have little or no effect upon the outcome of the host-parasite relationship. Whereas this can be investigated systematically in laboratory models, it is difficult, if not impossible, to do so in human infections, where the evidence for protective immunity is at best circumstantial.

4.1 Antigens and Immunomodulators.

The ability of intestinal worms to generate host-protective responses is, by definition, related to the production of potent immunogens. Polymorphism in these may therefore influence the efficacy of host response. In a few cases the identity of these immunogens is known in sufficient detail to draw conclusions about the possible consequences of variation in molecular structure. This can be illustrated by data for the two model parasites discussed later. *Trichinella spiralis* releases a 43kDa glycoprotein molecule from its stichocytes which is known to play a major role in eliciting host immunity. This has been amply demonstrated in mice by vaccination studies using purified or recombinant antigen (Silberstein & Despommier, 1984; Robinson, Bellaby & Wakelin, 1995). The molecule's immunogenicity is determined by its protein component (Jarvis & Pritchard, 1992) and it can therefore be predicted that polymorphism affecting glycosylation would have little or no effect upon protection in mice, although it might well affect antibody responses, as the carbohydrate component (tyvelose) is immunodominant (Denkers, Wassom & Hayes, 1990). In contrast, immunity against infection in newborn rats can be transferred with antibodies against tyvelose (Ellis *et al.*, 1994) and polymorphism could therefore affect the level of immunity expressed. Whereas *T. spiralis* is a potently immunogenic species, *T. pseudospiralis* appears to exert an immunomodulatory effect on the host (Stewart, 1989), although this has not yet been correlated with a particular molecular component. Polymorphism in such molecules may reduce modulation and therefore result in enhanced immunogenicity. Like *T spiralis, Trichuris muris* also produces a major immunogen from stichocytes which similarly can elicit host responses that bring about premature worm loss from the intestine (Jenkins & Wakelin, 1983). However, it appears that the later developmental stages of this species release a modulatory factor (Grencis & Entwistle, 1997 – see below). As with *T. spiralis* and *T pseudospiralis,* therefore, polymorphism in the molecules secreted by *T. muris* may have differing effects on the host-parasite relationship by altering either immunogenicity or immunomodulation.

4.2 Phenotypic variation and host immune responses.

There are very few studies relating phenotypic variation in a human geohelminth with specific host immune responses. Fraser & Kennedy (1991) examined heterogeneity in expression of surface antigens of *A. lumbricoides* infective larvae by assessing variation in binding of antibodies taken from the population from whom the infective eggs were also taken (eggs were hatched *in vitro* and the larvae cultured for 48h for maximal antigen expression). There was considerable heterogeneity between larvae in the degree of binding of antibody from a given donor, and it was suggested that this might reflect quantitative differences arising from polymorphism in surface antigens or qualitiative effects from differential gene expression. It is possible that the striking differences in fecundity of *A. lumbricoides* from Nigeria and Bangladesh described by Hall & Holland (2000) might also reflect parasite differences in immunogenicity.

Serologically detectable antigenic variation in adult *Trichuris trichiura* was demonstrated by Currie *et al.* (1998), both at the level of worm populations taken from individual hosts and at the level of individual worms. In these experiments variation was detected by immunoblotting using plasma taken from the worm donors.

Some evidence of variation in a major cryptic antigen of *H. contortus* was suggested by difference in responses of Australian lambs to vaccination with H11 (a gut membrane-derived protective antigen) taken from British and Australian parasite sources followed by challenge with Australian larvae (Newton *et al.* 1995). Both antigens protected well, but the Australian H11 was marginally better (protection from intramuscular or subcutaneous vaccination was 75.5% and 87.7%, respectively for the Australian antigen and 60% and 55.9% for the British antigen). No major differences between the antigens were detected by SDS-PAGE.

Tang, Dobson & McManus (1995) found that the protein and antigen profiles of *H. polygyrus* worms that had been selected by repeated passage through immune mice showed differences from those of worms maintained in naïve mice. Immune-adapted worms survived better in immune mice, but there were no strong correlations with the molecular differences described. A subsequent study (Su & Dobson, 1997) showed that immune-adapted worms appeared to have reduced immunogenicity, eliciting lower antibody (IgG) and cellular (lymphocyte and eosinophil) responses; again, the extent of these changes varied with the genotype of the host used, implying a complex host-parasite interaction at both the phenotypic and genotypic levels.

The strain of *H polygyrus* used in most laboratory studies with mice was derived from the deermouse *Peromyscus maniculatus* and is now designated as *H. polygyrus bakeri.* This is one of four subspecies, which, although very similar, show phenotypic differences in morphology and host range. Comparative studies have been made between *H. p. polygyrus* and *H. p. bakeri* in terms of their ability to infect laboratory mice and the field mouse *Apodemus sylvaticus*, a natural host for *H. p. polygyrus* (Quinnell, Behnke & Keymer, 1991). Overall *H. p. bakeri* survived well in laboratory mice but poorly in field mice. Time course studies showed that, whereas even initial establishment of *H. p. polygyrus* was very limited in field mice, *H. p. bakeri* could establish well in *Apodemus* initially, but numbers of adult worms then declined, suggesting the operation of anti-worm immune responses. This was confirmed by immunesuppression studies, treatment with cortisone acetate enabling not only adult *H. p. bakeri* to survive in *Apodemus*, but also *H. p. polygyrus* to establish and survive in laboratory mice. This is a clear illustration that the phenotypic differences between subspecies are accompanied by genotypic differences affecting the production of protective immunogens that have a direct influence on the host-parasite relationship.

5. *TRICHINELLA*

The genus *Trichinella* comprises a complex of some ten genotypes, eight of which are currently designated as separate species (Murrell *et al.* 2000), although the precise taxonomic status of *Trichinella* species is not entirely clear, all being morphologically similar, but with distinctive biological characteristics. Among these are elements such as duration of the intestinal phase, and level of reproductive output (together determining 'infectivity') as well as location in the intestine. However, such characteristics are known to be influenced by components of the host's immune response and will therefore vary with changes in the host's response capacity. Many workers have shown that it is possible to create hybrids between genotypes now generally recognized as distinct species (reviewed in Dick & Chadee, 1983). The hybrid muscle larvae have not always been tested for 'viability' – i.e. their capacity to infect another host - but where this has been done they have sometimes proved non-infective (e.g. Martinez-Fernandez *et al.,* 1988; Wu *et al.* 2000) implying that species identity is correct. Nevertheless, the overall similarity of members of this genus, and the ease of establishing laboratory infections make it a valuable resource for studies concerned with the relationships between variation and immunity.

Members of the genus *Trichinella* hare a similar and distinctive life cycle. Infection is initiated when hosts ingest infective larvae contained in the muscles of other infected animals. After digestion in the stomach, the larvae are released, pass into the small intestine and then penetrate into the epithelial cells of the intestinal mucosa. After a short period of rapid growth and development, in which the larvae undergo four moults in as many days, the sexually mature adults mate, and the females then begin to release newborn larvae into the mucosa. These travel, via the blood and lymph, to striated muscles, where they penetrate and invade muscle cells. The muscle cells are transformed into nurse cells, with a host-derived collagenous capsule which provide a niche that supports development to the infective stage and allows them to survive for prolonged periods (months to years). Two species (*T. pseudospiralis* and the recently described *T. papuae*) do not induce nurse cell formation, the larvae continuing to migrate freely within the muscle tissue. Although predation is probably the commonest route by which infection is acquired, it is well established that larvae of the capsule-forming species can survive for several days after the death of their host, thus allowing infection through scavenging and carrion feeding. Both the intestinal and the muscle phase of the life cycle can be pathogenic, and moderately heavily infected hosts may also show behavioural changes that increase the likelihood that they will be predated.

5.1 Immunity

The species *T. spiralis* has been extensively used as a laboratory model for studies of immunity in rodent hosts, primarily mice and rats, in consequence there is an extensive literature describing the responses induced by infection. Primary infections are terminated by responses that result in the expulsion of adult worms from the intestine, but in most cases immunity has little effect on the survival of the larvae that have developed in the muscles. Subsequent infections are eliminated much more rapidly, and there may be marked reductions in parasite growth and fecundity. In rats, challenge infections may be eliminated very rapidly – within 24 hours – but in mice the process normally takes longer. Immunity is induced by the release of stichosomal and cuticular antigens. Although a complex of antigens is released, one component, a glycoprotein with a MW of 43kDa is immunodominant and capable by itself of inducing protective immunity (see above).

The precise mechanisms that result in loss of worms from the intestine remain controversial. Data on immunity to different genotypes of *Trichinella* comes almost exclusively from work in mice, although there have been some papers relating to infections in pigs. The focus here is on data obtained from the mouse model. Loss of worms during an initial infection is dependent on responses generated by Th2 cells and is associated with profound inflammatory changes in the intestine. Although it has for many years been assumed that worm loss is a direct consequence of these changes it has now been shown that it can occur in the absence of many of these changes (Garside *et al.* 2000). It does seem clear, however, that, in most circumstances, the activity of mucosal mast cells is necessary for expulsion to occur (Donaldson *et al.* 1996). Antibody responses during primary infections are low level, and serum transfer experiments have shown little or no effect on worm survival, although growth and fecundity may be affected. Stronger antibody responses are made after the completion of the intestinal phase, and are prolonged (presumably) by release of antigen from the muscle larvae. Where they have been studied, the overall pattern of responses to infection with the different genotypes is similar to those described for *T. spiralis,* with the exception of *T. pseudospiralis,* which exerts a marked immunomodulatory effect on the host (Stewart, 1989).

5.2 Genetic variation

Differences in the host-parasite relationships established in laboratory mice have been reported both within and between particular genotypes (reviewed by Wakelin & Goyal, 1996). For example, Bolas-Fernandez & Wakelin (1989) and Goyal & Wakelin (1993a) described variations in infectivity between isolates of *T. spiralis sensu stricto* obtained from five different geographical regions when these were used to infect a single strain of mouse. Experiments in which mice given a primary infection with one isolate were challenged with the homologous or heterologous isolate showed inter-isolate differences in immunizing capacity (Goyal & Wakelin 1993b). Isolate-specific differences were also seen when levels of serum IgG1, IgG2a and IgE, mucosal mastocytosis and peripheral blood eosinophilia were measured in infected mice. Some evidence was obtained that linked these differences to differential cytokine responses, isolates generating the largest antibody, mast cell and eosinophil responses showing the earliest switch from a type 1 (Th1) to a type 2 (Th2) cytokine profile (Goyal, Hermanek & Wakelin, 1994).

Some of the most striking differences in host reponse to *Trichinella* genotypes were observed in comparative studies using *T. spiralis* and *T. nativa* (a genotype associated with infections wild animals in northern rather than temperate latitudes and which does not have the domestic cycle typical of *T. spiralis*). When comparable infections were established in the same strain of mice, *T. nativa* was expelled very much more rapidly than was *T. spiralis.* Treating mice with a corticosteroid immunosuppressive drug showed that this difference was largely determined by the host's immune response (Figure 12.1) (Bolas-Fernandez & Wakelin, 1989). In treated mice the kinetics of intestinal infection were similar with both genotypes. Although *T. nativa* can be distinguished easily from *T. spiralis* using a variety of molecular techniques, there is no information about the nature of antigenic differences between the two. However, such differences must exist, because not only are the kinetics of infection different in immunologically competent mice, but infected animals also show differences in the level and specificity of their antibody response (Bolas-Fernandez & Wakelin 1989). Differences in the anti-carbohydrate (IgG3) responses of mice infected with six genotypes, including *T. spiralis* and *T. nativa,* were also described by Dea-Ayuela *et al.* (2000).

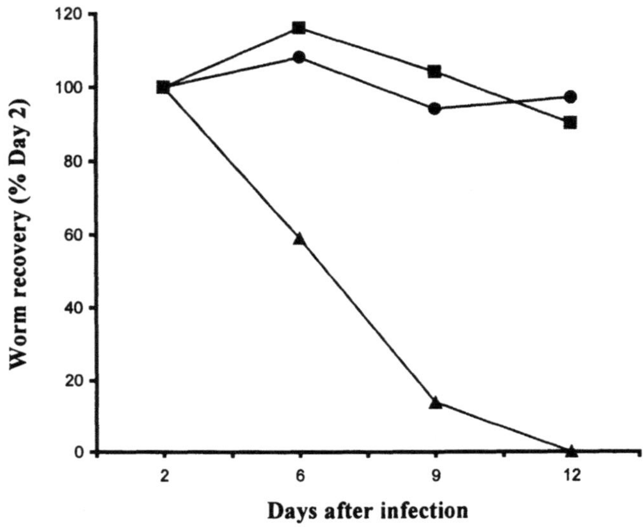

Figure 12.1. Survival of *Trichinella spiralis* and *Trichinella nativa* in untreated and immunosuppressed NIH mice (—■— =*T. spiralis*; —▲— = *T. nativa* in untreated mice; —●— = *T. nativa* in immunosuppressed mice). All mice were infected with 300 larvae on day 0, cortisone treatment was given before infection (Data from Bolas-Fernandez & Wakelin, 1989).

T. pseudospiralis is one of the most distinctive of all the genotypes within the genus because of the fact that its does not encapsulate within the host's muscles (Jasmer 1995). There can be little doubt, therefore, that it is a distinct species, although there is considerable molecular and genetic variation between worms from different geographical regions (La Rosa *et al.* 2001) and this extends to aspects of the host-parasite relationship (Alford *et al.* 1998). However, in its intestinal phase it is morphologically and behaviourally very similar to all the other members of the genus. It has been know for many years that *T. pseudospiralis* has an immunosuppressive influence upon its hosts (Stewart, 1989) and this is reflected in a prolonged intestinal phase and reduced intestinal inflammatory responses. Interestingly the latter, and specifically reduced mucosal mastocytosis (Wakelin *et al.* 1994), occur despite an earlier switch during infection to a type 2 cytokine profile, implying that the anti-inflammatory effects of *T. pseudospiralis* may operate through another mechanism, possibly through elevation of plasma corticosterone (Stewart *et al.* 1988).

Genetic and molecular differences between *T. spiralis* and *T. pseudospiralis* that may account for the distinctive features of their muscle phases have been described. There is differential expression of two genes - *tsJ5* and *tsmyd-1* - in the two species (Kuratli *et al.* 1999; 2001). *tsJ5* is developmentally regulated in *T .spiralis* and is down-regulated in *T. pseudospiralis*. *tsmyd-1* is not developmentally regulated in *T. spiralis* and its expression is slightly increased in *T. pseudospiralis*. The gene products of both are released in excretory-secretory (ES) material from the muscle larvae and that of *tsJ5* is present on the cuticular surface. This suggests that these molecules could therefore be available to the immune system during the intestinal as well as the muscle phase. The morphology of stichocyte granules (the source of most ES antigens) varies between the two species, as do the protein profiles of ES material (Wu, Nagano, & Takahashi 1998). These authors found that mRNA for the immunodominant 43kDa glycoprotein was present in both, but mRNA for a 53kDa component was found only in *T. pseudospiralis*.

There are few comparative studies of responses to *Trichinella* genotypes in hosts other than mice. Bolas-Fernandez *et al.* (1993) found differences in antibody responses to Spanish-origin *T. spiralis* and *T. britovi* in pigs exposed to experimental infection. A more detailed study of porcine responses against eight distinct genotypes was made by Kapel & Gamble (2000), whose data show significant differences in infectivity, *T. spiralis* being most infective, giving a mean larval burden of 427/g, *T. nativa, T. murrelli* and the genotype T6 being least infective, with a maximum of five larvae/g. Antibody responses against ES antigens differed significantly in level and kinetics between the genotypes and were, in general, highest against homologous ES antigen. Kapel (2001) made a similar study of with nine genotypes (including three of *T pseudospiralis*) in wild boars and also found marked differences in antibody response, although, again, these were most obvious during the post-intestinal period.

6. *TRICHURIS MURIS*

Members of the genus *Trichuris* occur widely in many species of mammals. The worms live in the large intestine and the females release eggs which pass out with faecal material and embryonate to the infective first stage larvae in the outside world. The life cycle is simple and direct, infection occurring when fully developed eggs are ingested by a suitable host. Eggs hatch in the small intestine and the larvae pass into the large intestine, where they penetrate into the epithelial layer of the mucosa. Like *Trichinella,* therefore, they are intracellular, but whereas *Trichinella* is a small worm (2-3mm) and can remain wholly within the epithelial layer *Trichuris* worms are much longer (2-3cm) and the larger posterior end eventually breaks free of the epithelial tunnel, leaving the anterior end in the intra-multicellular position. Development to the sexually mature adult stages is prolonged, taking several weeks.

A number of species occur in rodents, and one - *T. muris*, which is a natural parasite *of Mus musculus domesticus* - has been extensively used as an experimental model in laboratory mice (see also chapter 8). The few studies that have described this species in natural populations have reported relatively small numbers of worms per host (Behnke & Wakelin, 1973). It is possible to establish small infections (~ 10 worms) in laboratory mice with single pulse infections (Wakelin, 1973) but, in the majority of mouse strains, infection levels greater than this elicit an immune response that results in the loss of worms from the intestine before they reach sexual maturity, i.e. most strains become resistant, and this resistance operates rapidly against challenge infection. Certain strains, however, are susceptible to infection, worms establishing in large numbers and reaching maturity.

6.1 Immunity

The immunological basis of resistance and susceptibility to *T. muris* has been well-explored. Immunity is T cell dependent and can be transferred from resistant to susceptible mice with antibody and with lymphocytes. Recent work with a variety of mutant mice such as SCID mice (deficient in both T and B cells - Else & Grencis, 1996) and μMT mice (deficient in B cells – Blackwell & Else, 2001) has shown that each component can be effective separately. Although there is still no clear picture of the precise mechanisms that leads to worm loss, the use of genetically-defined strains of mice that show disparate phenotypes in terms of resistance to *T. muris* has provided much additional detail. It is now clear that strains of mouse that express resistance to infection mount a Th2-dominated response whereas susceptible strains show a Th1 response. Most strains of mouse fall into the first category, although there is considerable variation in the time at which they are able to expel infection. Some expel the parasite within two weeks, others within three to four weeks and others later still. Resistant strains of mice make anti-parasite antibody responses that are dominated by the IgG1 isotype, and their cytokine responses are dominated by

IL-4 and IL-13. At the other end of the spectrum the susceptible mouse strains which fail to expel parasites prior to patency produce IgG2a responses, with a bias to IFN-γ. The phenotype of a given mouse strain can be altered by appropriate treatment with anti-cytokine (or anti-cytokine receptor) antibodies, or with recombinant cytokines. An intriguing suggestion is that the later larval stages of worms can release molecules that also have the effect of altering the hosts' cytokine response from Th2 to Th1 (Grencis & Entwistle, 1997). If the host's response is too slow, or too inefficient, to bring about worm loss before these stages develop, then the host will become susceptible and allow adult worms to mature.

These data come from experiments using one particular laboratory-maintained isolate of *T. muris* (the Edinburgh (E or E/N) isolate). This was originally obtained from wild mice (*Mus musculus*) by Dr R.C. Rayski at Edinburgh Zoo in 1954 and has been passaged by D. Wakelin in immunosuppressed mice since 1963. There have recently become available two other isolates of *T. muris* with which it has been possible to examine phenotypic and genotypic variation in relation to host immunity. An isolate designated J (or E/K) derived from the same original stock as E, was sent to Dr E Pike in the USA some time in the mid 1960s and has been passaged in Japan since 1971 by Prof Y. Ito (Kitasato University School of Medicine, Kitasato, Japan), again in immunosuppressed mice. A third isolate (S) was obtained from *Mus spretus* in Portugal in 1992 by Prof. J.M. Behnke and has been maintained in Nottingham since that date in immunosuppressed mice. Assuming an average rate of three passages each year the E and J isolates as held in Nottingham are separated by more than 100 passage generations.

The responses of mice to infection with the E isolate and with the other two isolates show marked phenotypic differences (Figure 13.2) and these are reflected in the host's immune responses. Both aspects have been analysed in a series of papers (Bellaby *et al.* 1995; Bellaby, Robinson & Wakelin, 1996; Koyama & Ito; 1996, 2001) and the essential points can be summarized as follows:-

- The J isolate is the most protectively immunogenic and in consequence is more rapidly expelled by resistant strains of mice with a tendency to Th 2 phenotype. It is also expelled from strains that are susceptible to the E and S isolates, reversing their normal Th 1 response. It seems, therefore, that laboratory passage has selected for a phenotype which, under conditions of natural infection in the field, would be detrimental to parasite survival.

- The S isolate has a phenotype distinct from both E and J isolates and appears the least protectively immunogenic. Chronic infections are maintained even in mouse strains resistant to E and J, and infected mice generate a Th1 rather than a Th2 response.

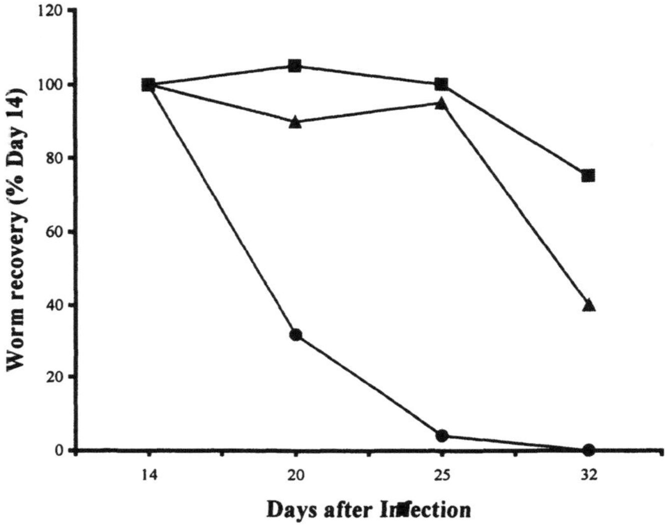

Figure 12.2. Survival of isolates of *Trichuris muris* in susceptible B10.BR mice (= ■ S isolate; ▲ = E isolate; ● = J isolate). Mice were infected with 400 eggs of each isolate on day 0. (Data from Bellaby et al., (1995).

These phenotypic differences must reflect underlying differences in molecules that elicit or, alternatively, divert or suppress immune responses. Accordingly, we have begun to examine the antigenic profile of the three isolates. Immunoblot analyses of the antibody responses of resistant and susceptible mouse strains to whole worm extracts and ES products of adult worms of all three isolates have failed to reveal any obvious quantitative or qualitative differences. It is possible, however, that functionally important antigenic differences may not be seen in the mature adult stages of the parasite, and it will be necessary to examine the larval stages, which may be critical in determining the establishment of the parasite. An alternative possibility is that the immunogenic or immunomodulatory molecules concerned may not be detectable with antibodies.

The Random Amplified Polymorphic DNA (RAPD) PCR technique has been used to differentiate between the three isolates, providing us with powerful markers for studies on the genetic basis of the phenotypic differences (Figure 12.3). The markers will make it possible to perform segregation analysis by interbreeding the isolates at either end of the virulence spectrum (J and S). Eggs from the hybrids can then be used to infect mice so that we can look for coordinated segregation of phenotype and genetic markers. This approach would not only be of value for theoretical studies of responses to selection pressures operating on parasite populations, but ultimately should allow us to identify the gene or genes responsible for parasite virulence and the mechanisms through which these genes act.

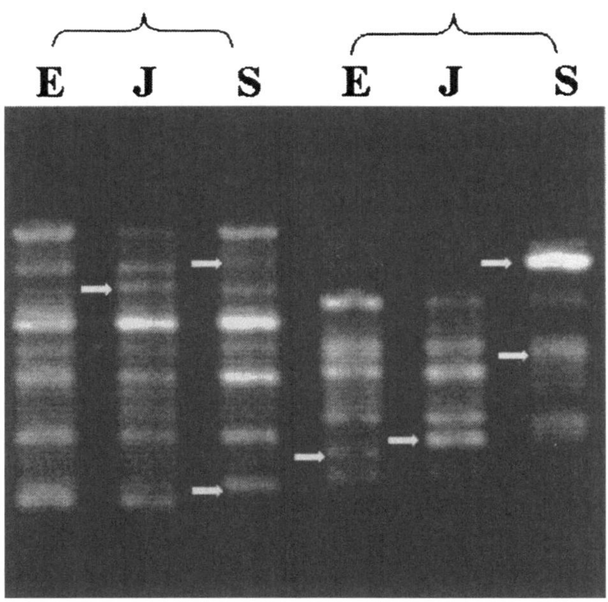

Figure 12.3. RAPD-DNA amplified from genomic DNA of *Trichuris muris* isolates E. J and S. The amplication was carried out with two different primer sets (bracketed). Arrows indicate bands unique to each isolate.

7. CONCLUSIONS

It is clear that the considerable genetic diversity found within nematode populations is reflected in phenotypic differences that influence the outcome of the host-parasite relationship. Perhaps the most dramatic example is seen in the spread of the genes responsible for anthelmintic resistance in trichostrongyles of ruminants, where the presence or absence of a point mutation determines whether or not worms can survive in benzimidazole-treated animals. A significant aspect of this example is the speed with which the mutant gene, once selected, has spread through nematode populations. Laboratory studies on host adaptation confirm the capacity of nematodes to respond rapidly to selection. As this review has shown, there is abundant evidence that genetic diversity in nematodes can influence the immunological outcome of the host parasite relationship in ways that alter resistance and susceptibility or increase or decrease pathogenicity. A possible consequence is that vaccination strategies may select for genes in nematode populations that reduce their immunogenicity and therefore decrease their susceptibility to vaccine-induced immunity.

Knowledge of the ways in which genetic diversity influences the outcome of nematode infections is relevant to the development of concepts relating to the evolution of the host-parasite relationship. Each partner in the relationship must optimize fitness, and this is not necessarily achieved simply by evolving maximum resistance (host) or minimum immunogenicity (parasite). For the host there must be a trade-off between the loss of resources caused by parasitism (particularly acute for intestinal infections that disrupt the digestive/absorptive function) and the loss of resources associated with the expression of protective immunity. For the parasite there must be a balance between being highly immunogenic, with the result that the host rapidly becomes resistant, or being insufficiently immunogenic so that the host is overwhelmed. For intestinal host-parasite relationships there is the added complication that levels of immunogenicity are positively correlated with levels of immunopathogenicity. Host fitness is reduced if enhanced immunity leads to greater pathology and parasite fitness will be reduced whether the host becomes resistant or if it suffers so much pathology that it dies. Knowledge of the contribution of specific genes in nematode populations to these questions of host-parasite balance will be crucial to a fuller understanding of the evolutionary pressures acting on the two partners.

REFERENCES

ALFORD K., OBENDORF D.L., FREDEKING T.M., HAEHLING E. & STEWART G.L. (1998). Comparison of the inflammatory responses of mice infected with American and Australian *Trichinella pseudospiralis* or *Trichinella spiralis*. *International Journal for Parasitology* **28,** 343-8.

ANDERSON T.J., BLOUIN M.S. & BEECH R.N. (1998). Population biology of parasitic nematodes: applications of genetic markers. *Advances in Parasitology* **41,** 219-83.

ANDERSON T.J., ROMERO-ABAL M.E. & JAENIKE J. (1995). Mitochondrial DNA and *Ascaris* microepidemiology: the composition of parasite populations from individual hosts, families and villages. *Parasitology* **110,** 221-9.

BARRETT F., JACKSON F. & HUNTLEY J.F. (1998). Pathogenicity and immunogenicity of different isolates of *Teladorsagia circumcincta*. *Veterinary Parasitology* **76,** 95-104.

BEHNKE J.M. & WAKELIN D. (1973). The survival of *Trichuris muris* in wild populations of its natural hosts. *Parasitology* **67,** 157-64.

BEHNKE J.M., PAUL V. & RAJASEKARIAH G.R. (1986). The growth and migration of *Necator americanus* following infection of neonatal hamsters. *Transactions of the Royal Society of Tropical Medicine and Hygiene* **80,** 146-9.

BELLABY T., ROBINSON K. & WAKELIN D. (1996). Induction of differential T-helper-cell responses in mice infected with variants of the parasitic nematode *Trichuris muris*. *Infection and Immunity* **64,** 791-5

BELLABY T., ROBINSON K., WAKELIN D. & BEHNKE J.M. (1995). Isolates of *Trichuris muris* vary in their ability to elicit protective immune responses to infection in mice. *Parasitology* **111,** 353-7

BLACKWELL, N. M. & ELSE, K.J. (2001). B cells and antibodies are required for resistance to the parasitic gastrointestinal nematode parasite *Trichuris muris*. *Infection and Immunity* **69,** 3860-68.

BOLAS-FERNANDEZ F. & WAKELIN D. (1989). Infectivity of *Trichinella* isolates in mice is determined by host immune responsiveness. *Parasitology* **99,** 83-8.

BOLAS-FERNANDEZ F., ALBARRAN-GOMEZ E., NAVARRETE I. & MARTINEZ-FERNANDEZ A.R. (1993). Dynamics of porcine humoral responses to experimental infections by Spanish *Trichinella* isolates: comparison of three larval antigens in ELISA. *Journal of Veterinary Medicine Series B* **40,** 223-9.

CURRIE R.M., NEEDHAM C.S., DRAKE L.J., COOPER E.S. & BUNDY D.A. (1998). Antigenic variability in *Trichuris trichiura* populations. *Parasitology* **117,** 347-53.

DEA-AYUELA M.A., MARTINEZ-FERNANDEZ A.R. & BOLAS-FERNANDEZ F. (2000). Comparison of IgG3 responses to carbohydrates following mouse infection or immunization with six species of *Trichinella*. *Journal of Helminthology* **74,** 215-23.

DENKERS E.Y., WASSOM D.L. & HAYES C.E. (1990). Characterization of *Trichinella spiralis* antigens sharing an immunodominant, carbohydrate-associated determinant distinct from phosphorylcholine. *Molecular and Biochemical Parasitology* **41,** 241-50.

DICK T.A. & CHADEE K. (1983). Interbreeding and gene flow in the genus *Trichinella*. *Journal of Parasitology* **69,**176-80.

DOBSON C. & OWEN M. E. (1977). Influence of serial passage on the infectivity and immunogenicity of *Nematospiroides dubius* in mice. *International Journal for Parasitology* **7,** 463-6.

DOBSON C. & TANG J.M. (1991). Genetic variation and host-parasite relations: *Nematospiroides dubius* in mice. *Journal of Parasitology* **77**, 884-9.

DONALDSON L.E., SCHMITT E., HUNTLEY J.F., NEWLANDS G.F.J. & GRENCIS R.K. (1996). A critical role for stem cell factor and c-kit in host protective immunity to an intestinal helminth. *International Immunology* **8**, 559-67.

EDWARDS A.J., BURT J.S. & OGILVIE B.M. (1971). The effect of immunity upon some enzymes of the parasitic nematode *Nippostrongylus brasiliensis*. *Parasitology* **62**, 339-47.

ELLIS L.A., REASON A.J., MORRIS H.R., DELL A., IGLESIAS R., UBEIRA F.M. & APPLETON J.A. (1994). Glycans as targets for mononclonal antibodies that protect rats against *Trichinella spiralis*. *Glycobiology* **4**, 585-92.

ELSE K. J. & GRENCIS R.K. (1996). Antibody-independent effector mechanisms in resistance to the intestinal nematode parasite *Trichuris muris*. *Infection and Immunity* **64** 2950-4

FRASER E.M. & KENNEDY M.W. (1991). Heterogeneity in the expression of surface-exposed epitopes among larvae of *Ascaris lumbricoides*. *Parasite Immunol*ogy **13**, 219-25.

GARSIDE P., KENNEDY M.W., WAKELIN D. & LAWRENCE C.E. (2000). Immunopathology of intestinal helminth infection. *Parasite Immunology* **22**, 605-12.

GASSER R.B.& NEWTON S.E. (2000). Genomic and genetic research on bursate nematodes: significance, implications and prospects. *International Journal for Parasitology* **30**, 509-34.

GEMMILL A.W., VINEY M.E. & READ AF. (2000). The evolutionary ecology of host-specificity: experimental studies with *Strongyloides ratti*. *Parasitology* **120**, 429-37.

GOYAL P.K. & WAKELIN D. (1993a). Influence of variation in host strain and parasite isolate on inflammatory and antibody responses to *Trichinella spiralis* in mice. *Parasitology* **106**, 371-8.

GOYAL P.K. & WAKELIN D. (1993b). Vaccination against *Trichinella spiralis* in mice using antigens from different isolates. *Parasitology* **107**, 311-7.

GOYAL P.K., HERMANEK J. & WAKELIN D. (1994). Lymphocyte proliferation and cytokine production in mice infected with different geographical isolates of *Trichinella spiralis*. *Parasite Immunology* **16**, 105-10.

GRANT W.N & MASCORD L.J. (1996). Beta-tubulin gene polymorphism and benzimidazole resistance in *Trichostrongylus colubriformis*. *International Journal for Parasitology* **26**, 71-7.

GRENCIS R.K. & ENTWISTLE G.M. (1997). Production of an interferon-gamma homologue by an intestinal nematode: functionally significant or interesting artefact? *Parasitology* **115**, S101-6.

HALEY J.A. (1966). Biology of the rat nematode *Nippostrongylus brasiliensis* (Travassos, 1914). III. Characteristics of *N. brasiliensis* after 30-120 serial passages in the Syrian hamster. *Journal of Parasitology* **52**, 98-108.

HALL A. & HOLLAND C. (2000). Geographical variation in *Ascaris lumbricoides* fecundity and its implications for helminth control. *Parasitology Today* **16**, 540-4.

HAWDON J.M., LI T., ZHAN B. & BLOUIN M.S. (2001). Genetic structure of populations of the human hookworm, *Necator americanus*, in China. *Molecular Ecology* **10**, 1433-7.

JARVIS L.M. & PRITCHARD D.I. (1992). An evaluation of the role of carbohydrate epitopes in immunity to *Trichinella spiralis*. *Parasite Immunology* **14**, 489-501.

JASMER D.P. (1995). *Trichinella spiralis*: subversion of differentiated mammalian skeletal muscle cells. *Parasitology Today* **11**, 185-8.

JENKINS D.C.. & PHILLIPSON R.F. (1972). Evidence that the nematode *Nippostrongylus brasiliensis* can adapt to and overcome the effects of host immunity. *International Journal for Parasitology* **2**, 353-9.

JENKINS S.N. & WAKELIN D. (1983). Functional antigens of *Trichuris muris* released during *in vitro* maintenance: their immunogenicity and partial purification. *Parasitology* **86**, 73-82.

KAPEL C.M. (2001). Sylvatic and domestic *Trichinella* spp. in wild boars; infectivity, muscle larvae distribution, and antibody response. *Journal of Parasitology* **87,** 309-14.

KAPEL C.M. & GAMBLE H.R. (2000). Infectivity, persistence, and antibody response to domestic and sylvatic *Trichinella* spp. in experimentally infected pigs. *International Journal for Parasitology* **30,** 215-21.

KOYAMA K. & ITO Y. (1996). Comparative studies on immune responses to infection in susceptible B10.BR mice infected with different strains of the murine nematode parasite *Trichuris muris*. *Parasite Immunology* **18,** 257-63

KOYAMA K. & ITO Y. (2001). Comparative studies on the levels of serum IgG1 and IgG2a in susceptible B10.BR mice infected with different strains of the intestinal nematode parasite *Trichuris muris*. *Parasitology Research.* In press.

KURATLI S., HEMPHILL A., LINDH J., SMITH D.F. & CONNOLLY B. (2001). Secretion of the novel *Trichinella* protein TSJ5 by *T. spiralis* and *T. pseudospiralis* muscle larvae. *Molecular and Biochemical Parasitol*ogy **115,** 199-208.

KURATLI S., LINDH J., GOTTSTEIN B, SMITH D.F. & CONNOLLY B. (1999). *Trichinella* spp.: differential expression of two genes in the muscle larva of encapsulating and non-encapsulating species. *Experimental Parasitology* **93,** 153-9.

LA ROSA G. & POZIO E. (2000). Molecular investigation of African isolates of *Trichinella* reveals genetic polymorphism in *Trichinella nelsoni. International Journal for Parasitology* **30,** 663-7.

LA ROSA G., MARUCCI G., ZARLENGA D.S. & POZIO E. (2001). *Trichinella pseudospiralis* populations of the Palearctic region and their relationship with populations of the Nearctic and Australian regions *International Journal for Parasitology* **31,** 297-305.

MAINGI N., SCOTT M.E. & PRICHARD R.K. (1990). Effect of selection pressure for thiabendazole resistance on fitness of *Haemonchus contortus* in sheep. *Parasitology,* **100,** 327-35.

MALLET S. & HOSTE H. (1995). Physiology of two strains of *Trichostrongylus colubriformis* resistant and susceptible to thiabendazole and mucosal response of experimentally infected rabbits. *International Journal for Parasitology* **25,** 23-7.

MARTINEZ-FERNANDEZ A.R., ARMAS-SERRA, C. de, GOMEZ-BARRIO A. & BOLAS-FERNANDEZ F. (1988). Single-pair cross hybridization test among Spanish *Trichinella* isolations. *Proceedings of the Seventh International Conference on Trichinellosis* 96-101.

MURRELL K.D., LICHTENFELS R.J., ZARLENGA D.S. & POZIO E. (2000). The systematics of the genus *Trichinella* with a key to species. *Veterinary Parasitology* **93,** 293-307.

NAGANO I., WU Z., MATSUO A., POZIO E. & TAKAHASHI Y. (1999.) Identification of *Trichinella* isolates by polymerase chain reaction--restriction fragment length polymorphism of the mitochondrial cytochrome c-oxidase subunit I gene. *International Journal of Parasitology* **29,** 1113-20.

NEWTON S.E., MORRISH L.E., MARTIN P.J., MONTAGUE P.E. & ROLPH T.P. (1995). Protection against multiply drug-resistant and geographically distant strains of *Haemonchus contortus* by vaccination with H11, a gut membrane-derived protective antigen. *International Journal for Parasitology* **25,** 511-21.

OGILVIE B.M. (1972). Protective immunity to *Nippostrongylus brasiliensis* in the rat. II. Adaptation by worms. *Immunology* **22,** 111-8.

QUINELL, R.J.,BEHNKE, J.M. & KEYMER, A. (1991) Host specificity of and cross-immunity between two strains of *Heligmosomoides polygyrus. Parasitology,* **102,** 419-27

ROBINSON K., BELLABY T. & WAKELIN D. (1995). High levels of protection induced by a 40-mer synthetic peptide vaccine against the intestinal nematode parasite *Trichinella spiralis. Immunology* **86,** 495-8.

SANGSTER N.C. (1999). Anthelmintic resistance: past, present and future. *International Journal of Parasitology* **29,** 115-24.

SEN H.G. & SETH D. (1967).Complete development of the human hookworm, *Necator americanus* in golden hamsters, *Mesocricetus auratus*. *Nature, Lond.* **214,** 609-10.

SILBERSTEIN D.S. & DESPOMMIER D.D. (1984). Antigens from *Trichinella spiralis* that induce a protective response in the mouse. *Journal of Immunology* **132**, 898-904.

SOLOMON M.S. & HALEY J.A. (1966). Biology of the rat nematode *Nippostrongylus brasiliensis* (Travassos, 1914). V. Characteristics of *N. brasiliensis* after serial passage in the laboratory mouse. *Journal of Parasitology* **52,** 237-41.

STEWART G.L. (1989). Biological and immunological characteristics of *Trichinella pseudospiralis*. *Parasitology Today* **5**, 344-9.

STEWART G.L., MANN M.A., UBELAKER J.E., MCCARTHY J.L. & WOOD B.G. (1988). A role for elevated plasma corticosterone in modulation of host response during infection with *Trichinella pseudospiralis*. *Parasite Immunology* **10,** 139-50.

SU Z, & DOBSON C. (1997). Genetic and immunological adaptation of *Heligmosomoides polygyrus* in mice. *International Journal for Parasitology* **27,** 653-63.

TANG J., DOBSON C. & McMANUS D.P. (1995). Antigens in phenotypes of *Heligmosomoides polygyrus* raised selectively from different strains of mice. *International Journal for Parasitology* **25,** 847-52.

WAKELIN D. (1973). The stimulation of immunity to *Trichuris muris* in mice exposed to low-level infections. *Parasitology* **66,** 181-9.

WAKELIN D. & GOYAL P.K. (1996). *Trichinella* isolates: parasite variability and host responses. *International Journal for Parasitology* **26**, 471-81.

WAKELIN D., GOYAL P.K., DEHLAWI M.S. & HERMANEK J. (1994). Immune responses to *Trichinella spiralis* and *T. pseudospiralis* in mice. *Immunology* **81,** 475-9.

WU Z., NAGANO I. & TAKAHASHI Y. (1998). Differences and similarities between *Trichinella spiralis* and *T. pseudospiralis* in morphology of stichocyte granules, peptide maps of excretory and secretory (E-S) products and messenger RNA of stichosomal glycoproteins. *Parasitology* **116,** 61-6.

WU Z., NAGANO I., MATSUO A. & TAKAHASHI Y. (2000). The genetic analysis of F1 hybrid larvae between female *Trichinella spiralis* and male *Trichinella britovi*. *Parasitology International* **48,** 289-95.

WU Z., NAGANO I., POZIO E. & TAKAHASHI Y. (1999). Polymerase chain reaction-restriction fragment length polymorphism (PCR-RFLP) for the identification of *Trichinella* isolates. *Parasitology,* **118,** 211-8.

Chapter 13

THE VALUE OF MUTATION SCANNING APPROACHES FOR DETECTING GENETIC VARIATION - IMPLICATIONS FOR STUDYING INTESTINAL NEMATODES OF HUMANS

Robin B. Gasser, Xingquan Zhu and Neil B. Chilton
Department of Veterinary Science, The University of Melbourne, Princes Highway, Werribee, Victoria 3030, Australia.
e-mail: robinbg@unimelb.edu.au

1. INTRODUCTION

Investigating genetic variation in parasites has significant implications for various areas of research, including epidemiology, population genetics, systematics and macromolecular evolution. Various DNA approaches have been applied to study these fields and have provided a great deal of information (McManus & Bowles, 1996; Gasser & Newton, 2000). Particularly, polymerase chain reaction (PCR)-based techniques have found wide-spread use because of their ability to specifically amplify genes from small amounts of DNA. However, little attention has been paid to the capacity of some analytical approaches to resolve sequence variation in (e.g., multi-copy genes) in individual organisms. For instance, in PCR-based restriction fragment length polymorphism (RFLP) analysis, sequence variation in an individual parasite may go undetected if a small panel of restriction enzymes scans a subset of putatively variable sites. Sequencing of PCR products (= amplicons) allows polymorphisms to be detected, but does not allow different sequence types (= paralogues) to be separated and characterised (Gasser, 1997; Gasser & Zhu, 1999). Also, there are limitations in determining sequences from chromatograms or gels when significant sequence heterogeneity (e.g., polymorphisms, indels or microsatellites) exists within an amplicon of a particular size. Attempting to circumvent such limitations by cloning of amplicons for subsequent sequencing can also lead to a loss of sequence types or to erroneous data relating to PCR-induced artefacts (Gasser, 1997). PCR-based mutation scanning techniques, such as single strand conformation polymorphism (SSCP), represent useful and cost effective alternatives for the direct analysis of genetic variation (Cotton, 1997; Gasser, 1997; Kristensen *et al.* 2001), particularly when large numbers of samples require analysis.

The principle of SSCP (Figure 13.1) is that the electrophoretic

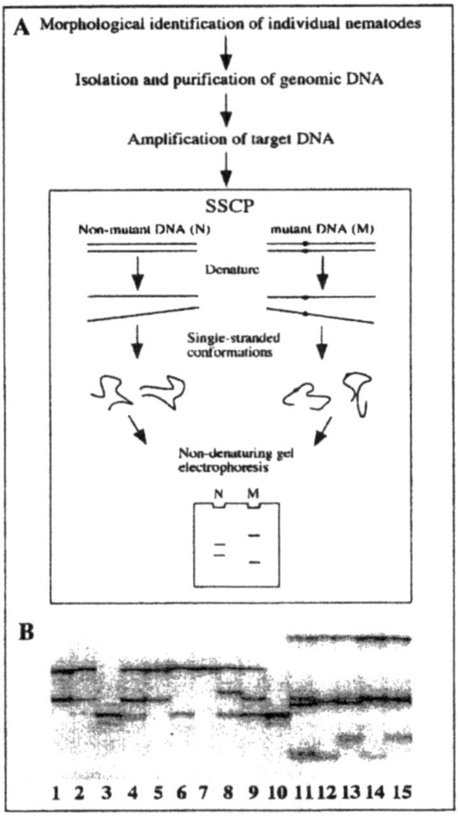

Figure 13.1. Polymerase chain reaction (PCR)-based single-strand conformation polymorphism (SSCP) analysis of genetic variation in nematodes. (A) Schematic representation of the principle of SSCP: individual nematodes are identified morphologically to species, and genomic DNA is isolated and column-purified. The target DNA region (e.g. the second internal transcribed spacer of ribosomal DNA, ITS-2) is amplified by PCR using radioactively-labelled oligonucleotide primers, heat-denatured, snap-cooled and then subjected to non-denaturing gel electrophoresis. A mutation (represented by a dot on the DNA strands) leads to the formation of different single-stranded conformations of the mutant DNA molecule (M) as compared with non-mutant DNA (N), consequently resulting in differential mobilities in the gel. (B) Example of an SSCP gel displaying sequence variability within and among individuals representing two very closely-related species of *Zoniolaimus* (Strongylida) (lanes 1-10 *versus* lanes 11-15). The existence of >2 bands indicates the formation of >1 conformation per single-stranded molecule or the existence of multiple sequence types within a PCR product.

mobility of a single-stranded DNA molecule in a non-denaturing polyacrylamide gel depends on its size and structure (Orita *et al.* 1989; Hayashi, 1991). Single-stranded molecules take on structures (secondary and tertiary conformations, or conformers) in solution, caused by base pairing among nucleotides within each strand. These conformations are highly dependent on the primary sequence and length of the molecule, and location and number of regions of base pairing. Hence, molecules differing in sequence (e.g., by a single base) can be separated in a non-denaturing polyacrylamide gel because of mobility difference(s) between different conformers. The SSCP method has been used to display point mutations for small (100-300 bp) amplicons (Cotton, 1997). However, it has been demonstrated that such mutations can be resolved for amplicons of up to 530 bp (Gasser & Zhu, 1999). Given its high mutation detection rate and technical simplicity, SSCP represents a sensitive analytical and diagnostic tool for some molecular investigations into parasitic nematodes. This chapter reviews some applications of SSCP to nematode taxonomy and population genetics and to investigate aspects of molecular evolution and structure, and proposes applications to important intestinal nematodes of humans, such as *Ascaris*, *Trichuris* and hookworms.

2. SSCP AS A DIAGNOSTIC AND TAXONOMIC TOOL

Specific identification of nematodes at any stage of development is central to diagnosis, epidemiology and disease control. Individual nematodes are usually identified and distinguished on the basis of morphological features, the host they infect, their pathological effect on the host or/and their geographical origin (Grove, 1990). However, these criteria are sometimes inadequate for identification (Andrews & Chilton 1999), which can seriously affect diagnosis. Substantial progress has been made in developing biochemical and molecular approaches for nematode identification (Andrews & Chilton, 1999; Gasser, 1999; Gasser & Newton, 2000). Central to developing PCR-based diagnostic approaches has been the choice of appropriate genetic markers in nuclear or mitochondrial DNA (mtDNA). As different genes evolve at different rates, the markers chosen should provide sufficient nucleotide sequence variation to allow identification of parasites at the taxonomic level required. For species identification, adequate DNA sequence differences should exist between species, with no or only low-level sequence variation within a species. In contrast, for the purpose of 'strain typing', a significant level of sequence variation should exist within the species under study. For example, repetitive (e.g., satellite) DNA (e.g., Christensen *et al.* 1994; Gasser, Nanson & Bøgh, 1995; Grenier, Gastagnone-Sereno & Abad, 1997), mtDNA (e.g., Anderson, Blouin & Beech, 1998; Blouin, 1998; Blouin *et al.* 1995, 1997, 1998; Viney, 1998; Zhu *et al.* 2000c) and nuclear ribosomal

DNA (rDNA) (reviewed in Gasser, 1999) have been employed as genetic markers to achieve the identification of parasites to species or strains.

In particular, the sequences of the first and second internal transcribed spacers (ITS-1 and ITS-2 = ITS) of rDNA have been shown to represent reliable genetic markers for PCR-based identification of strongylid and ascaridoid nematodes to species, irrespective of life cycle stage or sex (e.g., Chilton, Gasser & Beveridge, 1995; Epe, Von Samson-Himmelstjerna & Schneider, 1997; Jacobs *et al.* 1997; Romstad *et al.* 1997a,b; Monti *et al.* 1998; Newton *et al.* 1998a-c; Romstad *et al.*,1998; Zarlenga *et al.* 1998; Zhu *et al.* 1998a,b; Chilton & Gasser, 1999; Gasser *et al.* 1999a,b; Hung *et al.* 1999b, 2000; Zhu *et al.* 1999b; Verweij *et al.* 2000, 2001; Huby-Chilton *et al.* 2001a.b), because intraspecific variation in these sequences is mostly low (<1%) compared with higher levels of interspecific difference. However, a higher degree of intraindividual or intraspecific variation has been shown to occur within some parasitic nematodes, mainly due to microsatellites and/or indels in different ITS paralogues (Conole *et al.* 2001; Gasser *et al.* 2001).

Being able to accurately characterise species by their ITS sequences has enabled a number of taxonomic problems to be addressed (e.g., Gasser & Monti, 1997; Gasser *et al.* 1998a-d; Zhu & Gasser, 1998; Zhu *et al.* 1998a,b; 2000a,b). For instance, in a recent study, we tested the hypothesis that *Ascaris lumbricoides* represents a distinct species to *Ascaris suum* (see Zhu *et al.* 1999b). Sequence analysis of the ITS of *Ascaris* individuals (n=13) obtained from different geographical regions and countries revealed that all *Ascaris* individuals from humans differed from those from pigs by six nucleotide differences in the ITS-1 (Zhu *et al.* 1999). This result provided support for the hypothesis that they represent different species, although the lack of any difference in the ITS-2 sequence did not fully support this. Nevertheless, exploiting the nucleotide differences in the ITS-1, an SSCP approach was established for the unequivocal differentiation of *Ascaris* individuals from pigs from those from humans by differences in their profiles. The findings of Zhu *et al.* (1999b) were similar to those of Peng *et al.* (1998), but distinct from those of Anderson, Romero-Abal & Jaenike (1995), who demonstrated that some *Ascaris* individuals from endemic regions possessed 'mixed' RFLP patterns, suggesting that hybridisation may occur between human and pig *Ascaris* in sympatric zones. Using the genetic markers in the ITS-1 (Zhu *et al.* 1999b), it may be possible to test experimentally in pigs whether porcine and human *Ascaris* are capable of interbreeding and producing viable offspring. This would have significant epidemiological implications by providing insights into transmission patterns (cf. Anderson, 2001).

Analysis of ITS rDNA sequences has also enabled the detection of cryptic species of intestinal nematodes (Chilton *et al.* 1995; Hung *et al.* 1999a; Zhu *et al.* 2000a, 2001b; Gasser *et al.* 2001). For instance, Chilton *et al.* (1995) detected significant sequence differences (12-25%) in the ITS-2 among three members of the *Hypodontus macropi* (Strongylida) species complex, hookworm-like nematodes from Australian macropodid marsupials. Of particular significance in this study was that morphologically indistinguishable, but genetically distinct (i.e. cryptic) species identified by multilocus enzyme electrophoresis (Chilton *et al.* 1992) were compared and characterised by their ITS-2 sequences. The enzyme electrophoretic data also formed the basis for a detailed mutation scanning study of *H. macropi* individuals representing different populations (Gasser *et al.* 2001).

In another recent study (Huby-Chilton *et al.* 2001b), SSCP and DNA sequencing of the ITS-2 were used to genetically compare individual nematodes belonging to two sympatric species of *Zoniolaimus* (previously thought to represent a single species). Based on their SSCP profiles, the two species were readily distinguishable. Importantly, female nematodes, which were more difficult to identify morphologically compared with males, were readily identified by SSCP. This study highlights the value of SSCP to genetically identify individual nematodes for which very few and minor morphological characters are available for identification. Other recent studies have used SSCP for the detection and characterisation of cryptic species of ascaridoid (e.g., Zhu *et al.* 1998b; Hu *et al.* 2001; Zhu *et al.* 2001a,b), such as *Toxocara malaysiensis* n.sp. (see Zhu *et al.* 1998b; Gibbons *et al.* 2001), which could differ in their biology and transmission patterns, which may have implications with respect to control and human health. Hence, mutation scanning provides a highly sensitive analytical tool to address fundamental questions relating to the population biology and transmission of such cryptic species.

There are also considerable problems associated with the specific identification of some developmental stages (in particular eggs and larvae) of hookworms to species (Nelson, 1990), which can impact negatively on diagnosis and epidemiological studies (Schad & Warren, 1990). In order to overcome this limitation, PCR-based SSCP analysis of ITS rDNA regions has been employed effectively to identify and distinguish between seven species of hookworm (*Necator americanus, Ancylostoma duodenale, A. caninum, A. ceylanicum, A. tubaeforme, Uncinaria stenocephala* and *Bunostomum trigonocephalum*) (Gasser *et al.* 1998a). The method also allowed the direct display of sequence microheterogeneity between individuals of the same species, thus providing a valuable means of studying population variation. In another study of strongylid nematodes, Gasser *et al.* (1998b) employed SSCP (utilising the ITS-2) to overcome the limitation of not being able to distinguish morphologically (at the third larval stage) between the two species of nodule worm, *Oesophagostomum dentatum* and *O. quadrispinulatum*. This approach was applied to 'quality control' the purity of laboratory-maintained, monospecific lines of parasite (Gasser *et al.* 1998b), and also formed the basis for molecular studies of the prevalence and population biology of the human nodule worm,

Oesophagostomum bifurcum, in parts of Africa (Gasser *et al.* 1999a; Verweij *et al.* 2000, 2001).

3. STUDYING GENETIC DIVERSITY AND POPULATION GENETIC STRUCTURES

To date, there has been limited research on the population genetics of nematode species (Anderson *et al.* 1998). Most population genetic investigations have employed sequencing or RFLP approaches (e.g. Anderson *et al.*, 1993, 1995; Dame, Blouin & Courtney, 1993; Blouin *et al.* 1997). Given its technical simplicity, high resolving capacity and cost-effectiveness, SSCP provides a useful complementary tool for population studies, where large sample sizes are required for analysis.

Recent studies have used SSCP approaches to investigate mtDNA diversity in a range of parasitic helminths (e.g., Bøgh *et al.* 1999; Zhang *et al.* 1999; Zhu, Bøgh & Gasser, 1999a), including members of the order Enoplida, to which the genera *Capillaria, Trichinella* and *Trichuris* belong. For example, Zhu *et al.* (2000c) studied nucleotide variation in mtDNA within and among species of *Capillaria sensu lato* from Australian rodents and marsupials. A portion of the cytochrome *c* oxidase subunit I gene (p*cox*1) was enzymatically amplified from total genomic DNA from individual nematodes and analysed by SSCP, and representative samples with differing SSCP profiles were subjected to sequencing. While minor variation in SSCP profiles was displayed within a morphospecies from a particular host species, significant genetic variation was detected among morphospecies of *Capillaria* from different host species. The same morphospecies was shown to occur in different tissue habitats within one host individual or within different individuals of a particular species of host from the same or different geographical areas, and the morphospecies appeared to be relatively host specific at the generic level. These findings suggested that the members of *Capillaria* examined (although very variable in their host and tissue specificities) may exhibit greatest specificity at the level of host genus. Given that SSCP analysis of the expansion segment 5 of the large subunit of rDNA has also been used effectively to genotype other enoplids, such as members of the genus *Trichinella* (Gasser *et al.* 1998c), similar approaches could be employed to test the hypothesis that *Trichuris (suis)* of pigs is a different species to *Trichuris (trichiura)* from humans (cf. Grove, 1990; Oliveros *et al.* 2000).

Mutation scanning combined with 'selective' DNA sequencing has also been used to characterise sequence heterogeneity in the p*cox*1 of the hookworms *A. duodenale* from China, *A. caninum* from Australia, and *N. americanus* from China and from Togo (Hu *et al.* submitted). Haplotype diversity was found to differ markedly among the two *N. americanus* populations. For individual nematodes displaying genetic variation in SSCP within each of the three species, the p*cox*1 sequences were determined, and these were then compared with the p*cox*1 sequences of four heterologous species of hookworm. While intraspecific variation in the p*cox*1 sequence ranged from 0.3-3.5% for *A. duodenale,* 0.5-8.5% for *A. caninum* and 0.3-4.3% for *N. americanus*, interspecific differences varied from 5-13%.

The sequence data obtained also provided useful information on substitution patterns, nucleotide composition and DNA saturation, and indicated that the p*cox*1 had not reached saturation for the seven species of hookworm examined. Genetically distinct subpopulations were detected within *A. caninum* and *A. duodenale*, indicating significant population substructuring within each of these two species. Also, all *N. americanus* individuals from China differed from those from Togo at four nucleotide positions, supporting a previous proposal based on ITS rDNA sequence data (Romstad *et al.* 1998) that *N. americanus* may represent a species complex. Overall, the findings indicated the value of the SSCP approach and the p*cox*1 sequence data for studying the structure of hookworm populations, which may have important epidemiological implications. For instance, the genetic substructuring within both *A. duodenale* and *N. americanus* may relate to within-species variation in transmission and biology (e.g., migratory routes, prepatent periods and/or hypobiosis). Since this study (Hu *et al.* submitted), we have determined the entire mitochondrial genome sequence for both *A. duodenale* and *N. americanus* from China, which will provide a foundation for detailed population genetic studies of hookworms using a mutation scanning approach.

SSCP has also shown excellent promise for population genetic studies of nematodes within hybrid zones. In a multilocus enzyme electrophoresis study, Chilton *et al.* (1997) demonstrated the existence of hybrid individuals between *Paramacropostrongylus iugalis* and *P. typicus* (stomach-dwelling strongylid nematodes of western and eastern grey kangaroos) in a zone of host sympatry. Given that there were fixed differences in the ITS-1 and ITS-2 sequences between *P. iugalis* and *P. typicus* (Chilton *et al. unpublished observations*), SSCP should be a useful analytical tool for investigating the genetics of these nematodes (previously genotyped by multilocus enzyme electrophoresis) within the hybrid zone. Such an approach (using a range of genetic markers) is applicable to population genetic structure studies of any species of intestinal nematode infecting humans.

4. ANALYSIS OF MOLECULAR EVOLUTION AND STRUCTURE

Mutation scanning approaches provide a means of studying the evolution of genes, irrespective of intraindividual or intraspecific sequence heterogeneity (Gasser, 1997). Ribosomal DNA exhibits patterns of 'concerted evolution', leading to greater sequence similarity ('homogeneity') within a species than between species (Elder & Turner, 1995). Homogenisation of rDNA sequences is thought to take place by the mechanisms of 'molecular drive' which influence the turnover in DNA (Gerbi, 1986). Various processes (such as slippage, gene conversion, transposition and unequal crossing-over) are proposed to achieve and maintain sequence homogeneity in an rDNA array of repeats within individuals and species, but the relative contribution of each process remains unclear (Elder & Turner, 1995).

Little is known about the homogenisation process in rDNA sequences of nematodes and functional constraints on evolutionary divergence. This appears mainly to relate to technical limitations associated with analysing nucleotide variations in single organisms. Using an SSCP-based approach, Gasser *et al.* (1999a) showed that individuals of populations of *O. bifurcum* from human and Mona monkey hosts in Africa possessed ITS-2 rDNA arrays which were partially or fully homogenised for different sequence variants, a finding consistent with that achieved by denaturing gradient gel electrophoresis (DGGE) for individuals representing populations of *Haemonchus contortus* from different countries (Gasser *et al.* 1998d). This finding was also concordant with a study of the ITS-2 of *Drosophila melanogaster*, indicating that the homogenisation in the ITS rDNA is driven mainly by intra-chromosomal exchange (Schlötterer & Tautz, 1994). By contrast, some species of nematode exhibit relatively high levels of intraspecific sequence heterogeneity in the ITS-2 rDNA (Epe *et al.* 1997; Leignel, Humbert & Elard, 1997; Conole *et al.* 2001; Gasser *et al.* 2001), indicating that the homogenisation processes may differ from species to species.

Gasser *et al.* (2001) employed SSCP for a detailed analysis of sequence heterogeneity in the ITS-2 within and among individuals representing three operational taxonomic units (OTUs) of the *H. macropi* species complex from different species of Australian macropodid marsupial. Of the 96 nematodes analysed, three (OTU1 from *Petrogale persephone*), ten (OTU2 from *Macropus robustus robustus*) and seven (OTU9 from *Macropus rufus*) representative individuals were selected for DNA sequencing to characterise and estimate the magnitude of nucleotide variation in the ITS-2. While no unequivocal nucleotide difference in the ITS-2 was detectable within OTU1, most sequence variation ($\leq$ 4.7%) detected within OTU2 and OTU9 was related largely to dinucleotide (CA, TA, or a combination of both) differences. This microsatellite variability in some *H. macropi* OTUs suggests that the ITS-2 rDNA may be subjected to slippage events during DNA replication, resulting in a dispersal of short dinucleotide repeat tracts throughout ITS-2 lineages, or possibly transposition and/or crossing-over events (cf. Elder & Turner, 1995). These findings should have implications for studying speciation events and population differentiation in nematodes at the molecular level. The nucleotide variation in the ITS-2 of individual OTUs of *H. macropi* was also related to the predicted secondary structure of the precursor (pre-) rRNA. Most of the sequence heterogeneity or polymorphism within OTU2 and OTU9 occurred in unpaired regions (i.e. loops or bulges) of the structure, which appear not to be under functional constraint.

Interestingly, the ITS-2 pre-rRNA secondary structure model for *H. macropi* OTUs has essentially the same shape as for a range of nematodes, including hookworms (genus *Ancylostoma*) (Chilton & Gasser, 1999), nodule worms (genus *Oesophagostomum)* (Newton *et al.* 1998c) and trichostrongyloids (Chilton *et al.* 1998; Gasser *et al.* 1998d; Chilton *et al.* 2001) of the gastro-intestinal tract. Given that these parasites belong to different superfamilies within the order Strongylida, the pre-rRNA model may be applicable to a wide range of members within this order, with the exception of lungworms which lack one of the stems characteristic for gastro-intestinal strongylid nematodes (cf. Conole *et al.* 2001).

Although no studies have yet examined the function(s) of ITS-2 pre-rRNA for parasitic nematodes, the relevance of the conserved regions in the predicted structure may be inferred only from studies of other eukaryotic organisms, such as yeast. For *Saccharomyces cerevisiae,* it has been shown that regions in the ITS-2 with the highest degree of sequence conservation are crucial for pre-rRNA processing (Musters *et al.* 1990; van der Sande *et al.* 1992; van Nues *et al.* 1995), which suggests that such regions in the structure of strongylid nematodes may be associated with processing and/or binding to other RNA molecules or ribosomal proteins (cf. Peculis & Greeg, 1998; Michot *et al.* 1999). Thus, an SSCP-sequencing approach should provide a sensitive tool for 'pin-pointing' mutations to specific parts of the ITS-2, which has implications for investigating molecular evolutionary mechanisms, modes of inheritance as well as pre-rRNA structure and function.

5. CONCLUSION

Measuring genetic variation is important for studying the epidemiology, systematics and population genetics of parasitic nematodes as well as for their diagnosis and control. Technological advances pave the way for rapid progress in gene discovery and analysis. In particular, mutation scanning allows high-resolution and high-throughput analysis of sequence or allelic variation between and within individual parasitic nematodes and their populations. This chapter has highlighted a range of applications of SSCP (combined with selective DNA sequencing) to parasitic nematodes for the purposes of species identification or delineation, detection of cryptic species and diagnosis of infections. Importantly, it proposes future applications of the approach to population genetic and molecular evolutionary studies, and indicates its attributes for investigating the ecology and epidemiology of intestinal nematodes of humans.

ACKNOWLEDGEMENTS

The authors acknowledge contributions made by colleagues and students with whom they have published previously. NBC's research is currently supported by the Australian Research Council (ARC). RBG's research has been supported mainly through grants from the ARC, Melbourne Water Corporation, the Department of Industry, Science and Tourism, the Melbourne University Equine Research Fund, the Rural Industries Research and Development Corporation, the Collaborative Research Program of the University of Melbourne, the Canine Research Foundation and the Australian Companion Animal Health Foundation.

REFERENCES

ANDERSON, T. J. C. (2001). The dangers of using single locus markers in parasite epidemiology: *Ascaris* as a case study. *Trends in Parasitology* **17**, 183-188.

ANDERSON, T. J. C., ROMERO-ABAL, M. E. & JAENIKE, J. (1993). Genetic structure and epidemiology of *Ascaris* populations: patterns of host affiliation in Guatemala. *Parasitology* **107**, 319-334.

ANDERSON, T. J. C., ROMERO-ABAL, M. E. & JAENIKE, J. (1995). Mitochondrial DNA and *Ascaris* microepidemiology: the composition of parasite populations from individual hosts, families and villages. *Parasitology* **110**, 221-229.

ANDERSON, T. J. C., BLOUIN, M. S. & BEECH, R. N. (1998). Population biology of parasitic nematodes: applications of genetic markers. *Advances in Parasitology* **41**, 219-283.

ANDREWS, R. H. & CHILTON, N. B. (1999). Multilocus enzyme electrophoresis: a valuable technique for providing answers to problems in parasite systematics. *International Journal for Parasitology* **29**, 213-253.

BLOUIN, M. S. (1998). Mitochondrial DNA diversity in nematodes. *Journal of Helminthology* **72**, 285-289.

BLOUIN, M. S., YOWELL, C. A., COURTNEY, C. H. & DAME, J. B. (1995). Host movement and the genetic structure of populations of parasitic nematodes. *Genetics* **141**, 1007-1014.

BLOUIN, M. S., YOWELL, C. A., COURTNEY, C. H. & DAME, J. B. (1997). *Haemonchus placei* and *Haemonchus contortus* are distinct species based on mtDNA evidence. *International Journal for Parasitology* **27**, 1383-1387.

BLOUIN, M. S., YOWELL, C. A., COURTNEY, C. H. & DAME, J. B. (1998). Substitution bias, rapid saturation, and the use of mitochondrial DNA for nematode systematics. *Molecular Biology and Evolution* **15**, 1719-1727.

BØGH, H. O., ZHU, X. Q., QIAN, B-Z. & GASSER, R. B. (1999). Scanning for nucleotide variations in mitochondrial DNA fragments of *Schistosoma japonicum* by single-strand conformation polymorphism. *Parasitology* **118**, 73-82.

CHILTON, N. B., BEVERIDGE, I. & ANDREWS, R. H. (1992). Detection by allozyme electrophoresis of cryptic species of *Hypodontus macropi* (Nematoda: Strongyloidea) from macropodid marsupials. *International Journal for Parasitology* **22**, 271-279.

CHILTON, N. B., GASSER, R. B. & BEVERIDGE, I. (1995). Differences in a ribosomal DNA sequence of morphologically indistinguishable species within the *Hypodontus macropi* complex (Nematoda: Strongyloidea). *International Journal for Parasitology* **25**, 647-651.

CHILTON, N. B., BEVERIDGE, I., HOSTE, H. & GASSER, R. B. (1997). Evidence for hybridisation between *Paramacropostrongylus iugalis* and *P. typicus* (Nematoda: Strongyloidea) in grey kangaroos, *Macropus fuliginosus* and *M. giganteus* in eastern Australia. *International Journal for Parasitology* **27**, 475-482.

CHILTON, N. B., HOSTE, H., NEWTON, L. A., BEVERIDGE, I. & GASSER, R. B. (1998). Common secondary structures for the second internal transcribed spacer pre-rRNA of two subfamilies of trichostrongylid nematodes. *International Journal for Parasitology* **28**, 1765-1773.

CHILTON, N. B. & GASSER, R. B. (1999). Sequence differences in the internal transcribed spacers of rDNA among four species of hookworm (Ancylostomatoidea: *Ancylostoma*). *International Journal for Parasitology* **29**, 1971-1977.

CHILTON, N. B., NEWTON, L. A., BEVERIDGE, I. & GASSER, R. B. (2001). Evolutionary relationships of trichostrongyloid nematodes (Strongylida) inferred from ribosomal DNA sequence data. *Molecular Phylogenetics and Evolution* **19**, 367-386.

CHRISTENSEN, C. M., ZARLENGA, D. S. & GASBARRE L. C. (1994). *Ostertagia*, *Haemonchus*, *Cooperia*, and *Oesophagostomum*: construction and characterization of genus-specific DNA probes to differentiate important parasites of cattle. *Experimental Parasitology* **78**, 93-100.

CONOLE, J. C., CHILTON, N. B., JARVIS, T. & GASSER, R. B. (2001). Mutation scanning analysis of microsatellite variability in the ITS-2 (precursor rRNA) for three species of *Metastrongylus* (Nematoda: Metastrongyloidea). *Parasitology* **122**, 195-206.

COTTON, R. G. H. (1997). *Mutation Detection*. Oxford University Press, Oxford.

DAME, J. B., BLOUIN, M. S. & COURTNEY, C. H. (1993). Genetic structure of populations of *Ostertagia ostertagi*. *Veterinary Parasitology* **46**, 55-62.

ELDER, J. F. & TURNER, B. J. (1995). Concerted evolution of repetitive DNA sequences in eukaryotes. *The Quarterly Review of Biology* **70**, 297-320.

EPE, C., VON SAMSON-HIMMELSTJERNA, G. & SCHNIEDER, T. (1997). Differences in a ribosomal DNA sequence of lungworm species (Nematoda: Dictyocaulidae) from fallow deer, cattle, sheep and donkeys. *Research in Veterinary Science* **62**, 17-21.

GASSER, R. B, NANSEN, P. & BØGH, H. O. (1995). Specific fingerprinting of parasites by PCR with single primers to defined repetitive elements. *Acta Tropica* **60**, 127-131.

GASSER, R. B. (1997). Mutation scanning methods for the analysis of parasite genes. *International Journal for Parasitology* **27**, 1449-1463.

GASSER, R. B. (1999). PCR-based technology in veterinary parasitology. *Veterinary Parasitology* **84**, 229-258.

GASSER, R. B. & MONTI, J. R. (1997). Identification of parasitic nematodes by PCR-SSCP. *Molecular and Cellular Probes* **11**, 201-209.

GASSER, R. B. & ZHU, X. Q. (1999). Sequence-based analysis of DNA fragments by mutation detection techniques. *Parasitology Today* **15**, 427-468.

GASSER, R. B. & NEWTON, S. E. (2000). Genomic and genetic research on bursate nematodes: significance, implications and prospects. *International Journal for Parasitology* **30**, 509-534.

GASSER, R. B., MONTI, J. R., QIAN, B.-Z., POLDERMAN, A. M., NANSEN, P. & CHILTON, N. B. (1998a). A mutation scanning approach for the identification of hookworm species and analysis of population variation. *Molecular and Biochemical Parasitology* **92**, 303-312.

GASSER, R. B., WOODS, W. G. & BJØRN, H. (1998b). PCR-based SSCP to distinguish *Oesophagostomum dentatum* from *O. quadrispinulatum* developmental stages.

International Journal for Parasitology **28**, 1903-1909.
GASSER, R. B., ZHU, X. Q., MONTI, J. R., DOU, L., CAI, X. & POZIO, E. (1998c). PCR-SSCP of rDNA for the identification of *Trichinella* isolates from mainland China. *Molecular and Cellular Probes* **12**, 27-34.
GASSER, R. B., ZHU, X. Q., CHILTON, N. B., NEWTON, L. A., NEDERGAARD, T. & GULDBERG, P. (1998d). Analysis of sequence homogeneity in rDNA arrays of *Haemonchus contortus* by denaturing gradient gel electrophoresis. *Electrophoresis* **19**, 2391-2395.
GASSER, R. B., WOODS, W. G., BLOTKAMP, J., VERWEIJ, J., STOREY, P. A. & POLDERMAN, A. M. (1999a). Screening for nucleotide variations in ribosomal DNA arrays of *Oesophagostomum bifurcum* by PCR-coupled single-strand conformation polymorphism. *Electrophoresis* **20**, 1486-1491.
GASSER, R. B., WOODS, W. G., HUFFMAN, M. A., BLOTKAMP, J. & POLDERMAN, A. M. (1999b). Molecular separation of *Oesophagostomum stephanostomum* and *Oesophagostomum bifurcum* (Nematoda: Strongyloidea) from non-human primates. *International Journal for Parasitology* **29**, 1087-1091.
GASSER, R. B., ZHU, X. Q., BEVERIDGE, I. & CHILTON, N. B. (2001). Mutation scanning analysis of sequence heterogeneity in the second internal transcribed spacer (rDNA) within some members of the *Hypodontus macropi* (Nematoda: Strongyloidea) complex. *Electrophoresis* **22**, 1076-1085.
GERBI, S. A. (1986). The evolution of eukaryotic ribosomal DNA. *BioSystems* **19**, 247-258.
GIBBONS, L. M., JACOBS, D. E. & SANI, R. A. (2001). *Toxocara malaysiensis* n. sp. (Nematoda: Ascaridoidea) from the domestic cat (*Felis catus* Linnaeus, 1758). *Journal of Parasitology*, In press.
GRENIER, E., CASTAGNONE-SERENO, P. & ABAD, P. (1997). Satellite DNA sequences as taxonomic markers in nematodes of agronomic interest. *Parasitology Today* **13**, 398-401.
GROVE, D. I. (1990). *A History of Human Helminthology*. CAB International, Wallingford, UK.
HAYASHI, K. (1991). PCR-SSCP: a simple and sensitive method for detection of mutations in the genomic DNA. *PCR Methods and Applications* **1**, 34-38.
HU, M., D'AMELIO, S., ZHU, X. Q., PAGGI, L. & GASSER, R. B. (2001). Mutation scanning analysis of sequence variability in three mitochondrial DNA regions for members of the *Contracaecum osculatum* (Nematoda: Ascaridoidea) complex. *Electrophoresis* **22**, 1069-1075.
HU, M., CHILTON, N. B., ZHU, X. Q. & GASSER, R. B. - Evidence of population substructuring in some hookworm species based on variation in the cytochrome *c* subunit oxidase subunit 1 gene. *Molecular Ecology*, Submitted.
HUBY-CHILTON, F., BEVERIDGE, I., GASSER, R. B. & CHILTON, N. B. (2001a). Single-strand conformation polymorphism analysis of genetic variation in *Labiostrongylus longispicularis* from kangaroos. *Electrophoresis* **22**,1925-1929.
HUBY-CHILTON, F., BEVERIDGE, I., GASSER, R. B. & CHILTON, N. B. (2001b). Redescription of *Zoniolaimus mawsonae* (Nematoda: Strongyloidea) and the description of *Z. latebrosus* n. sp. from the red kangaroo, *Macropus rufus* (Marsupialia: Macropodidae) based on morphological and molecular data. *Systematic Parasitology*, In press.
HUNG, G.-C., CHILTON, N. B., BEVERIDGE, I. & GASSER, R. B. (1999a). Molecular evidence for cryptic species within *Cylicostephanus minutus* (Nematoda: Strongylidae). *International Journal for Parasitology* **29**, 285-291.
HUNG, G.-C., CHILTON, N. B., BEVERIDGE, I. & GASSER, R. B. (1999b). Secondary structure model for the ITS-2 precursor rRNA of strongylid nematodes of equids: implications for phylogenetic inference. *International Journal for Parasitology* **29**, 1949-1964.
HUNG, G-C., CHILTON, N. B. BEVERIDGE, I. & GASSER, R. B. (2000). Molecular systematic framework for equine strongyles based on ribosomal DNA sequence data. *International Journal for Parasitology* **30**, 95-103.

JACOBS, D. E., ZHU, X. Q., GASSER, R. B. & CHILTON, N. B. (1997). PCR-based methods for identification of potentially zoonotic ascaridoid parasites of the dog, fox and cat. *Acta Tropica* **68**, 191-200.

KRISTENSEN, V. N., KELEFIOTIS, D., KRISTENSEN, T. & BØRRENSEN-DALE, L. (2001). High-throughput methods for detection of genetic variation. *BioTechniques* **30**, 318-332.

LEIGNEL, V., HUMBERT, J. F. & ELARD, L. (1997). Study by ribosomal DNA ITS 2 sequencing and RAPD analysis on the systematics of four *Metastrongylus* species (Nematoda: Metastrongyloidea). *Journal of Parasitology* **83**, 606-611.

McMANUS, D. P. & BOWLES, J. (1996). Molecular genetic approaches to parasite identification: their value in diagnostic parasitology and systematics. *International Journal for Parasitology* **26**, 687-704.

MICHOT, B., JOSEPH, N., MAZAN, S. & BACHELLERIE, J. P. (1999). Evolutionarily conserved structural features in the ITS2 of mammalian pre-rRNAs and potential interactions with the snoRNA U8 detected by comparative sequence analysis of new mouse sequences. *Nucleic Acids Research* **27**, 2271-2282.

MONTI, J. R, CHILTON, N. B., QIAN, B.-Z. & GASSER, R. B. (1998). PCR-based differentiation of *Necator americanus* from *Ancylostoma duodenale* using specific markers in ITS-1 rDNA. *Molecular and Cellular Probes.* **12**, 71-78.

MUSTERS, W., BOON, K., VAN DER SANDE, C. A., VAN HEERIKHUIZEN, H. & PLANTA, R. J. (1990). Functional analysis of transcribed spacers of yeast ribosomal DNA. *EMBO Journal* **9**, 3989-3996.

NELSON, G. S. (1990). Hookworms in perspective. In *Hookworm Disease: Current Status and New Directions*, (ed. Schad, G. & Warren, K.S.), pp. 417-430. Taylor & Francis, London.

NEWTON, L. A., CHILTON, N. B., BEVERIDGE, I. & GASSER, R. B. (1998a). Differences in the second internal transcribed spacer of four species of *Nematodirus* (Nematoda: Molineidae). *International Journal for Parasitology* **28**, 337-341.

NEWTON, L. A., CHILTON, N. B., BEVERIDGE, I. & GASSER, R. B. (1998b). Genetic evidence indicating that *Cooperia surnabada* and *Cooperia oncophora* are one species. *International Journal for Parasitology* **28**, 331-336.

NEWTON, L. A., CHILTON, N. B., BEVERIDGE, I. & GASSER, R. B. (1998c). Systematic relationships of some members of the genera *Oesophagostomum* and *Chabertia* (Nematoda: Chabertiidae) based on ribosomal DNA sequence data. *International Journal for Parasitology* **28**, 1781-1789.

OLIVEROS, R., CUTILLAS, C., DE ROJAS, M. & ARIAS, P. (2000). Characterization of four species of *Trichuris* (Nematoda: Enoplida) by their second internal transcribed spacer ribosomal DNA sequence. *Parasitology Research* **86**, 1008-1013.

ORITA, M., SUZUKI, Y., SEKIYA, T. & HAYASHI, K. (1989). Rapid and sensitive detection of point mutations and DNA polymorphisms using the polymerase chain reaction. *Genomics* **5**, 874-879.

PECULIS, B. A. & GREEG, C. L. (1998). The structure of the ITS2-proximal stem is required for pre-rRNA processing in yeast. *RNA* **4**, 1610-1622.

PENG, W., ANDERSON T. J. C., ZHOU, X. & KENNEDY, M. (1998). Genetic variation in sympatric *Ascaris* populations from humans and pigs in China. *Parasitology* **117**, 355-361.

ROMSTAD, A., GASSER, R. B., POLDERMAN, A. M., NANSEN, P., PIT, D. S. S. & CHILTON, N. B. (1997a). Differentiation of *Oesophagostomum bifurcum* from *Necator americanus* by PCR using genetic markers in spacer ribosomal DNA. *Molecular and Cellular Probes* **11**, 169-176.

ROMSTAD, A, GASSER, R. B., NANSEN, P., POLDERMAN, A. M., MONTI, J. R. & CHILTON, N. B. (1997b). Characterization of *Oesophagostomum bifurcum* and *Necator americanus* by PCR-RFLP of rDNA. *Journal of Parasitology* **83**, 963-966.

ROMSTAD, A., GASSER, R. B., NANSEN, P., POLDERMAN, A. M. & CHILTON, N. B. (1998). *Necator americanus* (Strongylida: Ancylostomatidae) from Africa and Malaysia have different ITS-2 rDNA sequences. *International Journal for*

Parasitology **28**, 611-615.
SCHAD, G. & WARREN, K. S. (1990). *Hookworm Disease: Current Status and New Directions*. Taylor & Francis, London.
SCHLÖTTERER, C. & TAUTZ, D. (1994). Chromosomal homogeneity of *Drosophila* ribosomal DNA arrays suggests intrachromosomal exchanges drive concerted evolution. *Current Biology* **4**, 777-783.
VAN NUES, R. W., RIENTJES, J. M. J., MORRE, S. A., MOLLEE, E., PLANTA, R. J., VENEMA, J. & RAUE, H. A. (1995). Evolutionarily conserved structural elements are critical for processing of internal transcribed spacer 2 from *Saccharomyces cerevisiae* precursor ribosomal RNA. *Journal of Molecular Biology* **250**, 24-36.
VAN DER SANDE, C. A., KWA, M., VAN NUES, R. W., VAN HEERIKHUIZEN, H., RAUE, H. A. & PLANTA, R. J. (1992). Functional analysis of internal transcribed spacer 2 of *Saccharomyces cerevisiae* ribosomal DNA. *Journal of Molecular Biology* **223**, 899-910.
VERWEIJ, J., POLDERMAN, A. M., WIMMENHOVE, M. C. & GASSER, R. B. (2000). PCR assay for the specific amplification of *Oesophagostomum bifurcum* DNA from human faeces. *International Journal for Parasitology* **30**, 137-142.
VERWEIJ, J., PIT, D. S. S., VAN LIESHOUT, L., BAETA, S. M., DERY, G. D., GASSER, R. B. & POLDERMAN, A. M. (2001). Prevalence of *Oesophagostomum bifurcum* and *Necator americanus* infections using specific PCR amplification of DNA from faecal samples. *Tropical Medicine and International Health*, In press.
VINEY, M. E. (1998). Nematode population genetics. *Journal of Helminthology* **72**, 281-283.
ZARLENGA, D. S. GASBARRE, L. C., BOYD, P., LEIGHTON, E. & LICHTENFELS, J. R. (1998). Identification and semi-quantitation of *Ostertagia ostertagi* eggs by enzymatic amplification of ITS-1 sequences. *Veterinary Parasitology* **77**, 245-257.
ZHANG, L.-H., GASSER, R. B., ZHU, X. Q. & McMANUS, D. P. (1999). Screening for different genotypes of *Echinococcus granulosus* within China and Argentina by SSCP-sequencing. *Transactions of the Royal Society of Tropical Medicine and Hygiene* **93**, 329-334.
ZHU, X. Q. & GASSER, R. B. (1998). Single-strand conformation polymorphism (SSCP)-based mutation scanning approaches to fingerprint sequence variation in ribosomal DNA of ascaridoid nematodes. *Electrophoresis* **19**, 1366-1373.
ZHU, X. Q., GASSER, R. B., PODOLSKA, M. & CHILTON, N. B. (1998a). Characterisation of anisakid nematodes with zoonotic potential by ribosomal DNA sequences. *International Journal for Parasitology* **28**, 1911-1921.
ZHU, X. Q., JACOBS, D. E., CHILTON, N. B., SANI, R. A., CHENG, N. A. B. Y. & GASSER, R. B. (1998b). Molecular characterization of a *Toxocara* variant from cats in Kuala Lumpur, Malaysia. *Parasitology* **117**, 155-164.
ZHU, X. Q., BØGH, H. O. & GASSER, R. B. (1999a). Dideoxy fingerprinting of low-level nucleotide variation in mitochondrial DNA of the human blood fluke, *Schistosoma japonicum*. *Electrophoresis* **20**, 2830-2833.
ZHU, X. Q., CHILTON, N. B., JACOBS, D. E., BOES, J. & GASSER, R. B. (1999b). Characterisation of *Ascaris* from human and pig hosts by nuclear ribosomal DNA sequences. *International Journal for Parasitology* **29**, 469-478.
ZHU, X. Q., D'AMELIO, S., PAGGI, L. & GASSER, R. B. (2000a). Assessing sequence variation in the internal transcribed spacers of ribosomal DNA within and among members of the *Contracaecum osculatum* complex (Nematoda: Ascaridoidea: Anisakidae). *Parasitology Research* **86**, 677-683.
ZHU, X. Q., GASSER, R. B., JACOBS, D. E., HUNG, G.-C. & CHILTON, N. B. (2000b). Relationships among some ascaridoid nematodes based on ribosomal DNA sequence data. *Parasitology Research* **86**, 738-744.
ZHU, X. Q., SPRATT, D. M., BEVERIDGE, I., HAYCOCK, P. & GASSER, R. B. (2000c). Analysis of mitochondrial DNA polymorphism within and among species of *Capillaria sensu lato* from Australian marsupials and rodents by single-strand conformation polymorphism of mitochondrial DNA. *International Journal for*

Parasitology **30**, 933-938.
ZHU, X. Q., D'AMELIO, S., HU, M., PAGGI, L. & GASSER, R. B. (2001a). Electrophoretic detection of population variation within *Contracaecum ogmorhini* (Nematoda: Ascaridoidea: Anisakidae). *Electrophoresis* **22**, 1930-1934.
ZHU, X. Q., GASSER, R. B., CHILTON, N. B. & JACOBS, D. E. (2001b). Molecular approaches for studying ascaridoid nematodes with zoonotic potential, with an emphasis on *Toxocara* species. *Journal of Helminthology* **75**, 101-108.

Chapter 14

OPPORTUNITIES AND PROSPECTS FOR INVESTIGATING DEVELOPMENTALLY REGULATED AND SEX-SPECIFIC GENES AND THEIR EXPRESSION IN INTESTINAL NEMATODES OF HUMANS

Susan E. Newton[1], Peter R. Boag[1,2] and Robin B. Gasser[2]
[1] *Victorian Institute of Animal Science, Attwood, Victoria 3049, Australia,*
[2] *Department of Veterinary Science, The University of Melbourne, Werribee, Victoria 3030, Australia*
e-mail: Sue.Newton@nre.vic.gov.au

1. INTRODUCTION

Surprisingly little is known about the molecular aspects of development and reproduction in intestinal nematodes, particularly those of humans. Study of stage-specific, and the subset of sex-specific, molecules and their expression will provide an improved understanding of the molecular biology and physiology of nematode moulting, invasion of and development in the host, hypobiosis, as well as sexual differentiation, maturation and behaviour. Such knowledge has the potential to lead to novel means of parasite control by disrupting one or more of these processes at the molecular level.

Traditional studies of developmentally regulated molecules of parasitic nematodes typically involved characterisation of proteins expressed by different stages using techniques, such as metabolic labelling and/or immunochemical analysis by one- or two-dimensional gel electrophoresis. For example, Northern & Grove (1990) compared the profiles of proteins expressed by the infective larval and adult stages of the human intestinal nematode, *Strongyloides stercoralis*. The introduction and adoption of molecular biology techniques has allowed subsequent cloning of cDNAs, and genes encoding stage- and sex-specific molecules, and analysis of the regulation of their transcription and of their relationship to homologous genes in other species. For example, PcDNA clones of ES proteins from larval canine and human hookworms have been isolated (ASPs), revealing their relationship to a family of molecules from a range of organisms (Hawdon *et al.* 1996; Bin *et al.* 1999; Hawdon, Narasimhan & Hotez, 1999).

The advent of rapid DNA sequencing now enables genome-wide approaches for the analysis of gene expression by the isolation of partial cDNA sequences, termed expressed sequence tags (ESTs). For example, Blaxter *et al.* (1996) surveyed genes expressed in third-stage larvae (L3) of the human filarial parasite, *Brugia malayi*, and Hoekstra *et al.* (2000) used this approach to examine changes in gene expression in the gastric (i.e. abomasal) nematode of sheep or goats, *Haemonchus contortus*, upon transition from the pre-parasitic to the parasitic stage. EST sequencing is well suited for automation and for 'electronic subtraction' (e.g. of adult from L3 EST data, and *vice versa*). The use of this approach is becoming common for the identification of stage-specific genes from nematodes of human and animal health significance, particularly for drug discovery and vaccine development.

Table 14.1 lists current EST projects for intestinal nematodes of humans and animals, and filarioid nematodes of humans. Despite the usefulness of EST sequencing and the relative ease with which large amounts of data can be produced, the approach has the disadvantage that many of the sequences obtained are 'house-keeping' genes, present in all developmental stages, and that abundant genes are highly represented, thus generating redundant sequence information.

An improved approach to identifying stage-specific genes is by differential display (Liang & Pardee, 1992), a PCR-based method which has been used, for example, for the identification of stage-specifically expressed cDNAs of the rat lungworm, *Angiostrongylus cantonensis* (see Joshua & Hsieh, 1995) and of *H. contortus* (see Hartman *et al.* 2001). In the latter study, the advantage of this approach was demonstrated by the identification of adult-specific genes which had not been identified by EST sequencing of conventional cDNA libraries prepared from adult *H. contortus* (see Hoekstra *et al.* 2000). However, differential display can also have some disadvantages, particularly in the generation of 'false positive display products'.

Table 14.1. Current expressed sequence tag (EST) sequencing projects for parasitic nematodes of animals and humans

Site of EST project or web network	Parasite
Genome Sequencing Centre Washington University School of Medicine (http://www.nematode.net/)	**Human intestinal nematodes:** *Ancylostoma duodenale* *Ascaris* sp. *Strongyloides stercoralis*
	Animal intestinal nematodes: *Ancylostoma caninum* *Ascaris* sp. *Strongyloides ratti* *Toxocara canis*
The Blaxter Nematode Genetics Lab. Institute of Cell, Animal and Population Biology, The University of Edinburgh (http://nema.cap.ed.ac.uk/nematodeESTs/Nest.html) (http://nema.cap.ed.ac.uk/fgn/filgen1.html)	**Human intestinal nematodes:** *Necator americanus*
	Animal intestinal nematodes: *Haemonchus contortus* *Trichuris muris* *Ascaris* sp. *Toxocara canis* *Nippostrongylus brasiliensis* *Teladorsagia circumcincta*
	Human filarial nematodes: *Loa loa* *Brugia malayi*
William's Laboratory, Clark Science Centre, Smith College (http://www.math.smith.edu/~sawlab/OnchoNet/OnchoNet.html)	**Human filarial nematodes:** *Onchocerca volvulus*

The recent development of another PCR-based method, suppression subtractive hybridisation (SSH), allows the effective removal of common, house-keeping genes from the mRNA population of interest prior to library construction (Diatchenko *el al.* 1996). This method has the further advantage that rare transcripts are efficiently amplified. Although this relatively new technique has not yet found widespread application to parasites, its utility has been demonstrated by the identification of genes from *Plasmodium berghei* (a malarial parasite) expressed specifically in the mosquito mid-gut (Dessens *et al.* 2000).

Given the rapid growth of knowledge in genomics of the free-living nematode, *Caenorhabditis elegans*, and of several parasitic nematodes of human health significance (Williams & Johnston, 1999; Blaxter, 2000; Daub *et al.* 2000; Maizels, Tetteh & Loukas, 2000), this chapter provides a timely review of current information on developmentally regulated and sex-specific genes and/or their expression in nematodes of socio-economic importance, describes advanced approaches for the characterisation and analysis of such genes, and indicates the opportunities and prospects for research on molecular aspects of growth, development and reproduction in intestinal nematodes of humans.

2. DEVELOPMENTALLY REGULATED GENES IN NEMATODES

2.1 Genes triggered during infection of the host and transition to parasitism

During invasion of their hosts, infective larvae of nematodes encounter signals which initiate developmental events, including exsheathment, expression of genes associated with development and cell differentiation, and the release of excretory-secretory (ES) products. During development in the host, the repertoire of molecules expressed by a parasite changes, presumably in response to particular requirements for its survival and/or the host-parasite relationship. A number of developmentally regulated ES molecules have been characterised due to their potential as vaccine or diagnostic antigens, particularly for intestinal nematodes of veterinary importance and, to a lesser extent, for nematodes of humans, including filarial nematodes (see Table 14.2).

ES products from nematodes can play important roles related to infection of the host, such as exsheathment, tissue invasion and/or immune modulation (Riffkin *et al.* 1996). It has been shown that secreted aspartyl proteases are important in skin penetration by *Necator americanus* (Brown *et al.* 1999), while a 47 kDa ES pore-forming protein involved in the initiation and maintenance of infection has been described from the human intestinal whipworm, *Trichuris trichiura* (see Drake *et al.* 1994). This protein is also found in extracts of the stichosome (which encloses the oesophagus) and is encoded by a small gene family (Barker & Bundy, 1999). Recombinant or native porin is capable of inducing pore formation in planar lipid bilayers (Drake *et al.* 1998). It has been proposed that this 'porin' relates to the parasite's ability to bury its anterior end into the large intestinal wall and remain attached, despite continuous sloughing of the epithelial layers and the host's immune response (Drake *et al.* 1994; Barker & Bundy, 1999).

Two ES proteins (termed ASPs) from the L3 stage of the dog hookworm, *Ancylostoma caninum*, have been cDNA cloned (Hawdon *et al.* 1996; 1999) and shown to belong to a family of molecules present also in other parasitic nematodes (see Table 14.2), and are related to venom allergens of wasps and proteins from mammals and *C. elegans*. cDNAs encoding similar proteins have been cloned recently from the human hookworms, *Ancylostoma duodenale* and *N. americanus* (see Bin *et al.* 1999). Since the ASPs of hookworms are released during activation of the L3, it has been suggested that they play a significant role in the initial phase of host infection, and that interference with ASP function may prevent infection (Hawdon *et al.* 1999). Experiments using a mouse model of *A. caninum* infection (Ghosh, Hawdon & Hotez, 1996) have indicated the potential of ASPs as a vaccine against dog and human hookworm infection.

'Transition to parasitism' induces a number of molecular and biochemical changes due to the new environment in the host, in particular an increase to body temperature (~ 37°C) and, in the case of gastrointestinal and tissue-dwelling nematodes, a switch from aerobic to anaerobic metabolism. For the filarial nematode, *Brugia pahangi*, it has been shown that the temperature shift is a key factor in the regulation of two genes expressed in the parasitic stages (Hunter *et al.* 2001). Interestingly, small

Table 14.2. Examples of developmentally regulated molecules produced by parasitic nematodes

Nematode	Molecule	Type/role in parasite	Reference
Human intestinal nematodes:			
Ancylostoma duodenale/ A. ceylanicum	*Ancylostoma*-secreted protein (ASP)	ES, infection?	Bin *et al.* (1999)
	Kunitz-type serpin	ES, inhibition of host proteases?	Milstone *et al.* (2000)
Necator americanus	ASP homologue	ES, infection?	Bin *et al.* (1999)
	Acetylcholinesterase (AChE)	ES, immune evasion?	Pritchard *et al.* (1991)
	Metalloprotease which cleaves eotaxin	ES, immune evasion?	Culley *et al.* (2000)
	Glutathione-*S*-transferase (GST)	ES, antioxidant	Brophy *et al.* (1995)
	Superoxide dismutase (SOD)	ES, antioxidant	Taiwo *et al.* (1999)
	Aspartyl protease	ES, infection	Brown *et al.* (1999)
	Calreticulin-like	ES, immune evasion?	Kaspar *et al.* (2001)
	Neutrophil inhibitory factor (NIF)	ES, immune evasion?	Ali *et al.* (2001)
Trichuris trichiura	Porin-like proteins	ES, and stichosome, Attachment	Barker & Bundy (1999)
Strongyloides stercoralis	G-protein α subunit	Arrested development?	Massey *et al.* (2001)
Animal intestinal nematodes:			
Ancylostoma caninum	Anti-coagulant serpin	ES, anti-coagulant	Stanssens *et al.* (1996)
	NIF	ES, immune evasion?	Moyle *et al.* (1994)

Nematode	Molecule	Type/role in parasite	Reference
Animal intestinal nematodes (continued)			
Haemonchus contortus	ASP family (HES-15, HES-24)	ES, unknown	Schallig *et al.* (1997)
	Microsomal aminopeptidase (H11)	Gut membrane, digestive	Smith *et al.* (1997)
	Aspartate proteinase	Gut membrane, digestive	Longbottom *et al.* (1997)
	Cysteine proteases (cathepsin B)	Gut secreted, digestive	Pratt *et al.* (1992a)
	Metallopeptidase	Gut membrane, digestive	Redmond *et al.* (1997)
	Cystatin	Cytoplasmic, unknown	Newlands *et al.* (2001)
	Hyaluronidase	ES, unknown	Rhoads *et al.* (2000b)
	Succinate dehydogenase isoforms	Metabolic	Roos & Tielens (1994)
	Glutamate dehydrogenase	Metabolic	Skuce *et al.* (1999)
	SOD	ES and cytoplasmic, antioxidant	Liddell & Knox (1998)
	Galectin	Unknown	Greenhalgh *et al.* (2000)
Trichostrongylus spp.	AChE	ES, immune evasion?	Griffiths & Pritchard (1994)
	Globin-like protein	Oxygen supply?	Frenkel *et al.* (1992)
Ostertagia ostertagi	Cysteine proteases (cathepsin B)	Gut secreted, digestive	Pratt *et al.* (1992b)
	Globin-like protein	Oxygen supply?	de Graaf *et al.* (1996)
	GTP cyclohydrolase	Biosynthetic	Moore *et al.* (2000)

	Molecule	Type/role in parasite	Reference
Animal intestinal nematodes (continued)			
Dictyocaulus viviparus	AChE	ES, immune evasion?	McKeand *et al*. (1994)
	SOD isoforms	ES and cytoplasmic, antioxidant	Britton *et al*. (1994)
Toxocara canis	Mucin-like protein TES-120	Surface, immune evasion?	Gems & Maizels (1996)
	Prohibitin	Arrested development?	Loukas & Maizels (1998
	C-type lectin	Immune evasion?	Loukas *et al*. (2000)
Ascaris sp.	Pyruvate dehydrogenase isoforms	Metabolic	Huang *et al*. (1998)
	Catalase	Antioxidant	Eckelt *et al*. (1998)
	Peroxiredoxin	Antioxidant	Tsuji *et al*. (2000)
Trichuris muris	Interferon-γ homologue	ES, immune evasion?	Grencis & Entwhistle (1997)
Nippostrongylus brasiliensis	Small heat shock protein (sHSP)	Chaperone	Tweedie *et al*. (1993)
Human filarial nematodes:			
Brugia malayi/B. pahangi	Thioredoxin peroxidase	Antioxidant	Ghosh *et al*. (1998)
	Glutathione peroxidase	Cuticular and ES, antioxidant immune evasion?	Cookson *et al*. (1992)
	Chitinase	ES, unknown	Fuhrman *et al*. (1992)
	Serpin	ES, immune evasion?	Yenbutr & Scott (1995) Zang *et al*. (1999)
	Transforming growth factor-β (TGF-β) homologues	ES? development? immune evasion?	Gomez-Escobar *et al*. (1998;2000)

	Molecule	Type/role in parasite	Reference
Human filarial nematodes (continued):			
Brugia malayi/B. pahangi (continued)	TGF-β receptor	Development?	Gomez-Escobar *et al.* (1997)
	Macrophage migration inhibitory factor (MIF)	ES, immune evasion?	Pastrana *et al.* (1998)
	sHSP	Chaperone	Raghaven *et al.* (1999)
	HSP 70	Chaperone	Rothstein & Rajan (1991
	Nuclear receptors	Unknown	Moore & Devaney (1999
Onchocerca volvulus	Zinc finger protein	ES, unknown	Holst & Zipfel (1996)
	Galectin	Unknown	Joseph *et al.* (2000)
	Chitinase	ES, unknown	Wu *et al.* (1996)
Wuchereria bancrofti	Glutathione peroxidase	Antioxidant	Cookson *et al.* (1992)
	Chitinase	Unknown	Raghaven *et al.* (1994)

heat shock protein (sHSP) genes have been shown to be 'switched on' at the transition to parasitism in *Brugia malayi* and in the rat intestinal nematode, *Nippostrongylus brasiliensis* (see Tweedie *et al.* 1993; Raghaven *et al.* 1999). sHSPs are known to function as molecular chaperones, protecting proteins from heat-induced aggregation and denaturation, a role consistent with their upregulation when the parasite infects the mammalian host. However, in the case of *B. malayi* and *N. brasiliensis*, *in vitro* studies (Tweedie *et al.* 1993; Raghaven *et al.* 1999) suggest that the sHSPs characterised to date do not perform this function. The role of HSPs in protecting nematode parasites from heat shock damage is not yet clear, and no heat shock/heat regulated genes have yet been characterised for human intestinal nematodes, although ESTs for HSPs from *N. americanus* and *S. stercoralis* have been deposited in the databases.

After infection of the host's intestinal tract, changes are induced in biochemical pathways of nematode parasites due to the anaerobic environment (reviewed by Bryant, 1975; Kita, Hirawake & Takamija, 1997). For example, there is a change from succinate oxidation via succinate dehydrogenase (SDH) in the Kreb's cycle in the pre-parasitic stages to the reverse reaction of reduction of fumarate to succinate in parasitic stages. The pathways operating vary according to the micro- or macro-niche(s) occupied by the parasite *in vivo*, and thus the relative availability of O_2 and glucose. However, in completely anaerobic environments, the phosphoenolpyruvate carboxykinase (PEPCK)-succinate pathway operates effectively (Kita *et al.* 1997). The intestinal hookworm of humans and carnivorous animals, *Ancylostoma ceylanicum*, has been shown to have both NADH oxidase and fumarate reductase activities at the adult stage, enabling utilisation of both aerobic and anaerobic pathways (Goyal *et al.* 1991). Some enzymes involved in aerobic *versus* anaerobic metabolism have been cDNA cloned and characterised for *H. contortus*, revealing that, in this nematode, SDH may perform both succinate oxidation and reduction via different isoforms (Roos & Tielens, 1994). Surprisingly, no genes associated with anaerobic metabolism have yet been characterised for intestinal nematodes of humans (although there are ESTs for SDH and PEPCK in the databases), despite their importance in parasite survival in the mammalian host. Since biochemical and metabolic pathways in nematodes differ from those of their vertebrate hosts, molecules related to these pathways are potential targets for new anthelmintic compounds.

2.2 Surface molecules

Some of the earliest studies of developmentally regulated molecules of nematode parasites examined 'surface' or coat proteins. Nematodes must synthesise a new cuticle prior to each moult. In the case of intestinal nematodes, the surface of each developmental stage has been demonstrated to differ significantly from the preceding one (e.g. Maizels, Meghji & Ogilvie, 1983). In particular, the cuticle differs markedly between the pre-parasitic and the parasitic stages (Proudfoot *et al.* 1993), reflecting the different environments inhabited. The surface of the parasitic nematode is in intimate contact with its host, where it lives under harsh physical and immunological pressure/conditions. Experiments using *in vitro* cultured parasites have shown that infective L3 of some gastrointestinal nematodes of animals can actively shed their outer surface layer when complexed with antibodies (Ashman *et al.* 1995) and/or immune cells, such as eosinophils (Badley *et al.* 1987), and thus play a role in immune evasion.

In the case of the common roundworm of dogs, *Toxocara canis*, a protein involved in this surface shedding, TES-120, has been cDNA cloned and shown to encode a mucin-like protein (Gems & Maizels, 1996). Surface molecules have also been characterised for filarial nematodes. For example, Storey & Philipp (1992) studied expression patterns of surface antigens from infective L3 and the stages of *B. malayi* parasitising the mammalian host, and a microfilarial surface antigen from *Onchocerca volvulus* has been cDNA cloned (Lustigman *et al.* 1992). Surprisingly, there is little published information on developmentally regulated surface proteins of human intestinal nematodes, although non-surface cuticular proteins, such as collagen and cuticulin, have been studied (e.g. Winkfein *et al.* 1985; D'Auria *et al.* 1998), and there is a number of ESTs in current databases.

2.3 Gene expression related to parasite feeding within the host

Although molecules associated with parasite feeding have not yet been characterised for intestinal nematodes of humans, serpins (proposed to be associated with the ability of *A. ceylanicum* to evade the host response) have been demonstrated to play a functional role in feeding in some parasites. For example, a serpin from the canine hookworm, *A. caninum*, has been shown to be an anti-coagulant (Stanssens *et al.* 1996) and thus plays a role in blood-feeding. It is very likely that the closely-related human hookworms also possess serpins with anti-coagulant properties. Other molecules, such as proteases and peptidases involved in digesting the blood meal, have been well characterised for *H. contortus* (reviewed by Newton & Munn, 1999). Such molecules represent a range of digestive enzymes (see Table 14.2) from cysteine proteases (cathepsins), to aspartic and metalloproteases, to microsomal aminopeptidases (which cleave the dipeptides which are the final products of digestion), and are expressed from the onset of blood-feeding. However, there has been limited study of such proteases in nematodes which do not suck blood. Although it is not yet known whether intestinal nematodes of humans, including hookworms, possess a similar repertoire of these proteases, ESTs for cathepsins from *N. americanus*, *T. trichiura* and *S. stercoralis* have been deposited in the databases, and it is likely that there is significant similarity in digestive processes amongst intestinal nematodes.

2.4 Genes involved in evasion of host responses

Parasitic nematodes reside in 'hostile' environments within their mammalian hosts, and thus employ strategies to protect themselves. The strategies vary depending on the ecological niche occupied by the parasite, and include the synthesis of antioxidant, immune modulating and/or immune evasion molecules, including those which may downregulate or otherwise alter the immune system.

Parasites are exposed to free radicals produced both by their own metabolism and by the cellular immune responses in the host, and it has been proposed that parasite-expressed antioxidants have therefore evolved as a protective mechanism (Callahan, Crouch & James, 1988; Henkle-Dührsen & Kampkotter, 2001). Antioxidants, such as thioredoxin peroxidases, superoxide dismutases, glutathione-S-transferases, catalases and glutathione peroxidases have been cDNA cloned and characterised for filarial nematodes of humans and gastrointestinal nematodes of ruminants, where expression is restricted to parasitic stages (e.g. Cookson, Blaxter & Selkirk, 1992; Eckelt *et al.* 1998; Ghosh *et al.* 1998; Liddell & Knox, 1998; Lattemann, Matzen & Apfel, 1999). However, little is known about antioxidant enzymes produced by intestinal parasites of humans, although glutathione-S-transferase activity has been demonstrated for hookworms (Brophy *et al.* 1995). To date, no antioxidant enzymes have been fully characterised for any human intestinal nematode, but partial cDNA sequences encoding putative antioxidants have been deposited in the databases for *S. stercoralis* and *N. americanus*.

Developmentally regulated ES products have been proposed to play a role in protecting parasites from their hosts by immune evasion. An adult-specific, Kunitz-type serine protease inhibitor (serpin) has been cloned from *A. ceylanicum* (Milstone *et al.* 2000). The broad-spectrum activity of this inhibitor led to speculation that it may protect the parasite from attack by digestive proteases in the host intestine. Also, secreted serine protease inhibitors may modulate mucosal host immune responses to nematode infection in the intestine (Rhoads *et al.* 2000a), and it has been shown that a serpin from *B. malayi* can inhibit neutrophil serine protease activity (Zang *et al.* 1999). Moreover, acetylcholinesterases (AChEs) secreted from parasitic nematodes have also been proposed to play a role in immune evasion, for instance, via the inhibition of local cellular immune responses, and by decreasing gut peristalsis and mucus secretion (Rhoads, 1984). An AChE secreted by *N. americanus* has been characterised biochemically (Pritchard *et al.* 1991; Pritchard, Brown & Toutant, 1994), but has not yet been cDNA cloned. Immunoblotting has demonstrated low levels of AChE in L3, increasing amounts in L4, and maximum expression in adults (Pritchard *et al.* 1991). AChEs have been identified in a wide range of nematode parasites of animals (e.g. Table 14.2), and a corresponding EST from *S. stercoralis* is present in current sequence databases. Although their function in ES is unclear, they have been proposed as candidate vaccine antigens on the basis that immunisation with ES fractions containing AChE elicits some degree of protection against parasite challenge (Griffiths & Pritchard, 1994; McKeand *et al.* 1995). However, in a recent vaccination study of the bovine lungworm, *Dictyocaulus viviparus*, AChE failed to induce protective immunity (Matthews *et al.* 2001).

Recently, nematode parasites have been shown to specifically synthesize homologues of mammalian immune molecules, which may interfere with host protective immune responses. For instance, it has been demonstrated that adults of *Trichuris muris*, an intestinal nematode of the mouse, secrete an interferon-γ homologue in ES which could switch a 'protective' Th2 host response to a 'susceptible' Th1 response (Grencis & Entwhistle, 1997). *N. americanus* secretes a protease which inhibits eotaxin-mediated eosinophil recruitment (Culley *et al.* 2000) and a calreticulin-like molecule which binds to the complement component C1q (Kaspar *et al.* 2001). Macrophage inhibitory factors (MIFs) have been described from the filarial nematodes of humans, including *Brugia pahangi* (see Pennock *et al.* 1998), *B. malayi*, *O. volvulus* and *Wuchereria bancrofti* (see Pastrana *et al.* 1998), as well as from *T. muris* and *Trichinella spiralis* (see Pennock *et al.* 1998). In *B. malayi*, MIF was detected in infective, mosquito-derived L3, is upregulated in expression in all developmental stages occurring within the mammalian host, and is present in ES (Pastrana *et al.* 1998). Since MIF functions by attracting macrophages, it was postulated that the parasite attracts and then alters macrophage effector functions, perhaps by manipulating host cytokine expression profiles (Pastrana *et al.* 1998). Homologues of transforming growth factor-β (TGF-β) have been cloned from *B. malayi* and *B. pahangi* using a PCR approach (Gomez-Escobar, Lewis & Maizels, 1998). However, TGF-β expression was low or absent in developmentally arrested stages, and is highly expressed around the times of larval moults in the mammalian host, suggesting that its role is in developmental maturation of larval parasites rather than as a regulator of the immune response in the host. This is supported by its close genetic relationship with other members of the TGF-β subfamily, which include many key molecules associated with development (Gomez-Escobar *et al.* 1998), and by the fact that a TGF-β receptor has also been cloned from *B. pahangi* (see Gomez-Escobar, Van Den Biggelaar & Maizels, 1997). More recently, a second TGF-β from *B. malayi* has been cloned which may play a role in immune evasion (Gomez-Escobar, Gregory & Maizels, 2000). The most extensively characterised immuno-modulatory molecule from any nematode parasite is the neutrophil inhibitory factor (NIF) from the canine hookworm, *A. caninum* (see Moyle *et al.* 1994). This protein is a potent inhibitor of CD11/CD18-dependent neutrophil function *in vitro* (Moyle *et al.* 1994) and was suggested to play a key role in evading the host's inflammatory response (Rieu *et al.* 1994). Moreover, it has been shown recently that vaccination with recombinant NIF from *A. ceylanicum* reduces the fecundity of this parasite in the hamster model (Ali *et al.* 2001).

Parasite homologues of mammalian immuno-modulatory molecules have been proposed to play a role in evading host responses, but such molecules and their function(s) have not yet been adequately studied. ESTs are present in the databases for a putative MIF from *T. trichiura* and a NIF from *N. americanus*, but they have not yet been characterised. The nature and extent of immuno-modulatory molecules synthesised by intestinal nematodes of humans remain to be determined. Such information will contribute toward a better understanding of host-parasite relationships.

3. GENES EXPRESSED IN A SEX-SPECIFIC MANNER IN NEMATODES

3.1 Major sperm proteins

Amongst the first sex-specifically expressed proteins of nematodes described were those of the major sperm protein (MSP) family, originally isolated from *C. elegans* and *Ascaris* sp. (see Klass & Hirsh, 1981; Nelson & Ward, 1981). MSPs are small (~14 kDa), nematode-specific, cytoskeletal proteins, which account for ~10-15% of the total cellular protein in spermatozoa (Klass, Dow & Herndon, 1982) and are involved sperm motility (King *et al.* 1994) and in oocyte maturation and sheath contraction (Miller *et al.* 2001). Expression of *msp* genes in *Ascaris* and *C. elegans* is confined to the testes during the meiotic stages of spermatogenesis (Klass *et al.* 1982). In *C. elegans*, most of the large family of ~60 *msp* genes, located on chromosomes II and IV, appear to be transcribed, with each gene contributing ~1-3% to the total cellular mRNA, while in *Ascaris* only a single gene is transcribed (Nelson & Ward, 1981; Bennett & Ward, 1986; Ward *et al.* 1988).

Most of the parasitic nematodes investigated to date have 5-13 *msp* genes, while free-living nematodes usually have a higher number, ranging from 15-50 (Scott *et al.* 1989a). The high concentration of MSP in *C. elegans* sperm is likely to be due to simultaneous expression of the numerous *msp* genes, resulting in the availability of a large pool of *msp* mRNA for translation. It has been proposed that MSP production is a rate-limiting step in sperm production in *C. elegans*, and thus a relatively high copy number of *msp* genes is maintained (Scott *et al.* 1989a). Parasitic nematodes may not have the high rate of sperm production of the free-living nematode, due to longer life cycles and/or the presence of larger testes, and thus the required *msp* mRNA concentration can be achieved from fewer *msp* genes. Alternatively, *msp* genes from parasitic nematodes may be transcribed from a more efficient promoter or have increased mRNA stability.

Apart from *Ascaris* sp. (see Bennett & Ward, 1986), no other *msp*s from human intestinal nematodes have yet been cDNA cloned, and no ESTs representing *msp* are currently present in the databases. Moreover, the human filarioid, *O. volvulus* is presently the only parasitic nematode for which the *msp* genomic organisation has been determined (Scott *et al.* 1989b). Two *O. volvulus msp* genes, *Ovgs-1* and *Ovgs-2*, have been isolated and these show ~80% identity to *Ascaris msp* cDNA and 79% to the *C. elegans msp-3* cDNA sequence. However, there is limited DNA sequence similarity between the promoters of the *O. volvulus msp* genes and those of *C. elegans*, although two GATA binding motifs have also been identified for *O. volvulus*, suggesting that they may be important for *msp* gene expression in this parasite.

3.2 Vitellogenins

Vitellogenins represent another important family of sex-specifically expressed genes. These are large (170-700 kDa) phospho-glycolipoproteins occurring in a broad range of vertebrates and invertebrates, and are thought to provide a source of amino acids and lipids for embryos to consume during development (Chen, Sappington & Raikhel, 1997). Vitellogenins have the ability to non-covalently bind hormones, vitamins and/or metal ions in some species (Chen *et al.* 1997). In *C. elegans*, they are abundantly expressed in the intestine of the late, fourth-stage larva and adult hermaphrodite, secreted into the body cavity, taken up by the gonad and absorbed into the oocyte by receptor-mediated endocytosis (Grant & Hirsh, 1999). *C. elegans* has six vitellogenin genes (*vit-1* to *vit-6*). Interestingly, *vit-6* encodes a protein which is proteolytically cleaved into two parts (88 and 115 kDa) before being absorbed by the oocyte, whereas those of the *vit-2, vit-3* and *vit-5* are absorbed without prior cleavage; *vit-1* and *vit-4* appear to be pseudogenes (Spieth *et al.* 1991).

There is relatively little published information regarding vitellogenin genes of parasitic nematodes. However, numerous vitellogenin expressed sequence tags (ESTs) for a range of veterinary gastrointestinal nematodes, including *H. contortus* and *Oesophagostomum dentatum* (Strongylida), have recently been deposited in databases or published (Boag *et al.* 2000; Hartman *et al.* 2001). An adult female-specific cDNA clone from *H. contortus* with greatest similarity to *C. elegans vit-6* has been isolated, and expression of the *vit-6* transcript in the intestine confirmed by *in situ* hybridisation (Hartman *et al.* 2001). Although vitellogenins have been shown to exist in a number gastrointestinal nematodes of veterinary importance, database searches using *C. elegans vit-2*, *vit-5* and *vit-6* cDNA sequences, or keyword searches, have not identified any putative vitellogenin ESTs for human nematodes, thus providing opportunities for future research. Comparative analyses of the genomic organisation and structure of vitellogenin genes between *C. elegans* and a wide range of parasitic nematodes may assist in identifying promoters critical for regulating sex-specific gene expression (cf. MacMorris *et al.* 1994).

3.3 Other recently-characterised sex-specific genes

Other than *msp* and vitellogenin genes, a number of sex-specific genes have recently been isolated from parasitic nematodes of humans and animals (e.g. Bessarab & Joshua, 1997; Michalski & Weil, 1999; Boag *et al.*, 2000). However, other than *msp*, no sex-specific genes have been characterised from intestinal nematodes of humans, although there are ESTs from *S. stercoralis* in the databases with similarity to a sex-determining gene (*tra-2*) from *Drosophila* and a *C. elegans* gene (*gld-1*) essential for oogenesis. Michalski & Weil (1999) characterised sex-specific genes from the human filarial nematode, *B. malayi*, using differential display (Liang & Pardee, 1992) combined with database analysis. Of the 12 adult sex-specific genes isolated, only five had similarity to sequences contained within current databases. Some of the partial gene sequences encoded proteins with similarity to molecules expressed in a sex-specific manner in other nematode species. Joshua & Hsieh (1995) isolated a female-specific gene fragment from the strongylid nematode, *Angiostrongylus cantonensis*. The full-length cDNA (*Ac-fmp-1*) encoded a peptide of 417 amino acids, which was localised to the musculature adjacent to the pseudocoelomic space, although its role or function is currently unclear (Bessarab & Joshua, 1997).

The nodule worm of the large intestine of pigs, *O. dentatum*, provides a unique model system for investigating reproductive processes in parasitic nematodes (Christensen, 1997). It has a short, direct life-cycle (21 days) (Talvik *et al.* 1997), produces large numbers of off-spring and can easily be maintained as a laboratory line. Importantly, uni-sex or mixed-sex infections can be established by rectal transplantation to naïve pigs (Christensen, Grøndahl-Nielsen & Nansen, 1996), thus proving the opportunity for studying mating behaviour and sexual maturation *in vivo*. Also, the parasite may be maintained relatively effectively in cultures *in vitro* (Daugschies & Watzel, 1999). Recently, ten male- and two female-specific ESTs were isolated from *O. dentatum* by differential display analysis (Boag *et al.* 2000). Of these, six ESTs appeared to represent 'novel' genes, while the others had varying levels of sequence similarity to sequences in the databases (Boag *et al.* 2000). One male-specific EST encoded the catalytic subunit of a serine/threonine protein phosphatase, showing greatest similarity to a *C. elegans* protein implicated in sperm production (Reinke *et al.* 2000). Characterisation of these and other sex-specific genes should enable detailed investigations into the molecular aspects of reproduction in *O. dentatum,* both *in vivo* and/or *in vitro*, which should also have implications for study of intestinal nematodes of humans.

4. ADVANCED APPROACHES FOR THE MOLECULAR CHARACTERISATION OF GENES FROM PARASITIC NEMATODES USING *CAENORHABDITIS ELEGANS*

The free-living nematode, *C. elegans* represents one of the best characterised metazoan organisms (Riddle *et al.* 1997), and recently its near complete genome sequence was determined (The *C. elegans* Sequencing Consortium, 1998). Of the ~20,000 genes, ~ 58% appear to be nematode-specific (Blaxter, 1998). Given that a number of key biological processes appear to be conserved among a wide range of nematodes (e.g. Favre *et al.* 1998; Ashton, Li & Schad, 1999), *C. elegans* may be considered a model for parasitic nematodes (Blaxter, 1998; Bürglin, Lobos & Blaxter, 1998). Other features, including the nematode's short life cycle (three days at 25°C), ease of propagation of well-defined lines using a simple bacterial food source (*Escherichia coli*), relatively small genome size, ability to produce clonal progeny from hermaphrodites and to cross hermaphrodites with males, make *C. elegans* an attractive system (cf. Riddle *et al.* 1997). These attributes have allowed the acquisition of considerable knowledge and understanding of reproductive biology of the nematode (reviewed by Boag, Newton & Grasser, 2001) and have also enabled the development of functional genomics techniques, such as gene transformation, RNA-triggered gene silencing and global profiling of gene expression by microarray analysis. These approaches are likely to assist significantly in the study of developmentally regulated and sex-specific genes of intestinal nematodes of humans and other hosts.

4.1 Gene transformation

Transformation of *C. elegans* with homologous or heterologous genes represents a powerful tool for assessing gene function (Stinchcomb *et al.* 1985; Fire, 1986). Expression profiles for genes from parasitic nematodes can be studied using their promoter region to drive a reporter gene, such as the green fluorescent protein (GFP) or β-galactosidase. For example, Britton *et al.* (1999) produced transgenic *C. elegans* containing the promoter regions from two *H. contortus* protease gene (*pep-1* and *AC-2*) and the cuticle collagen gene *colost-1* from *Teladorsagia circumcincta.* Spatial expression of the reporter genes correlated with the expression profiles in the parasite, although there was a difference in temporal expression of the *colost-1* gene between the transformed *C. elegans* and *T. circumcincta.* Another application of transformation technology is 'genetic rescue' to demonstrate functional similarity among proteins (Fire & Waterston, 1989), although a possible limitation can be that transgenes are not usually well-expressed in germline tissues (Kelly *et al.* 1997). In such studies, mutant phenotypes are 'rescued' by introducing the wild-type or a homologous gene into *C. elegans*, with restoration of the mutant to wild-type, providing evidence of functional similarity of the homologue (Stinchcomb *et al.* 1985; Fire, 1986). The availability of a large number of *C. elegans* lines carrying mutations for defined genes provides opportunities for conducting such experiments using parasite genes (cf. Kwa *et al.* 1995). Such an approach could also be used to gain an understanding of molecular events relating to development, including the transition to parasitism and sexual differentiation and reproduction in intestinal nematodes of humans.

4.2 RNA-triggered gene silencing

Double-stranded RNA-mediated interference (RNAi) provides a powerful tool for the rapid analysis of gene function in *C. elegans* and other organisms (Fire *et al.* 1998; Tabara, Grishok & Mello, 1998; Fire, 1999). In brief, the principle of the method is that expression of a specific gene is reduced (silenced) by 'microinjecting', 'soaking' or 'feeding' the nematode with specific double-stranded RNA (dsRNA) (Carthew, 2001). The RNAi effect is systemic, with gene silencing occurring throughout the entire organism. Degradation of the target mRNA is usually specific and is passed on to the progeny in both *C. elegans* and *Drosophila.* Although some aspects of RNAi are still unclear, the current consensus view is that the introduction of dsRNA into an organism leads to a targeted degradation of the homologous mRNA, in many cases producing a null mutant phenotype (Carthew, 2001).

The ability to assess the possible functions of EST sequences from parasitic nematodes by RNAi would improve our understanding of their developmental and reproductive biology and may also, in the long term, aid in identifying targets for anti-parasitic drugs or vaccines (Kuwabara & Coulson, 2000). RNAi has been applied effectively to the protozoan parasite, *Trypanosoma brucei* (see Ngo *et al.* 1998; Shi *et al.* 2000), but the challenge now is to adapt RNAi to parasitic nematodes. RNAi feeding experiments could be carried out on developmental stages which feed on bacteria, such as first and second stage larvae of strongylid nematodes, although this may only be applicable to studying genes related to early developmental processes. Therefore, the development of an effective *in vitro* culture system for intestinal nematodes (which allows access to all developmental stages) is needed for RNAi studies in the parasite itself (see Eckert, 1997). Nonetheless, if a high level of sequence identity exists between a gene from a parasitic nematode and its *C. elegans* homologue, the function of the former can be inferred from RNAi experiments in *C. elegans*. For example, Boag *et al.* (2000) recently cloned a male-specific serine/threonine phosphatase from the intestinal nematode, *O. dentatum*, which shares ~90% similarity with its *C. elegans* homologue. RNAi experiments in *C. elegans* have indicated that this phosphatase is involved in reproduction, as a reduction in the number of offspring in the F1 generation was observed in dsRNA-treated hermaphrodites (Boag, P. *et al.* unpublished observations). This result may have important implications for testing in *C. elegans* the function of genes of intestinal nematodes, including those of human health importance.

4.3 Global profiling of gene expression by microarray analysis

The availability of the complete *C. elegans* genome sequence has facilitated the development of DNA microarray technology for studying differential gene expression during key developmental and reproductive phases in this nematode. Microarray analysis permits a 'global perspective' of gene expression (Gutierrez, 2000; Lockhart & Winzeler, 2000; Jiang *et al.* 2001). In brief, the technique involves the automated 'spotting out' of oligonucleotides, cDNAs or genomic DNA (usually corresponding to previously characterised genes) onto glass slides in precise positions. mRNAs from two different stages or tissue origins are labelled with different fluorescent markers and hybridised to the array. Then, the relative abundance of individual transcripts in each mRNA population is determined by comparing the relative signal intensity of each marker.

For example, Reinke *et al.* (2000) used a DNA microarray representing 11,917 (~63%) genes of *C. elegans* to analyse gene expression during the development of germline tissues. A total of 1,416 genes with enriched expression in the germline tissues was identified, and these were divided into three classes. The first class, termed 'germline intrinsic', comprising 508 genes expressed in the germline of hermaphrodites producing either sperm or oocytes, were predicted to have common functions in germline cells, such as meiosis and recombination, stem cell proliferation and germline development. The second class comprised 258 genes expressed at elevated levels in *C. elegans* hermaphrodites producing oocytes only (Reinke *et al.* 2000). The third class contained 650 genes with elevated or exclusive expression in hermaphrodites producing sperm compared with males (Reinke *et al.* 2000). The large number of differentially regulated genes identified using this microarray approach has provided broad insights into germline development and allowed the prediction of gene functions.

The availability of microarray and electronic subtraction techniques as well as the functional genomics capacity of the *C. elegans* system, provides an exciting opportunity to enhance our understanding of molecular developmental and reproductive processes in nematodes of humans. The apparent abundance of sex-specific genes expressed in germline tissues, particularly those related to sperm development and maturation, may represent targets for developing new prophylactic or therapeutic agents.

5. CONCLUSION

A range of intestinal nematodes which parasitise humans are of socio-economic importance because of the diseases they cause, particularly in children (see Chapters 3 and 4). However, due to difficulties in working with the parasitic stages of nematodes infecting humans (unless a suitable animal model is available), there has been limited study of processes of parasite growth and development, and/as well as parasite-host interaction(s). Recent technological advances now provide a unique opportunity to investigate the molecular basis of developmental and reproductive processes in parasitic nematodes. Large-scale EST sequencing programs (Blaxter & Ivens, 1999; Blaxter, 2000), together with PCR-based techniques which allow identification of genes which differ between stages and sexes (Diatchenko *el al.* 1996; Liang and Pardee, 1992), is enabling the identification of genes related to development (both in pre-parasitic and parasitic stages), sexual differentiation and maturation. Although relatively few developmentally regulated, and even fewer sex-specific, molecules from human intestinal nematodes have been characterised, a significant number have been identified from intestinal nematodes of animals and from human filarial nematodes (Table 14.2). It is expected that human intestinal nematodes also produce some of these molecules.

EST data sets will provide a foundation for gene expression profiling using DNA microarrays, and for gene deletion studies and/or gene silencing in *C. elegans*. Importantly, the availability of *C. elegans* and its complete genome sequence, and emerging information on gene function should provide a platform for testing the function of gene orthologues from parasitic nematodes, particularly given that most of these parasites are difficult to propagate and maintain in culture *in vitro*. Another technological revolution is now occurring in the field of proteomics (Lee, 2001). Recently, array techniques for mass spectroscopic analysis of differentially expressed proteins have been developed (von Eggeling *et al.* 2000), which will allow large scale analysis of expressed proteins from small amounts of parasite material. This will provide a link between the regulation of transcription and translation and, importantly for the study of parasites, allow the analysis of proteins expressed within short time-frames, or within organs or micro-environments of a parasite, such as ES molecules (Barrett, Jeffries & Brophy, 2000).

Hence, the use of genomics and proteomics to investigate molecular developmental and reproductive processes in intestinal nematodes will yield valuable information of fundamental biological significance, which should also have implications for developing novel ways of treating, controlling or preventing the diseases they cause, by blocking or disrupting key biological pathways.

ACKNOWLEDGEMENTS

Project support through the Australian Research Council, Novartis Animal Health Australia, the Department of Natural Resources and Environment Victoria, Agriculture Victoria and the Danish Centre for Experimental Parasitology is gratefully acknowledged. P.R.B. is the recipient of a scholarship from The University of Melbourne and Novartis Animal Health Australia.

REFERENCES

ALI, F., BROWN, A., STANSSENS, P., TIMOTHY, L. M., SOULE, H. R. & PRITCHARD, D. I. (2001). Vaccination with neutrophil inhibitory factor reduces the fecundity of the hookworm *Ancylostoma ceylanicum*. *Parasite Immunology* **23**, 237-249.

ASHMAN, K., MATHER, J., WILTSHIRE, C., JACOBS, H. J. & MEEUSEN, E. (1995). Isolation of a larval surface glycoprotein from *Haemonchus contortus* and its possible role in evading host immunity. *Molecular and Biochemical Parasitology* **70**, 175-179.

ASHTON, F. T., LI, J. & SCHAD, G. A. (1999). Chemo- and thermosensory neurons: structure and function in animal parasitic nematodes. *Veterinary Parasitology* **84**, 297-316.

BADLEY, J. E., GRIEVE, R. B., ROCKEY, J. H. & GLICKMAN, L. T. (1987). Immune-mediated adherence of eosinophils to *Toxocara canis* infective larvae: the role of excretory-secretory antigens. *Parasite Immunology* **9**, 133-143.

BARKER, G. C. & BUNDY, D. A. P. (1999). Isolation of a gene family that encodes the porin-like proteins from the human parasitic nematode *Trichuris trichiura*. *Gene* **229**, 131-136.

BARRETT, J., JEFFRIES, J. R. & BROPHY, P. M. (2000). Parasite proteomics. *Parasitology Today* **16**, 400-403.

BENNETT, G. & WARD, S. (1986). Neither a germ line-specific nor several somatically expressed genes are lost or rearranged during embryonic chromatin diminution in the nematode *Ascaris lumbricoides* var. *suum*. *Developmental Biology* **118**, 141-147.

BESSARAB, I. N. & JOSHUA, G. W. (1997). Stage-specific gene expression in *Angiostrongylus cantonensis*: characterisation and expression of an adult-specific gene. *Molecular and Biochemical Parasitology* **88**, 73-84.

BIN, Z., HAWDON, J., QIANG, S., HAINAN, R., HUIQING, Q., WEI, H., SHU-HUA, X., TIEHUA, L., XING, G., ZHENG, F. & HOTEZ, P. (1999). *Ancylostoma* secreted protein 1 (ASP-1) homologues in human hookworms. *Molecular and Biochemical Parasitology* **98**, 143-149.

BLAXTER, M. (1998). *Caenorhabditis elegans* is a nematode. *Science* **282**, 2041-2046.

BLAXTER, M. (2000). Genes and genomes of *Necator americanus* and related hookworms. *International Journal for Parasitology* **30**, 347-355.

BLAXTER, M. & IVENS, A. (1999). Reports from the cutting edge of parasitic genome analysis. *Parasitology Today* **15**, 430-431.

BLAXTER, M. L., RAGHAVEN, N., GHOSH, I., GUILIANO, D., WENHONG, L., WILLIAMS, S. A., SLATKO, B. & SCOTT, A. L. (1996). Genes expressed in *Brugia malayi* infective third stage larvae. *Molecular and Biochemical Parasitology* **77**, 77-93.

BOAG, P. R., NEWTON, S. E., HANSEN, N., CHRISTENSEN, C. M., NANSEN, P. & GASSER, R. B. (2000). Isolation and characterisation of sex-specific transcripts from *Oesophagostomum dentatum* by RNA arbitrarily-primed PCR. *Molecular and Biochemical Parasitology* **108**, 217-224.

BOAG, P. R., NEWTON, S. E. & GASSER, R. B. (2001). Molecular aspects of sexual development and reproduction in nematodes and schistosomes. *Advances in Parasitology*, In press.

BRITTON, C., KNOX, D. P. & KENNEDY, M. W. (1994). Superoxide dismutase (SOD) activity of *Dictyocaulus viviparus* and its inhibition by antibody from infected and vaccinated bovine hosts. *Parasitology* **109**, 255-261.

BRITTON, C., REDMOND, D. L., KNOX, D. P., McKERROW, J. H. & BARRY, J. D. (1999). Identification of promoter elements of parasite nematode genes in transgenic *Caenorhabditis elegans*. *Molecular and Biochemical Parasitology* **103**, 171-181.

BROPHY, P. M., PATTERSON, L. H., BROWN, A. & PRITCHARD, D. I. (1995). Glutathione-*S*-transferase (GST) expression in the human hookworm *Necator americanus*: potential roles for excretory-secretory forms of GST. *Acta Tropica* **59**, 259-263.

BROWN, A., GIROD, N., BILLETT, E. E. & PRITCHARD, D. I. (1999). *Necator americanus* (human hookworm) aspartyl proteinases and digestion of skin macromolecules during skin penetration. *American Journal of Tropical Medicine and Hygiene* **60**, 840-847.

BRYANT, C. (1975). Carbon dioxide utilisation, and the regulation of respiratory metabolic pathways in parasitic helminths. *Advances in Parasitology* **13**, 35-69.

BÜRGLIN, T. R., LOBOS, E. & BLAXTER, M. L. (1998). *Caenorhabditis elegans* as a model for parasitic nematodes. *International Journal for Parasitology* **28**, 395-411.

CALLAHAN, H. L., CROUCH, R. K. & JAMES, E. R. (1988). Helminth anti-oxidant enzymes: a protective mechanism against host oxidants? *Parasitology Today* **4**, 218-225.

CARTHEW, R. W. (2001). Gene silencing by double-stranded RNA. *Current Opinion in Cell Biology* **13**, 244-248.

CHEN, J. S., SAPPINGTON, T. W. & RAIKHEL, A. S. (1997). Extensive sequence conservation among insect, nematode, and vertebrate vitellogenins reveals ancient common ancestry. *Journal of Molecular Evolution* **44**, 440-451.

CHRISTENSEN, C. M. (1997). The effect of three distinct sex ratios at two *Oesophagostomum dentatum* worm population densities. *Journal of Parasitology* **83**, 636-640.

CHRISTENSEN, C. M., GRØNDAHL-NIELSEN, C. & NANSEN, P. (1996). Non-surgical transplantation of *Oesophagostomum dentatum* to recipient pigs via rectal intubation. *Veterinary Parasitology* **65**, 139-145.

COOKSON, E., BLAXTER, M. L. & SELKIRK, M. E. (1992). Identification of the major soluble cuticular glyocprotein of lymphatic filarial nematode parasites (gp29) as a secretory homolog of glutathione peroxidase. *Proceedings of the National Academy of Sciences of the United States of America* **89**, 5837-5841.

CULLEY, F. J., BROWN, A., CONROY, D. M., SABROE, I., PRITCHARD, D. I. & WILLIAMS, T. J. (2000). Eotaxin is specifically cleaved by hookworm metalloproteases preventing its action *in vitro* and *in vivo*. *Journal of Immunology* **165**, 6447-6453.

DAUB, J., LOUKAS, A., PRITCHARD, D. I. & BLAXTER, M. (2000). A survey of genes expressed in adults of the human hookworm, *Necator americanus*. *Parasitology* **120**, 171-184.

DAUGSCHIES, A. & WATZEL, C. (1999). *In vitro* development of histotropic larvae of *Oesophagostomum dentatum* under various conditions of cultivation. *Parasitology Research* **85**, 158-161.

D'AURIA, S., ROSSI, M., TANFANI, F., BERTOLI, E. & BAZZICALUPO, P. (1998). Structural analysis of ASCUT-1, a protein component of the parasitic nematode *Ascaris lumbricoides*. *European Journal of Biochemistry* **255**, 588-594.

DE GRAAF, D. C., BERGHEN, P., MOENS, L., MAREZ, T. M. D., RAES, S., BLAXTER, M. L. & VERCRUYSSE, J. (1996). Isolation, characterization and immunolocalization of a globin-like antigen from *Ostertagia ostertagi* adults. *Parasitology* **113**, 63-69.

DESSENS, J. T., MARGOS, G., RODRIGUEZ, M. C. & SINDEN, R. E. (2000). Identification of differentially regulated genes of *Plasmodium* by suppression subtractive hybridization. *Parasitology Today* **16**, 354-365.

DIATCHENKO, L., LAU, Y. F., CAMPBELL, A. P., CHENCHIK, A., MOQADAM, F., HUANG, B., LUKYANOV, S., LUKYANOV, K., GURSKAYA, N., SVERDLOV, E. D. & SIEBERT, P. D. (1996). Suppression subtractive hybridization: a method for generating differentially regulated or tissue-specific cDNA probes and libraries. *Proceedings of the National Academy of Sciences of the United States of America* **93**, 6025-6030.

DRAKE, L., KORCHEV, Y., BASHFORD, L., DJAMGOZ, M., WAKELIN, D., ASHALL, F. & BUNDY, D. (1994). The major secreted product of the whipworm, *Trichuris*, is a pore-forming protein. *Proceedings of the Royal Society of London, B. Biological Sciences* **257**, 255-261.

DRAKE, L. J., BARKER, G. C., KORACHEV, Y., LAB, M., BROOKES, H. & BUNDY, D. A. P. (1998). Molecular and functional characterisation of a recombinant protein of *Trichuris Trichiura*. *Proceedings of the Royal Society of London, B. Biological Sciences* **265**, 1559-1565.

ECKELT, V. H., LIEBAU, E., WALTER, R. D. & HENKLE-DÜHRSEN, K. (1998). Primary sequence and activity analyses of a catalase from *Ascaris suum*. *Molecular and Biochemical Parasitology* **95**, 203-214.

ECKERT, J. (1997). Alternatives to animal experimentation in parasitology. *Veterinary Parasitology* **71**, 99-120.

FAVRE, R., CERMOLA, M., NUNES, C. P., HERMANN, R., MULLER, M. & BAZZICALUPO, P. (1998). Immuno-cross-reactivity of CUT-1 and cuticlin epitopes between *Ascaris lumbricoides, Caenorhabditis elegans* and *Heterorhabditis. Journal of Structural Biology* **123**, 1-7.

FIRE, A. (1986). Intergrative transformation of *Caenorhabditis elegans. EMBO Journal* **5**, 2673-2670.

FIRE, A. (1999). RNA-triggered gene silencing. *Trends in Genetics* **15**, 358-363.

FIRE, A. & WATERSTON, R. H. (1989). Proper expression of myosin genes in transgenic nematodes. *EMBO Journal* **8**, 3419-3428.

FIRE, A., XU, S., MONTGOMERY, M. K., KOSTAS, S. A., DRIVER, S. E. & MELLO, C. C. (1998). Potent and specific genetic interference by double-stranded RNA in *Caenorhabditis elegans. Nature*, **391**, 806-811.

FRENKEL, M. J., DOPHEIDE, T. A. A., WAGLAND, B. M. & WARD, C. W. (1992). The isolation, characterization and cloning of a globin-like, host-protective antigen from the excretory-secretory products of *Trichostrongylus colubriformis. Molecular and Biochemical Parasitology* **50**, 27-36.

FUHRMAN, J. A., LANE, W. S., SMITH, R. F., PIESSENS, W. F. & PERLER, F. B. (1992). Transmission-blocking antibodies recognize microfilarial chitinase in brugian lymphatic filariasis. *Proceedings of the National Academy of Sciences of the United States of America* **89**, 1548-1552.

GEMS, D. & MAIZELS, R. M. (1996). An abundantly expressed mucin-like protein from *Toxocara canis* infective larvae: The precursor of the larval surface coat glycoproteins. *Proceedings of the National Academy of Sciences of the United States of America* **93**, 1665-1670.

GHOSH, I., EISINGER, S. W., RAGHAVEN, N. & SCOTT, A. L. (1998). Thioredoxin peroxidases from *Brugia malayi. Molecular and Biochemical Parasitology* **91**, 207-220.

GHOSH, K., HAWDON, J. & HOTEZ, P. (1996). Vaccination with alum-precipitated recombinant *Ancylostoma*-secreted protein 1 protects mice against challenge infections with infective hookworm (*Ancylostoma caninum*) larvae. *Journal of Infectious Diseases* **174**, 1380-1383.

GOMEZ-ESCOBAR, N., GREGORY, W. F. & MAIZELS, R. M. (2000). Identification of *tgh-2*, a filarial nematode homolog of *Caenorhabditis elegans daf-7* and human transforming growth factor β, expressed in microfilarial and adult stages of *Brugia malayi. Infection and Immunity* **68**, 6402-6410.

GOMEZ-ESCOBAR, N., LEWIS, E. & MAIZELS, R. M. (1998). A novel member of the transforming growth factor-beta (TGF-β) superfamily from the filarial nematodes *Brugia malayi* and *Brugia pahangi. Experimental Parasitology* **88**, 200-209.

GOMEZ-ESCOBAR, N., VAN DEN BIGGELAAR, A. & MAIZELS, R. (1997). A member of the TGF-beta receptor gene family in the parasitic nematode *Brugia pahangi. Gene* **199**, 101-109.

GOYAL, N., GUPTA, S., KATIYAR, J. C. & SRIVASTAVA, V. M. L. (1991). NADH oxidase and fumarate reductase of *Ancylostoma ceylanicum. International Journal for Parasitology* **21**, 673-676.

GRANT, B. & HIRSH, D. (1999). Receptor-mediated endocytosis in the *Caenorhabditis elegans* oocyte. *Molecular Biology of the Cell* **10**, 4311-4326.

GREENHALGH, C. J., LOUKAS, A., DONALD, D., NIKOLAOU, S. & NEWTON, S. E. (2000). A family of galectins from *Haemonchus contortus*. *Molecular and Biochemical Parasitology* **107**, 117-121.

GRENCIS, R. K. & ENTWHISTLE, G. M. (1997). Production of an interferon-gamma homologue by an intestinal nematode: functionally significant or interesting artefact? *Parasitology* **115** (Suppl.) S101-S105.

GRIFFITHS, G. & PRITCHARD, D. I. (1994). Vaccination against gastrointestinal nematodes of sheep using purified secretory acetylcholinesterase from *Trichostrongylus colubriformis* - an initial pilot study. *Parasite Immunology* **16**, 507-510.

GUTIERREZ, J. A. (2000). Genomics: from novel genes to new therapeutics in parasitology. *International Journal for Parasitology* **30**, 247-252.

HARTMAN, D., DONALD, D.R., NIKALOAU, S., SAVIN, K.W., HASSE, D., PRESIDENTE, P.J.A. & NEWTON, S.E. (2001). Analysis of developmentally regulated genes of the parasite *Haemonchus contortus*. *International Journal for Parasitology*, In press.

HAWDON, J. M., JONES, B. F., HOFFMAN, D. R. & HOTEZ, P. J. (1996). Cloning and characterisation of *Ancylostoma*-secreted protein. *Journal of Biological Chemistry* **271**, 6672-6678.

HAWDON, J. M., NARASIMHAN, S. & HOTEZ, P. J. (1999). *Ancylostoma* secreted protein 2: cloning and characterisation of a second member of a family of nematode secreted proteins from *Ancylostoma caninum*. *Molecular and Biochemical Parasitology* **99**, 149-165.

HENKLE-DÜHRSEN, K. & KAMPKOTTER, A. (2001). Antioxidant enzyme families in parasitic nematodes. *Molecular and Biochemical Parasitology* **114**, 129-142.

HOEKSTRA, R., VISSER, A., OTSEN, M., TIBBEN, J., LENSTRA, J. A. & ROOS, M. H. (2000). EST sequencing of the parasitic nematode *Haemonchus contortus* suggests a shift in gene expression during transition to the parasitic stages. *Molecular and Biochemical Parasitology* **110**, 53-68.

HOLST, C. & ZIPFEL, P. F. (1996). A zinc finger gene from *Onchocerca volvulvus* encodes a protein with a functional signal peptide and an unusual Ser-His finger motif. *Journal of Biological Chemistry* **271**, 16725-16733.

HUANG, Y. J., WALKER, D., CHEN, W., KLINGBEIL, M. & KOMUNIECKI, R. (1998). Expression of pyruvate dehydrogenase isoforms during the aerobic/anaerobic transition in the development of the parasitic nematode *Ascaris suum*: altered stoichiometry of phosphorylation/inactivation. *Archives of Biochemistry and Biophysics* **352**, 263-270.

HUNTER, S. J., THOMPSON, F. J., TETLEY, L. & DEVANEY, E. (2001). Temperature is a cue for gene expression in the post-infective L3 of the parasitic nematode *Brugia pahangi*. *Molecular and Biochemical Parasitology* **112**, 1-9.

JIANG, M., RYU, J., KIRALY, M., DUKE, K., REINKE, V. & KIM, S. K. (2001). Genome-wide analysis of developmental and sex-regulated gene expression profiles in *Caenorhabditis elegans*. *Proceedings of the National Academy of Sciences of the United States of America* **98**, 218-223.

JOSEPH, G. T., HUIMA, T., KLION, A. & LUSTIGMAN, S. (2000). A novel developmentally regulated galectin of *Onchocerca volvulvus*. *Molecular and Biochemical Parasitology* **106**, 187-195.

JOSHUA, G. W. & HSIEH, C. Y. (1995). Stage-specifically expressed genes of *Angiostrongylus cantonensis*: identification by differential display. *Molecular and Biochemical Parasitology* **71**, 285-289.

KASPAR, G., BROWN, A., EBERL, M., VALLAR, L., KIEFFER, N., BERRY, C., GIRDWOOD, K., EGGLETON, P., QUINNELL, R. & PRITCHARD, D. I. (2001). A calreticulin-like molecule from the human hookworm *Necator americanus* interacts with C1q and the cytoplasmic signalling domains of some integrins. *Parasite Immunology* **23**, 141-152.

KELLY, W. G., XU, S., MONTGOMERY, M. K. & FIRE, A. (1997). Distinct requirements for somatic and germline expression of a generally expressed *Caernorhabditis elegans* gene. *Genetics* **146**, 227-238.

KING, K. L., ESSIG, J., ROBERTS, T. M. & MOERLAND, T. S. (1994). Regulation of the *Ascaris* major sperm protein (MSP) cytoskeleton by intracellular pH. *Cell Motility and the Cytoskeleton* **27**, 193-205.

KITA, K., HIRAWAKE, H. & TAKAMIYA, S. (1997). Cytochromes in the respiratoty chain of helminth mitochondria. *International Journal for Parasitology* **27**, 617-630.

KLASS, M. R. & HIRSH, D. (1981). Sperm isolation and biochemical analysis of the major sperm protein from *Caenorhabditis elegans*. *Developmental Biology* **84**, 299-312.

KLASS, M., DOW, B. & HERNDON, M. (1982). Cell-specific transcriptional regulation of the major sperm protein in *Caenorhabditis elegans*. *Developmental Biology* **93**, 152-164.

KUWABARA, P. E. & COULSON, A. (2000). RNAi - prospects for a general technique for determining gene function. *Parasitology Today* **16**, 347-349.

KWA, M. S., VEENSTRA, J. G., VAN DIJK, M. & ROOS, M. H. (1995). Beta-tubulin genes from the parasitic nematode *Haemonchus contortus* modulate drug resistance in *Caenorhabditis elegans*. *Journal of Molecular Biology* **246**, 500-510.

LATTEMANN, C. T., MATZEN, A. & APFEL, H. (1999). Up-regulation of extracellular copper/zinc superoxide dismutase mRNA after vertical transmission of the filarial parasite *Acanthocheilonema vitae* in the vertebrate host *Meriones unguiculatus*. *International Journal for Parasitology* **29**, 1437-1446.

LEE, K. H. (2001). Proteomics: a technology-driven and technology-limited discovery science. *Trends in Biotechnology* **19**, 217-222.

LIANG, P. & PARDEE, A. (1992). Differential display of eukaryotic messenger RNA by means of the polymerase chain reaction. *Science* **257**, 967-971.

LIDDELL, S. & KNOX, D. P. (1998). Extracellular and cytoplasmic Cu/Zn superoxide dismutases from *Haemonchus contortus*. *Parasitology* **116**, 383-394.

LOCKHART, D. J. & WINZELER, E. A. (2000). Genomics, gene expression and DNA arrays. *Nature* **405**, 827-836.

LONGBOTTOM, D., REDMOND, D. L., RUSSELL, M., LIDDELL, S., SMITH, W. D. & KNOX, D. P. (1997). Molecular cloning and characterisation of a putative aspartate proteinase associated with a gut membrane protein complex from adult *Haemonchus contortus*. *Molecular and Biochemical Parasitology* **88**, 63-72.

LOUKAS, A., DOEDENS, A., HINTZ, M. & MAIZELS, R. M. (2000). Identification of a new C-type lectin, TES-70, secreted by infective larvae of *Toxocara canis*, which binds to host ligands. *Parasitology* **121**, 545-554.

LOUKAS, A. & MAIZELS, R. M. (1998). Cloning and characterization of a prohibitin gene from infective larvae of the parasitic nematode *Toxocara canis*. *DNA Sequence* **9**, 323-328.

LUSTIGMAN, S., BROTMAN, B., JOHNSON, E. H., SMITH, A. B., HUIMA, T. & PRINCE, A. M. (1992). Identification and characterization of an *Onchocerca volvulus* cDNA clone encoding a microfilarial surface-associated antigen. *Molecular and Biochemical Parasitology* **50**, 79-93.

MacMORRIS, M., SPIETH, J., MADEJ, C., LEA, K. & BLUMENTHAL, T. (1994). Analysis of the VPE sequences in the *Caenorhabditis elegans vit-2* promoter with extrachromosomal tandem array-containing transgenic strains. *Molecular and Cell Biology* **14**, 484-491.

MAIZELS, R. M., MEGHJI, M. & OGILVIE, B. M. (1983). Restricted sets of parasite antigens from the different stages and sexes of the nematode parasite *Nippostrongylus brasiliensis*. *Immunology* **48**, 107-121.

MAIZELS, R. M., TETTEH, K. K. & LOUKAS, A. (2000). *Toxocara canis*: genes expressed by the arrested infective larval stage of a parasitic nematode. *International Journal for Parasitology* **30**, 495-508.

MASSEY, H. C. M. JR, BALL, C. C. & LOK, J. B. (2001). PCR amplification of putative *gpa-2* and *gpa-3* orthologs from the (A + T)-rich genome of *Strongyloides stercoralis*. *International Journal for Parasitology* **31**, 377-383.

MATTHEWS, J. B., DAVIDSON, A. J., FREEMAN, K. L. & FRENCH, N. P. (2001). Immunisation of cattle with recombinant acetylcholinesterase from *Dictyocaulus viviparus* and with adult worm ES products. *International Journal for Parasitology* **31**, 307-317.

McKEAND, J. B., KNOX, D. P., DUNCAN, J. L. & KENNEDY, M. W. (1994). The immunogenicity of the acetylcholinesterases of the cattle lungworm *Dictyocaulus viviparus*. *International Journal for Parasitology* **24**, 501-510.

McKEAND, J. B., KNOX, D. P., DUNCAN, J. L. & KENNEDY, M. W. (1995). Immunisation of guinea pigs against *Dictyocaulus viviparus* using adult ES products enriched for acetylcholinesterases. *International Journal for Parasitology* **25**, 829-837.

MICHALSKI, M. L. & WEIL, G. J. (1999). Gender-specific gene expression in *Brugia malayi*. *Molecular and Biochemical Parasitology* **104**, 247-257.

MILLER, M. A., NGUYEN, V. Q., LEE, M. H., KOSINSKI, M., SCHEDL, T., CAPRIOLI, R. M. & GREENSTEIN, D. (2001). A sperm cytoskeletal protein that signals oocyte meiotic maturation and ovulation. *Science* **291**, 2144-2147.

MILSTONE, A. M., HARRISON, L. M., BUNGIRO, R. D., KUZMIC, P. & CAPPELLO, M. (2000). A broad spectrum Kunitz type serine protease inhibitor secreted by the hookworm *Ancylostoma ceylanicum*. *Journal of Biological Chemistry* **275**, 29391-29399.

MOORE, J. & DEVANEY, E. (1999). Cloning and characterization of two nuclear receptors from the filarial nematode *Brugia pahangi*. *Biochemical Journal* **344**, 245-252.

MOORE, J., TETLEY, L. & DEVANEY, E. (2000). Identification of abundant mRNAs from the third stage larvae of the parasitic nematode, *Ostertagia ostertagi*. *Biochemical Journal* **347**, 763-770.

MOYLE, M., FOSTER, D. L., McGRATH, D. E., BROWN, S. M., LAROCHE, Y., MEUTTER, J. D., STANSSENS, P., BOGOWITZ, C. A., FRIED, V. A. & ELY, J. A. (1994). A hookworm glycoprotein that inhibits neutrophil function is a ligand of the integrin CD11b/CD18. *Journal of Biological Chemistry* **269**, 10008-10015.

NELSON, G. A. & WARD, S. (1981). Amoeboid motility and actin in *Ascaris lumbricoides* sperm. *Experimental Cell Research* **131**, 149-160.

NEWLANDS, G. F. J., SKUCE, P. J., KNOX, D. P. & SMITH, W. D. (2001). Cloning and expression of cystatin, a potent cysteine protease inhibitor from the gut of *Haemonchus contortus*. *Parasitology* **122**, 371-378.

NEWTON, S. E. & MUNN, E. A. (1999). The development of vaccines against gastrointestinal nematode parasites, particularly *Haemonchus contortus*. *Parasitology Today* **15**, 116-122.

NGO, H., TSCHUDI, C., GULL, K. & ULLU, E. (1998). Double-stranded RNA induces mRNA degradation in *Trypanosoma brucei*. *Proceedings of the National Academy of Sciences of the United States of America* **95**, 14687-14692.

NORTHERN, C. & GROVE, D. I. (1990). *Strongyloides stercoralis*: antigenic analysis of infective larvae and adult worms. *International Journal for Parasitology* **20**, 381-387.

PASTRANA, D. V., RAGHAVEN, N., FITZGERALD, P., EISINGER, S. W., METZ, C., BUCALA, R., SCLEIMER, R. P., BICKEL, C. & SCOTT, A. L. (1998). Filarial nematode parasites secrete a homologue of the human cytokine macrophage migration inhibitory factor. *Infection and Immunity* **66**, 5955-5963.

PENNOCK, J. L., BEHNKE, J. M., BICKLE, Q. D., DEVANEY, E., GRENCIS, R. K., ISAAC, R. E., JOSHUA, G. W. P., SELKIRK, M. E., ZHANG, Y. & MEYER, D. J. (1998). Rapid purification and characterization of L-dopachrome-methyl ester tautomerase (macrophage-migration-inhibitory factor) from *Trichinella spiralis*, *Trichuris muris* and *Brugia pahangi*. *Biochemical Journal* **335**, 495-498.

PRATT, D., ARMES, L. G., HAGEMAN, R., REYNOLDS, V., BOISVENUE, R. J. & COX, G. N. (1992a). Cloning and sequence comparisons of four distinct cysteine proteases expressed by *Haemonchus contortus* adult worms. *Molecular and Biochemical Parasitology* **51**, 209-218.

PRATT, D., BOISVENUE, R. J. & COX, G. N. (1992b). Isolation of putative cysteine protease genes of *Ostertagia ostertagi*. *Molecular and Biochemical Parasitology* **56**, 39-48.

PRITCHARD, D., LEGGETT, K. V., ROGAN, M. T., McKEAN, P. G. & BROWN, A. (1991). *Necator americanus* secretory acetylcholinesterase and its purification from excretory-secretory products by affinity chromatography. *Parasite Immunology* **13**, 187-199.

PRITCHARD, D. I., BROWN, A. & TOUTANT, J. P. (1994). The molecular forms of acetylcholinesterase from *Necator americanus* (Nematoda), a hookworm parasite of the human intestine. *European Journal of Biochemistry* **219**, 317-323.

PROUDFOOT, L., KUSEL, J. R., SMITH, H. V., HARNETT, W., WORMS, M. J. & KENNEDY, M. W. (1993). Rapid changes in the surface of parasitic nematodes during transition from pre- to post-parasitic forms. *Parasitology* **107**, 107-117.

RAGHAVEN, N., FREEDMAN, D. O., FITZGERALD, P. C., UNNASCH, T. R., OTTESEN, E. A. & NUTMAN, T. B. (1994). Cloning and characterization of a potentially protective chitinase-like recombinant antigen from *Wuchereria bancrofti*. *Infection and Immunity* **62**, 1901-1908.

RAGHAVEN, N., GHOSH, I., EISINGER, W. S., PASTRANA, D. & SCOTT, A. L. (1999). Developmentally regulated expression of a unique small heat shock protein in *Brugia malayi*. *Molecular and Biochemical Parasitology* **104**, 233-246.

REDMOND, D. L., KNOX, D. P., NEWLANDS, G. & SMITH, W. D. (1997). Molecular cloning and characterisation of a developmentally regulated putative metallopeptidase present in a host protective extract of *Haemonchus contortus*. *Molecular and Biochemical Parasitology* **85**, 77-87.

REINKE, V., SMITH, H. E., NANCE, J., WANG, J., VAN DOREN, C., BEGLEY, R., JONES, S. J., DAVIS, E. B., SCHERER, S., WARD, S. & KIM, S. K. (2000). A global profile of germline gene expression in *C. elegans*. *Molecular Cell* **6**, 605-616.

RHOADS, M. L. (1984). Secretory cholinesterases of nematodes: possible functions in the host-parasite relationship. *Tropical Veterinarian* **2**, 3-10.

RHOADS, M. L., FETTERER, R. H. HILL, D. E. &URBAN, J. F. JR. (2000a). *Trichuris suis*: a secretory chymotrypsin/elastase inhibitor with potential as an immunomodulator. *Experimental Parasitology* **95**, 36-44.

RHOADS, M. L., FETTERER, R. H. & ROMANOWSKI, R. D. (2000b). A developmentally regulated hyaluronidase of *Haemonchus contortus*. *Journal of Parasitology* **86**, 916-921.

RIDDLE, D. L., BLUMENTHAL, T. MEYER, B.J. & PRIESS, J. R.. (1997). *C. elegans II*. Cold Spring Harbor Laboratory Press, Cold Spring Harbor.

RIEU, P., UEDA, T., HARUTA, I., SHARMA, C. P. & ARNAOUT, M. A. (1994). The A-domain of beta 2 integrin CR3 (CD11b/CD18) is a receptor for the hookworm-derived neutrophil adhesion inhibitor NIF. *Journal of Cell Biology* **127**, 2081-2091.

RIFFKIN, M., SEOW, H. F., JACKSON, D., BROWN, L. & WOOD, P. (1996). Defence against the immune barrage: helminth survival strategies. *Immunology and Cell Biology* **74**, 564-574.

ROOS, M. H. & TIELENS, A. G. (1994). Differential expression of two succinate dehydogenase subunit-B genes and a transition in energy metabolism during the development of the parasitic nematode *Haemonchus contortus*. *Molecular and Biochemical Parasitology* **66**, 273-281.

ROTHSTEIN, N. & RAJAN, T. V. (1991). Characterization of an hsp70 gene from the human filarial parasite, *Brugia malayi* (Nematoda). *Molecular and Biochemical Parasitology* **49**, 229-237.

SCHALLIG, H. D. F. H., VAN LEEUWEN, M. A. W., VERSTREPEN, B. E. & CORNELISSEN, A. W. C. A. (1997). Molecular characterization and expression of two putative protective excretory secretory proteins of *Haemonchus contortus*. *Molecular and Biochemical Parasitology* **88**, 203-213.

SCOTT, A. L., DINMAN, J., SUSSMAN, D. J. & WARD, S. (1989a). Major sperm protein and actin genes in free-living and parasitic nematodes. *Parasitology* **98**, 471-478.

SCOTT, A. L., DINMAN, J., SUSSMAN, D. J., YENBUTR, P. & WARD, S. (1989b). Major sperm protein genes from *Onchocerca volvulus*. *Molecular and Biochemical Parasitology* **36**, 119-126.

SHI, H., DJIKENG, A., MARK, T., WIRTZ, E., TSCHUDI, C. & ULLU, E. (2000). Genetic interference in *Trypanosoma brucei* by heritable and inducible double-stranded RNA. *RNA* **6**, 1069-1076.

SKUCE, P. J., STEWART, E. M., SMITH, W. D. & KNOX, D. P. (1999). Cloning and characterization of glutamate dehydrogenase (GDH) from the gut of *Haemonchus contortus*. *Parasitology* **118**, 297-304.

SMITH, T. S., GRAHAM, M., MUNN, E. A., NEWTON, S. E., KNOX, D. P., COADWELL, W. J., McMICHAEL-PHILLIPS, D., SMITH, H., SMITH, W. D. & OLIVER, J. J. (1997). Cloning and characterization of a microsomal aminopeptidase from the intestine of the nematode *Haemonchus contortus*. *Biochimica et Biophysica Acta* **1338**, 295-306.

SPIETH, J., NETTLETON, M., ZUCKER-APRISON, E., LEA, K. & BLUMENTHAL, T. (1991). Vitellogenin motifs conserved in nematodes and vertebrates. *Journal of Molecular Evolution* **32**, 429-438.

STANSSENS, P., BERGUM, P. W., GANSEMANS, Y., JESPERS, L., LAROCHE, Y., HUANG, S., MAKI, S., MESSENS, J., LAUWEREYS, M., CAPPELLO, M., HOTEZ, P. J., LASTERS, I. & VLASUK, G. P. (1996). Anticoagulant repertoire of the hookworm *Ancylostoma caninum*. *Proceedings of the National Academy of Sciences of the United States of America* **93**, 2149-2154.

STINCHCOMB, D. T., SHAW, J. E., CARR, S. H. & HIRSH, D. (1985). Extrachromosomal DNA transformation of *Caenorhabditis elegans*. *Molecular and Cell Biology* **5**, 3484-3496.

STOREY, N. & PHILIPP, M. (1992). *Brugia malayi*: patterns of expression of surface-associated antigens. *Experimental Parasitology* **74**, 57-68.

TABARA, H., GRISHOK, A. & MELLO, C. C. (1998). RNAi in *C. elegans*: soaking in the genome sequence. *Science* **282**, 430-431.

TAIWO, F. A., BROPHY, P. M., PRITCHARD, D. I., BROWN, A., WARDLAW, A. & PATTERSON, L. H. (1999). Cu/Zn superoxide dismutase in excretory-secretory products of the human hookworm *Necator americanus*. An electron paramagnetic spectrometry study. *European Journal of Biochemistry* **264**, 434-438.

TALVIK, H., CHRISTENSEN, C. M., JOACHIM, A., ROEPSTORFF, A., BJORN, H. & NANSEN, P. (1997). Prepatent periods of different *Oesophagostomum* spp. isolates in experimentally infected pigs. *Parasitology Research* **83**, 563-568.

THE *C. ELEGANS* SEQUENCING CONSORTIUM. (1998). Genome sequence of the nematode *C. elegans*: a platform for investigating biology. *Science* **282**, 2012-2018.

TSUJI, N., KASUGA-AOKI, H., ISOBE, T. & YOSHIHARA, S. (2000). Cloning and characterization of a peroxiredoxin from the swine roundworm *Ascaris suum*. *International Journal for Parasitology* **30**, 125-128.

TWEEDIE, S., GRIGG, M. E., INGRAM, L. & SELKIRK, M. E. (1993). The expression of a small heat shock protein homologue is developmentally regulated in *Nippostrongylus brasiliensis*. *Molecular and Biochemical Parasitology* **61**, 149-154.

VON EGGELING, F., DAVIES, H., LOMAS, L., FIEDLER, W., JUNKER, K., CLAUSSEN, U. & ERNST, G. (2000). Tissue-specific microdissection coupled with ProteinChip® array technologies: applications in cancer research. *BioTechniques* **29**, 1066-1070.

WARD, S., BURKE, D. J., SULSTON, J. E., COULSON, A. R., ALBERTSON, D. G., AMMONS, D., KLASS, M. & HOGAN, E. (1988). Genomic organization of major sperm protein genes and pseudogenes in the nematode *Caenorhabditis elegans*. *Journal of Molecular Biology* **199**, 1-13.

WILLIAMS, S. A. & JOHNSTON, D. A. (1999). Helminth genome analysis: the current status of the filarial and schistosome genome projects. Filarial Genome Project. Schistosome Genome Project. *Parasitology* **118** (Suppl.) S19-S38.

WINKFEIN, R. J., PASTERNAK, J., MUDRY, T. & MARTIN, L. H. (1985). *Ascaris lumbricoides*: characterization of the collagenous components of the adult cuticle. *Experimental Parasitology* **59**, 197-203.

WU, Y., ADAM, R., WILLIAMS, S. A. & BIANCO, A. E. (1996). Chitinase genes expressed by infective larvae of the filarial nematodes, *Acanthocheilonema viteae* and *Onchocerca volvulus*. *Molecular and Biochemical Parasitology* **75**, 207-219.

YENBUTR, P. & SCOTT, A. L. (1995). Molecular cloning of a serine protease inhibitor from *Brugia malayi*. *Infection and Immunity* **63**, 1745-1753.

ZANG, X., YAZDANBAKHSH, M., JIANG, H., KANOST, M. R. & MAIZELS, R. M. (1999). A novel serpin expressed by blood-borne microfilariae of the parasitic nematode *Brugia malayi* inhibits human neutrophil serine proteinases. *Blood* **94**, 1418-1428.

Chapter 15
SCHISTOSOMIASIS AND REDUCED RISK OF ATOPIC DISEASES: NEW INSIGHTS AND POSSIBLE MECHANISMS

Anita H.J. van den Biggelaar and Maria Yazdanbakhsh, Department of Parasitology, Leiden University Medical Center, The Netherlands.
e-mail: A.H.J.van_den_Biggelaar@lumc.nl

1. INTRODUCTION

Infections with parasitic helminths and allergy are immunologically characterised by a skewing of the cellular immune response towards dominance by T helper type 2 (Th2) cells. The most evident sign of this is IgE antibody specific to the parasite or environmental allergen concerned, accompanied by high levels of apparently non-specific IgE. This has led to the seemingly antithetical ideas that helminth infection may exacerbate allergic reactions by enhancing IgE responses, or counter them by competition between total IgE and parasite-specific IgE with IgE specific for environmental allergens for mast cell activation. This area is currently under intense investigation as much for what it might tell us about the dramatic increase in atopic responses that has occurred over recent decades as for understanding the development of T cell immune biases in the human immune response as a whole. Despite the fact that schistosomes are neither nematodes nor intestinal and thus outside the scope of this book, new findings on schistosomiasis have greatly informed the helminth/allergy debate, and how Th2 biases arise in the situation for which they are assumed to have evolved resistance to macroparasites. With this in mind, therefore, we discuss immune responses in individuals infected with schistosomes, and the suppressive effect such infections may have on allergic diseases. We will discuss how Th2-skewed schistosome infected individuals not only do not develop, but even appear to have a reduced risk of, developing allergic diseases. We argue that a strong immunoregulatory network that develops upon persistent antigenic stimulation that helminths provide might play an important role in suppressing atopic reactions in individuals with chronic helminth infections.

2. THE ASSOCIATION BETWEEN PARASITES AND ALLERGY

It is a commonly held view that IgE and Th2 responses evolved to combat helminth infections. As allergic diseases are associated with the expression of Th2 immune responses, it is possible that these diseases are a negative consequence of an evolutionary benefit of an anti-helminth response. Allergic responses are induced, prolonged, and amplified by Th2 cells secreting IL-4, IL-5 and IL-13 (Robinson *et al.* 1992; Umetsu & DeKruyff, 1997, 1999). Inflammatory cells, particularly eosinophils, basophils and mast cells, which bind to specific IgE and respond to incoming environmental allergens, characterize the harmful responses in allergic diseases.

Although genetic factors must influence the development of atopy and allergic diseases, only environmental factors could explain the recent dramatic rise in these diseases. International studies have indicated that allergic diseases are increasing in industrialized countries (International Study of asthma and Allergies in Childhood, 1998). In addition, there are clear differences in the prevalence of allergies between rural and urban areas within one country (Yemaneberhan *et al.* 1997). To explain these observations, environmental factors associated with a more industrialized/urban living have been studied intensively. Possible candidates include the rise in air pollution, increased exposure to indoor allergens, changes in diet, and changed patterns of microbial exposure. An overview of literature covering such factors is given in Table 15.1.

Much attention has been paid to the association between the decline in infectious diseases, due to improved hygiene and successful vaccination programs, and the development of allergy. First evidence supporting this negative association has come from studies showing that allergic sensitization is more frequent in children from small families and is over represented among the first born (Strachan, 1989; Jarvis *et al.* 1997). This suggests that the frequent exchange of infections may have a protective effect. More recent studies have shown a negative association between the development of allergy and exposure to infectious diseases such as Hepatitis A (Matricardi *et al.* 1997), measles (Shaheen *et al.* 1996) and tuberculosis

(Shirakawa, *et al.* 1997). Since bacterial and viral infections are characterized by a skewing of the T-helper response toward type 1, the 'hygiene hypothesis' has been explained by the counterbalance between Th1 and Th2. It has been proposed that infections early in life that stimulate the Th1-arm of the immune system, suppress the development of Th2-immune responses and thus allergic disease later in life.

3. SCHISTOSOMIASIS

Schistosomiasis, is a parasitic disease caused by infection with trematodes belonging to the genus *Schistosoma*. The life cycle of the parasite is maintained in snails as intermediate hosts and in humans as definitive hosts. Infection with *S. haematobium* is clinically associated with urinary symptoms such as hematuria and obstructive uropathy, whereas infections with *S. mansoni* and *S. japonicum* often result in intestinal symptoms and hepatosplenic disease. The pathological consequences of infection are often not due to adult worms, but due to eggs that get trapped in tissues and cause inflammatory reactions, resulting in bladder wall irregularity and obstruction, and sometimes bladder cancer in urinary schistosomiasis, and liver fibrosis and portal hypertension in intestinal schistosomiasis (Gutierrez, 2000). Worldwide, about 200 million people are thought to be infected, of which an estimated 120 million are symptomatic. Approximately 20 million are considered to suffer from severe consequences of the infection (Report of WHO Informal Consultation on Schistosomiasis Control, 1999). Another current issue is that individuals infected with *S. haematobium* have been shown to have an increased risk of carrying HIV infection (Bichler *et al.* 2001).

4. ACQUIRED IMMUNITY: PARASITE CLEARANCE BY SPECIFIC IgE

From several population-based studies it has become clear that both the prevalence and intensity of infection rise during the first two decades of life, and thereafter decline in young adults. Since this pattern cannot be explained by changes in levels of exposure, it has been postulated that immunity to incoming parasites is acquired with increasing length of exposure. In search for immunological correlates of immunity, antibody isotypes have been

Table 15.1. Overview of studies describing different factors that might be associated with the rise in allergy.

	Factor	Outcome	Effect	Evoked response	(Hypothetical) involved mechanism	Study area
Diet	Vitamin C – E	BHR/Wheeze	- ve	Th1	Th1 / Th2	UK[1] Saudi-Arabia [2]
	Polyunsaturated fatty acids (ω-6)	BHR/Asthma	+ ve	Th2	Th1 / Th2	Australia[3,4] Europe [5] Saudi-Arabia [2]
	Sodium	BHR	+ ve	Hyperpolerization of smooth muscle	Na+/K+ -ATPase reactivity/ Na+ pump	Canada [6] UK [7]
Pollution	Airpollution (PM10;SO2;NO2)	Bronchitis/ Asthma/ Chronic cough	+ ve	Inflammation	Oxidative stress (IL8; RANTES)	Germany [8] Hong Kong [9] UK [10] Israel [11]
	Tobacco smoke	Asthma/ Wheeze/ Chronic bronchitis	+ ve			Switzerland[12] Kenya [13] 'Western countries'[14]
	Road traffic (DEP)	Wheeze/ Pollen-sensitisation	+ ve	Inflammation Increased sensitisation	Oxidative stress (RANTES;MIP1-α; MCP3) Allergen-carrier	Switzerland [15-17]
Lifestyle	Obesity	SPT/ BHR/ Wheeze/ Asthma	+ ve	Upregulated Th2/ dysregulated airway tone	IL-6 (85) / female hormone (leptin?)	Taiwan [18] Finland [19] UK [20. 21] US[22-25]

	Factor	Outcome	Effect	Evoked response	(Hypothetical) involved mechanism	Study area
Infection - directly	Schistosomiasis	SPT	- ve	Th2	IL-10 Total IgE	Gabon[26] Brazil [27]
	Ascaris	SPT/ IgE/ Asthma	- ve	Th2	Total IgE	Venezuela[28] Papua New Guinea[29]
	Measles	SPT/ Asthma/ Hay fever	- ve	Th1	Th1 / Th2	Guinea Bissau[30] UK[31]
		Eczema/Asthma/ Rhinitis	+ ve	Th1		Finland[32]
	Rubella	Asthma	- ve	Th1		New-Zealand[33]
Infection-serology/ Question-naires	*Ascaris*	IgE/ Asthma/ Rhinitis/	+ ve	Th2		East Germany [34]
		Asthma	- ve	Th2		Ethiopia[35]
	Necator	Asthma	- ve	Th2		Ethiopia[35]
	Toxocara	IgE/ Asthma/ Bronchitis	+ ve	Th2		Netherlands[36] Malaysia[37] Egypt[38]
	Myobacterium tuberculosis	IgE/ Asthma/ Rhinocon-junctivitis	- ve	Th1	Th1 / Th2	Japan[39] Finland[40] 'Industrialized areas'[41]

	Factor	Outcome	Effect	Evoked response	(Hypothetical) involved mechanism	Study area
Infection-serology/ Question-naires (continued)	'BCG'	SPT/ IgE	- ve	Th1	Th1 / Th2	Guinea Bissau [42] Norway[43]
			=			Sweden[44]
	rofecal Hepatitis A; *oxoplasma gondii; elicobacter pylori*)	SPT/ IgE/ Asthma/Rhino-conjunctivitis/ Chronic cough	- ve	Th1	Th1 / Th2	Italy[45, 46] UK[47]
	'Viral' infection (runny nose; herpes)	Asthma/BHR	- ve	Th1	Th1/Th2	Germany[48]
Indirectly infection	Family size	SPT/ IgE/ Hay fever/ Asthma/ Wheeze/Rhino-conjunctivitis	- ve	Possible Th1	Th1/ Th2	Italy[49,50] UK[51-53] New Zealand[33] USA [54]
	Childcare nursery	SPT/Asthma/ Wheeze	-ve	Possible Th1	Th1/Th2	USA [54] Germany[55]
	Anthroposophic	SPT/IgE	- ve	Possible Th1	Th1/ Th2	Sweden[56]
	Urbanization	Wheeze/ Asthma	+ ve	Possible Th1	Th1 / Th2	Ethiopia[57] Kenya[13] Tanzania[58] Australia-Nigeria[59] Hong Kong-China-Malaysia[60]

	Factor	Outcome	Effect	Evoked response	(Hypothetical) involved mechanism	Study area
Indirectly infection (continued)	Farming	SPT/ IgE/ Wheeze Hay fever/ Asthma/ Rhino-conjunctivitis	- ve	Possible Th1	Th1/ Th2	Finland[61] Germany[62] Austria[63] Switzerland [64]
	Antibiotic use	Asthma	+ ve	Reduced Th1/ Reduction gutmicroflora	Th1/Th2	New Zealand[65]
Bacteriae	Endotoxin	SPT	- ve	Th1	Th1 / Th2	US[66]
	Gut microflora (probiotics)	SPT/Atopic eczema	- ve	Th1/Anti-inflammatory	Th1/Th2 and/or TGFβ – IL10	Sweden[67] Finland[68]
Infections of lower respiratory tract	Bronchiolitis/ Pneumonia	Asthma/SPT/ IgE/Wheeze	+ ve	Reduced Th1	Th1/Th2	Canada[69] Sweden[70]
		Asthma/ Wheeze, But SPT	+ ve =		Tissue remodeling/ Predisposition	Germany[48] UK[50] US[51]
		Wheeze/SPT	=		Tissue remodeling/ Predisposition	US[52]

Overview of studies describing different factors that might be associated with the rise in allergy. The column entitled 'outcome' indicates which outcome of atopy and/or clinical allergy is studied. BHR is an abbreviation for bronchial hyperresponsiviness, IgE stands for levels of allergen-specific IgE, SPT for skin test reactivity. In the 'effect column' the indication '+ ve' stands for the studied factor having an exacerbating effect on the allergic outcome, '- ve' for the factor having a preventive influence, and = indicates that the studied factor had no significant effect on the outcome.

The numbers in the 'Study area' column refer to references, which are listed separately at the end of the main reference list.

analyzed and levels of parasite-specific IgE were found to increase with age, coinciding with the decline of infection during adolescence (Hussain *et al.* 1983; Ndhlovu *et al.* 1996). In humans, investigations on acquisition of immunity have been based on 'reinfection studies' which involve chemotherapy to clear resident worms, followed by comparative studies of immune responses in subjects who become reinfected ('susceptibles') and those who remain uninfected ('resistants'). In such studies high levels of parasite-specific IgE were found to be associated with low intensities of reinfection (Hagan *et al.* 1991; Dunne *et al.* 1992). Taken together, these findings have led to the conclusion that specific IgE antibodies are important mediators of immunity, and Th2-responses in schistosome infections should therefore be considered to be protective (Dunne *et al.* 1995).

In experimental models of schistosomiasis, when mice are infected with cercaria, the early responses to infection are predominantly of the Th1-type, but shift towards Th2 after maturation of the worms and onset of egg deposition (Sher *et al.* 1992; Sher & Coffman, 1992). The question of immunity in murine models has been addressed by vaccination and challenge experiments that show Th1 responses to be protective (Wilson *et al.* 1996). However, so far only one study has questioned concomitant immunity in the murine model, showing that IL-4 is a major player, which supports human immuno-epidemiological studies (Brunet, Kopf & Pearce, 1999).

The concept of immunity in human schistosomiasis has also been addressed in communities that have recently become exposed to this infection. Studies in Senegal and Kenya, in populations where both children and adults have been exposed to schistosomes for an equal length of time, have provided evidence that the acquisition of immunity is associated with an age-related factor rather than with the length of exposure to infection (Stelma *et al.* 1993; Ouma *et al.* 1998). So far the analysis of IgE antibodies in newly exposed communities has shown that levels of specific IgE increase with age and not with length of exposure, and are in fact associated with infection intensity, casting doubt on whether IgE antibodies play an active role in protective immunity (Van Dam *et al.* 1996; Naus *et al.* 1999).

5. SPECIFIC HYPORESPONSIVENESS IN CHRONIC INFECTIONS: IL-10 IN THE LIMELIGHT

Despite some evidence for Th2-related protective immune responses that are associated with parasite clearance, adult worms seem able to survive in the Th2-skewed human host for many years. Although the question of how schistosomes persist for many years in the face of often strong Th2 responses, has yet to be answered, an active suppression of the immune effectors needs to be considered. Indeed, it has been shown that in many, but not in all, schistosome infected humans, proliferation and production of cytokines by Th1- and Th2- cells may be depressed (De Jesus *et al.* 1993). The mechanism of loss of parasite-specific T-cell responses are not fully understood but as neutralization of regulatory cytokines can reverse some of the T cell reactivities, anergy rather than deletion may account for the observed hyporesponsiveness (Grogan *et al.* 1998a). In addition, removal of the parasites by chemotherapy often results in elevated T cell responses, vouching for active suppression of cellular immunity by parasites (Grogan *et al.* 1998b). With respect to suppression of Th1-responses in helminth infections, an important role has been attributed to the immunosuppressory cytokine IL-10 (Sher *et al.* 1991; King *et al.* 1993). Whether IL-10 can act directly on Th2 cells to suppress cytokine production is not clear yet. In murine models, Th2 responses (to *Aspergillus fumigatus* antigen) have indeed been shown to be prone to suppression by IL-10 (Grunig *et al.* 1997). In humans there is no direct evidence for such activity but it is possible that IL-10 interferes with mediators controlled by Th2 cells (Del Prete *et al.* 1993); suppression of eosinophil and mast cell function (Takanaski *et al.* 1994; Ohkawara *et al.* 1996; Royer *et al.* 2001) or inhibition of IgE production while stimulating IgG4 (Jeannin *et al.* 1998).

6. IgG4 ANTIBODIES IN SCHISTOSOMIASIS

T helper type-2 cells, through production of IL-4, stimulate B cells to produce not only IgE but also IgG4 (Lundgren *et al.* 1989; Armitage *et al.* 1993). The valency of IgG4 does not allow any Fc receptor cross-linking when this isotype binds antigen and therefore does not result in immune activation. The similarity in recognition of antigens by IgE and IgG4 implies that IgG4 can compete with IgE for binding to antigens and therefore can act as a blocking antibody for IgE function (Boctor & Peter, 1990). An inverse association between IgE and IgG4 has been found in schistosomiasis (Hagan *et al.* 1991; Demeure *et al.* 1993). Levels of IgE are found to be highest in young adults in whom intensities of infection are declining, whereas levels of parasite-specific IgG4 antibodies appear to peak in 10-14 years old subjects, an age group with the highest intensities of infection (Dunne *et al.* 1992; Hagan *et al.* 1991). In reinfection studies levels of specific IgG4 were found to be low in 'resistant' subjects, but high in 'susceptibles' (Demeure *et al.* 1993; Roberts *et al.* 1993). Although both IgG4 and IgE production are dependent on IL-4, the finding that these two isotypes are differentially expressed in schistosomiasis indicates that other factors may dissociate IgG4 from IgE. It was known that IL-12 and IFN-γ can suppress IgE production while leaving IgG4 unaffected (Kim *et al.* 1997; De Boer *et al.* 1997). However, more recently, IL-10 was shown to enhance IgG4 while suppressing IgE (Jeannin *et al.* 1998), which provides a plausible model whereby IL-10 elevated in chronic helminth infections can result in the observed amplification of the IgG4 response and allow parasite survival.

7. IgE, CYTOKINE RESPONSES AND CLINICAL MANIFESTATIONS: 'THE GOOD AND THE BAD'

In filariasis patients it has been observed that IgE antibodies are high in so-called 'endemic normals' who are resistant to the infections, but are also very high in elephantiasis patients who suffer from severe immunopathology (Yazdanbakhsh *et al.* 1993). In murine schistosomiasis it has been shown that Th2 cell responses are involved in granuloma formation and fibrosis (Wynn *et al.* 1995; Kaplan *et al.* 1998), and although clear evidence is lacking, it remains possible that in human schistosomiasis effector mechanisms associated with Th2 responses and especially IgE might be involved in pathogenesis. Therefore, immunosuppression during an infection may benefit not only the parasite, but also the host (Hoffmann Cheever & Wynn, 2000; Falcao *et al.* 1998).

This is exemplified in schistosomiasis when considering different manifestations of an infection in an endemic area. In parallel to filariasis, also in schistosomiasis, three clinical groups may be distinguished. The first group consists of individuals that are free of infection, despite living in an endemic area. It appears that such 'endemic normals' are capable of clearing parasites and resisting incoming infections. However, in comparison with filariasis, this group constitutes a very small percentage of the exposed population. Immunologically, these people are typified by having high proliferative T-helper cell responses to both egg- and worm antigens, producing high levels of IFN-γ and showing high levels of specific IgE but low IgG4 (Viana *et al.* 1994); this would predict that both Th1- and Th2-related effector mechanisms are involved.

Most people living in an endemic area belong to the second group that harbor (sometimes very high intensities of) infections, but do not show any severe clinical symptoms. In these chronically infected subgroups the production of both antigen-specific IL-5 and IFN-γ is decreased. Neutralization of IL-10 during *in vitro* culture of PBMC from chronically infected subjects restores the specific Th-cell proliferation and IFN-γ production (Grogan *et al.* 1998a; Sher *et al.* 1991; Araujo *et al.* 1996; Wynn *et al.* 1998; Montenegro *et al.* 1999). At serological level, these chronically infected individuals show relatively high levels of specific IgG4 and rather low levels of IgE, with high IgG4 presumptively 'blocking' harmful IgE-associated immune responses (Boctor & Peter, 1990).

The last subgroup consists of subjects that suffer from severe clinical symptoms. The immune response in these symptomatic patients is complex. Firstly, these individuals show high levels of IL-4 and IL-5, IL-13 and IgE, and high numbers of eosinophils, corresponding with their high capacity to clear parasites (Medhat *et al.* 1998). The pathologic response in this symptomatic subgroup is thus postulated to be associated with a failure to downregulate vigorous egg-induced Th2 responses (Williams *et al.* 1994). In baboons, however, it was found that granulomas were smaller after reinfection than during primary infection, and that this reduction was associated with an enhanced Th2 response (Mola *et al.* 1999). This finding argues against the association between Th2 responses and pathology, at least in baboons. Another study has shown that hepatosplenic patients produce low levels of IL-5 but high levels of IFN-γ and TNF-α (Mwatha *et al.* 1998), the latter suggesting that proinflammatory cytokines might be involved in pathogenesis (Allen & Maizels, 1996). Genetic studies of schistosomiasis patients showed that pathology was related to a major locus very close to the gene encoding the α-chain of the IFN-γ receptor, which again argues for the involvement of proinflammatory cytokines (Mohamed-Ali *et al.*1999; Dessein *et al.* 1999). Indeed, in murine studies it has been shown that granuloma formation is worsened in mice deficient for IL-10 (Wynn *et al.* 1998), suggesting that parasite induced IL-10 is likely to play an important role in controlling immunopathology in schistosomiasis (Falcao *et al.* 1998; Mola *et al.* 1999).

8. HELMINTH INFECTIONS REFUTE THE Th1/Th2 PARADIGM OF THE HYGIENE HYPOTHESIS

A number of observations do not support the immunological explanations of the hygiene hypothesis, which centers on the Th1/Th2 paradigm. First, there has been a number of recent studies which show that the prevalence of autoimmune diseases such as diabetes is also increasing in industrialized countries (Onkamo *et al.* 1999; Tedeschi & Airaghi, 2001). Given that diabetes is a Th1 disease, it has to be concluded that not only Th2 but also Th1 diseases are increasing in prevalence and that lack of childhood infections is unlikely to affect the immune system in terms of Th1/Th2 balance. Second, several studies show that Th1 responses may also be associated with airway inflammation. In animal models of asthma, adoptive transfer of Th1 cells exacerbates airway inflammation (Randolph *et al.* 1999a,b; Hansen *et al.* 1999). In asthmatic subjects, high levels of IFN-γ have been reported (Ten Hacken, 1998) which support the view that not only Th2 but also Th1 responses may be detrimental in allergic diseases that involve inflammation. Finally, evidence from the field of parasitology involving helminths, argues against a simple skewing of Th2 responses being responsible for increase in allergic diseases. Helminth infected subjects with prominent Th2 skewing, do not develop allergic diseases. In fact, a number of studies have shown that intestinal helminths as well as schistosomes can be protective (Araujo *et al.* 2000; Godfrey, 1975; Lynch *et al.* 1993; Van Den Biggelaar *et al.* 2001). Yet, when considering the hygiene hypothesis and the Th1/Th2 imbalance, one would expect that helminth infected subjects would be particularly at a high risk of developing allergy. As illustrated in Figure 15.1, helminth infected populations were found to produce high levels of IgE, not only to the parasite, but also to environmental allergens, indicating that these people are exposed and sensitized to allergens (Van den Biggelaar *et al.* 2001). The observed decrease of skin test reactivity against increasing levels of sensitization, would argue that suppression of allergic reactivity must act at a later point in the allergic cascade.

A suppressory role has been attributed to the high levels of polyclonal IgE produced during helminth infections. In the 'IgE blocking hypothesis' it is postulated that allergic reactivity in helminth-infected subjects is blocked due to the high levels of non-specific IgE antibodies competing with the low levels of allergen-specific IgE for binding to Fcε receptors on mast cells (Godfrey, 1975; Lynch *et al.* 1993). This would inhibit degranulation and immediate hypersensitivity responses to allergens. However, a number of

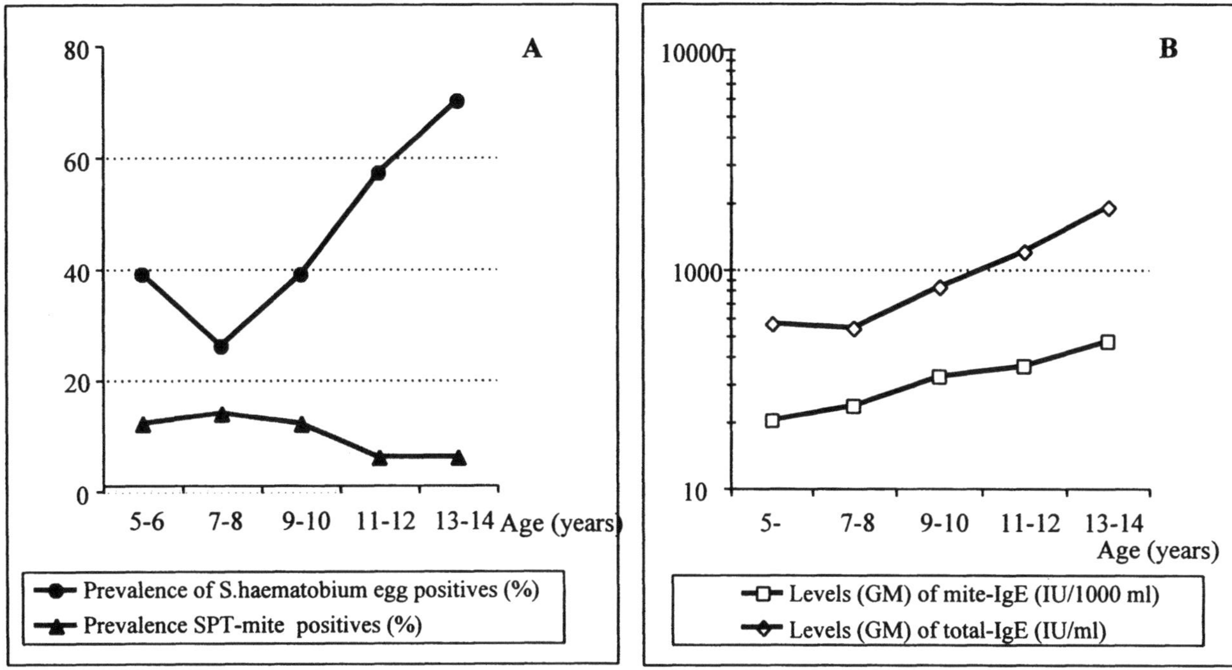

Figure 15.1. The inverse association between *S. haematobium* infection and SPT to house dust mite (A) and the production of total (IU/ml) and mite-specific IgE (IU/ml) (B), determined in 520 Gabonese schoolchildren and plotted against age. Levels of mite-specific IgE were found to rise with age whereas the percentage of children responding to mite in skin prick testing were decreased with increasing (Van den Biggelaar *et al* 2001).

arguments have been raised against this model. The number of Fcε receptor sites on mast cells appear to respond to the concentration of circulating IgE; direct evidence for this was obtained when clinical trials with anti-IgE antibodies were carried out (Saini *et al.* 1999). In addition, difficulties in functionally saturating Fcε receptors in *in vitro* experiments have been reported (Lynch *et al.* 1983). Moreover, in human studies, some studies have failed to demonstrate this negative association between allergy and total polyclonal IgE (Van den Biggelaar *et al.* 2000, 2001; Larrick *et al.* 1983).

9. SPILLOVER SUPPRESSION IN SCHISTOSOMIASIS

Chronic schistosome infections can induce a state of immune hyporesponsiveness, which is primarily directed at schistosome antigens but can also extend to non-parasite antigens. Early studies showed that delayed hypersensitivity to PPD was lower in schistosomiasis patients compared to control non-infected subjects (El-Kalouby *et al.* 1979) and more recent work on immunological responses to tetanus toxoid (TT) has shown that Th1 responses to TT are weaker in patients with schistosomiasis compared to healthy controls (Sabin *et al.* 1996). The suppressive effects of chronic helminth infections can even be transferred to a fetus in utero; children born to mothers infected with filarial worms or with schistosomes responded poorly to BCG vaccination (ten fold lower IFN-γ response to PPD) compared to children born to non infected mothers (Malhotra *et al.* 1999). These studies indicate that in chronic schistosomiasis suppression can spill over to unrelated antigens.

With this in mind, the question was asked whether schistosome-induced immunosuppression might also extend to responses mounted to allergens and therefore affect allergic diseases. In a study performed in an area in Gabon where *S. haematobium* is endemic, the first evidence to support such a proposition was obtained. Schoolchildren were screened for presence of helminth infections as well as atopic reactions to environmental allergens and cytokine responses were measured in PBMC stimulated with adult worm antigens. Antigen specific levels of IL-5, IL-10 and IL-13 were significantly elevated in schistosome-infected children, but multiple logistic regression analysis revealed that only the levels of parasite-specific IL-10 were significantly negatively associated with skin test reactivity to house dust mite (Van den Biggelaar *et al.* 2000): with increasing levels of parasite-specific IL-10 the risk of a positive skin test was reduced by almost 70%, independent of produced levels of mite-specific IgE. Thus, the amplification of the immunosuppressive cytokine in a chronic helminth infection which underlies hyporesponsiveness to parasite antigens, could also have a profound inhibitory effect on responses associated with allergy.

10. HOW PARASITE-INDUCED IL-10 MAY SUPPRESS ALLERGIC RESPONSES

It is already clear from immunotherapy studies that IL-10 is involved in downmodulating allergic responses. The administration of high doses of allergen at regular intervals was shown to result in an enhanced endogenous production of IL-10 by T-cells (Akdis *et al.* 1998). Immunotherapy was also shown to result in the increased production of allergen-specific IgG4 antibodies (Aalberse, van der Gaag & van Leeuwen, 1983) which, as discussed for parasite-specific IgG4 and IgE, may 'block' the IgE-mediated effector responses such as hypersensitivity reactions.

Interestingly, it was recently shown that intense exposure to cat allergens might be protective: children exposed to high levels of cat allergen produced increased levels of cat-specific IgG4 and slightly reduced IgE responses and were at a reduced risk for asthma (Platts-Mills *et al.* 2001). The same phenomenon was not observed for mite allergens; children exposed to increasing levels of mite were found to produce increasing levels of specific IgE. There is currently no explanation for this discrepancy. Nevertheless, the data on exposure to cat allergens indicate that with increasing allergen exposure a protective mechanism may be switched on. This so called 'alternative Th2' hypothesis which argues that under certain conditions Th2 responses will lead to the preferential production of IgG4 rather than IgE, may indeed involve IL-10, a cytokine that together with IL-4 enhances the production of IgG4 while suppressing IgE (Jeannin *et al.* 1998).

When considering the Gabon study, T cells may produce the parasite-induced IL-10. There is an increasing interest in the so-called regulatory T-cells, which release high levels of anti-inflammatory cytokines such as IL-10 and which are functionally characterized by their ability to suppress both Th1 and possibly Th2 cells (Jonuleit *et al.* 2000; Roncarolo, Levings & Traversari, 2001). Although currently not much is known about the involvement of dendritic cells (DC) on the outgrowth of regulatory T-cells, it has been suggested that a tolerogenic type of DC, the DC type 3, may play a mediating role (Jonuleit *et al.* 2001). Immature DC are polarized into type 1,

2 or 3, depending on their interaction with local concentrations of pathogen-derived or tissue-derived molecules. DC1 and DC2 are assumed to induce the polarization of naïve Th-cells into respectively Th1 and Th2 (Figure 15.2) (Kapsenbert *et al.* 1999, 2000; Kalinski *et al.* 1999) and a study with a well defined molecule derived from *A. vitaea* excretory-secretory antigens showed that helminth derived molecules can indeed act on DC and skew murine immune responses towards Th2 (Whelan *et al.* 2000). Considering schistosomes, it has recently been shown that immature DC cultured in the presence of soluble schistosome-egg antigens (SEA) gave rise to DC2 that subsequently polarize T cells into Th2 cells, indicating that these parasites carry molecules involved in the skewing of immune responses towards Th2 at the level of DC (De Jong *et al.* unpublished data). Since tolerance, and thus DC3, are supposedly the result of persistent antigenic challenge, it may be possible to find in the case of chronic schistosome infections parasite signature molecules that upon interaction with DC can program these cells to deliver signals to naïve T cells to differentiate into regulatory T cells. Such regulatory T cells may then possible also suppress IL-10 (Jonuleit *et al.* 2000). The expansion of this subset of T cells as a result of persistent antigenic challenge in schistosomiasis, may work to suppress allergen-specific Th1 and possibly Th2 cells. This might be due to a non-specific suppression by sheer abundance of such cells within a compartment where other antigens are encountered. Alternatively, schistosome antigens and allergens are known to share common epitopes (Thomas & Smith, 1999), therefore the effect could be specific involving regulatory T cells specific for cross-reactive epitopes.

The idea that IL-10 produced by T-cells is negatively associated with skin test reactivity is supported by a study showing that (in atopic Caucasian children) levels of mite-specific IL-10 produced by PBMC, were inversely associated with the wheal size in skin tests against mite (Macaubas *et al.* 1999).

Besides mechanisms involving regulatory T-cells, the local production of IL-10 in affected tissues should also be considered. With respect to allergic disease it has been shown that IL-10 is able to limit the survival of LPS-stimulated eosinophils and to reduce their cytokine production (Takanaski, *et al.* 1994). Moreover, lower levels of IL-10 were found in lungs of asthmatic patients compared to non-asthmatic controls (Takanashi, *et al.* 1999). Interesting new data have emerged that show IL-10 to be able to inhibit the release of histamine by activated human mast cells (Royer, *et al.* 2001), providing a plausible explanation for the observation that schistosome infected children, producing high levels of parasite-specific IL-10, react poorly in skin tests against mite despite the presence of mite-specific IgE (Figure 15.2). Moreover, the same study showed that IL-10 also inhibited the production and release of TNF-α and IL-8 from activated mast cells. TNF-α is an important inflammatory cytokine that is involved in mediating increased bronchial responsiveness. This cytokine enhances the synthesis of other pro-inflammatory cytokines, it upregulates adhesion molecules, increases the cytotoxic activity of eosinophils and amplifies monocyte activation (Galli & Costa, 1995). IL-8 released by mast cells is thought to recruit and induce the migration of other mast cells to the site of inflammation. Since the production of IL-5 by mast cells can not be inhibited by IL-10, it is postulated that IL-10 is involved in downregulating the early phase of asthma inflammation but not the late phase which is dependent on IL-5 (Keatings, *et al.* 1997). At the molecular level, in various cells, including mast cells, the translocation of NF-κB to the nucleus is important for the synthesis of TNF-α (Pelletier, *et al.* 1998) and IL-8 (Vlahopoulos, *et al.* 1999), whereas it is not involved in the transcription of the IL-5 gene (Tsuruta, *et al.* 1995). The ability of IL-10 to inhibit the NF-κB binding activity in human monocytes provides a possible mechanism whereby IL-10 can inhibit the production of certain number of pro-inflammatory cytokines (Schottelius, *et al.* 1999).

Taken together, anti-inflammatory cytokines may be the important key players that control allergic diseases by suppressing excessive inflammation. Childhood infections may be protective against allergic diseases by ensuring that immunostimulation occurs early and at high frequency, resulting in the establishment of a strong anti-inflammatory network that is a natural sequel of immune activation needed to resolve inflammation and limit damage to host tissues.

11. IMPLICATIONS OF THE 'ALTERNATIVE HYGIENE HYPOTHESIS'

The finding that in schistosome-infected individuals the risk of developing allergy is reduced by way of immuno-suppressory mechanisms, provides a new immunological way of looking at the hygiene hypothesis and has important implications for the development of therapeutic strategies. The established view assumes that inducing Th1 responses in allergic individuals

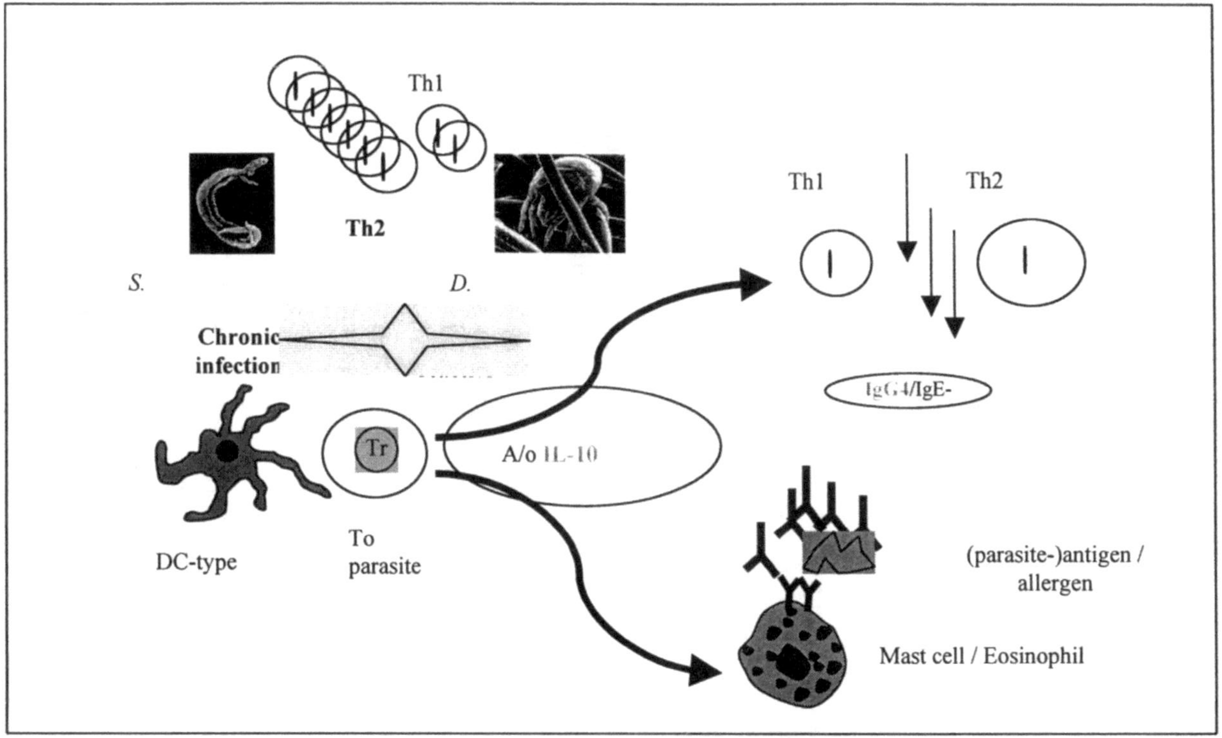

Figure 15.2. Mechanisms by which immunomodulation in chronic schistosomiasis might downregulate allergy. Chronic infection results in tolerance by inducing tolerogenic DC-type 3, regulatory T-cells (Tr) and high levels of IL-10. As a result, Th-cell responses are down-regulated. Possibly due to shared epitopes allergen-specific Th-cell responses might be down-regulated by parasite-specific Tr-cells. Also, high levels of IL-10 stimulate the production of IgG4 by B-cells; competition between IgG4 and IgE for parasite-Ag / allergen may lead to 'blocking' of effector responses. Moreover, IL-10 might act directly on effector cells and prevent the release of mediators such as histamine.

might oppose and thus suppress allergic-Th2 responses. The recent finding implies that it might be more appropriate to induce allergen-specific hyporesponsiveness by way of stimulating regulatory T-cells to produce high levels of suppressory cytokines such as IL-10.

REFERENCES

AALBERSE, R.C., VAN DER GAAG, R.. & VAN LEEUWEN, J. (1983). Serologic aspects of IgG4 antibodies. I. Prolonged immunization results in an IgG4-restricted response. *Journal of Immunology* **130**, 722-726.

AKDIS, C.A., BLESKEN, T., AKDIS, M., WUTHRICH, B. & BLASER, K. (1998). Role of interleukin 10 in specific immunotherapy. *Journal of Clinical Investigations* **102**, 98-106.

ALLEN, J.E. & MAIZELS, R.M. (1996). Immunology of human helminth infection. *International Archives of Allergy* **109**, 3-10.

ARAUJO, M.I., DE JESUS, A.R., BACELLAR, O., SABIN, E., PEARCE, E. & CARVALHO, E..M. (1996). Evidence of a T helper type 2 activation in human schistosomiasis. *European Journal of Immunology* **26**, 1399-1403.

ARAUJO, M.I., LOPES, A.A., MEDEIROS, M., CRUZ, A.A., SOUSA-ATTA, L., SOLE, D. & CARVALHO, E.M.. (2000). Inverse association between skin response to aeroallergens and schistosoma mansoni infection. *International archives of allergy and Immunology* **123**, 145-148.

ARMITAGE, R.J., MACDUFF, B.M., SPRIGGS, M.K. & FANSLOW, W.C. (1993). Human B cell proliferation and Ig secretion induced by recombinant CD40 ligand are modulated by soluble cytokines. *Journal of Immunology* **150**, 3671-3680.

BICHLER, K.H., FEIL, G., ZUMBRAGEL, A., EIPPER,, E. & DYBALLA, S. (2001). Schistosomiasis: a critical review. *Current Opinions in Urology* **11**, 97-101.

BOCTOR, F.N. & PETER, J,B,. (1990). IgG subclasses in human chronic schistosomiasis: over-production of schistosome-specific and non-specific IgG4.*Clinical and Experimental Immunology* **82**, 574-578.

BRUNET, L.R., KOPF, M.A. & PEARCE, E.J. (1999). Schistosoma mansoni: IL-4 is necessary for concomitant immunity in mice. *Journal of Parasitology* **85**, 734-736.

DE BOER, B.A., KRUIZE, Y.C., ROTMANS, P.J. & YAZDANBAKHSH, M. (1997) Interleukin-12 suppresses immunoglobulin E production but enhances immunoglobulin G4 production by human peripheral blood mononuclear cells. *Infection and Immunity* **65**, 1122-1125.

DE JESUS, A.M., ALMEIDA, R.P., BACELLAR, O., ARAUJO, M.I., DEMEURE, C., BINA, J.C., DESSEIN, A.J. & CARVALHO, E.M. (1993). Correlation between cell-mediated immunity and degree of infection in subjects living in an endemic area of schistosomiasis. *European Journal of Immunology* **23**, 152-158.

DE JONG, E.C., VIEIRA, P.L., KALINSKI, P., SCHUITEMAKER, J.H.N., TANAKA, Y., WIEREINGA, E.A., YAZDANBAKHSH, M. & KAPSENBERG, M. (2001). Microbial compounds selectively induce Th1 cell-promoting or Th2 cell-promoting dendretic cells through diverse Th cell-polarizing signals. Submitted.

DEL PRETE, G., DE CARLI, M., ALMERIGOGNA, F., GIUDIZI, M.G., BIAGIOTTI, R. & ROMAGNANI, S. (1993). Human IL-10 is produced by both type 1 helper (Th1) and type 2 helper (Th2) T cell clones and inhibits their antigen-specific proliferation and cytokine production, *Journal of Immunology* **150**, 353-360.

DEMEURE, C.E., RIHET, P., ABEL, L., OUATTARA, M., BOURGOIS, A. & DESSEIN, A.J. (1993). Resistance to Schistosoma mansoni in humans: influence of the IgE/IgG4 balance and IgG2 in immunity to reinfection after chemotherapy. *Journal of Infectious Diseases* **168**, 1000-1008.

DESSEIN, A.J., HILLAIRE, D., ELWALI, N..E., MARQUET, S., MOHAMED-ALI, Q., MIRGHANI, A., HENRI, S., ABDELHAMEED, A.A., SAEED, O.K., MAGZOUB, M.M. & ABEL, L. (1999). Severe hepatic fibrosis in *Schistosoma mansoni* infection is controlled by a major locus that is closely linked to the interferon-gamma receptor gene. *American Journal of Human Genetics* **65**, 709-721.

DUNNE, D.W., BUTTERWORTH, A.E., FULFORD, A.J., KARIUKI, H.C., LANGLEY, J.G., OUMA, J.H., CAPRON, A., PIERCE, R.J. & STURROCK, R.F. (1992). Immunity after treatment of human schistosomiasis: association between IgE antibodies to adult worm antigens and resistance to reinfection. *European Journal of Immunology* **22**, 1483-1494.

DUNNE, D.W., HAGAN, P. & ABATH, F.G. (1995). Prospects for immunological control of schistosomiasis. *Lancet* **345**, 1488-1491.

EL-KALOUBY, A.H., AMER, R., ABDEL-WAHAB, M.F. & EL-RAZIKY, E.H. (1979). Delayed hypersensitivity to specific antigen and hetrologous PPD antigen in patients infected with *Schistosoma mansoni* and/or *Schistosoma haematobium*. *Egyptian Journal of Bilharzia* **6**, 43-49.

FALCAO, P.L., MALAQUIAS, L.C., MARTINS-FILHO, O.A., SILVEIRA, A.M., PASSOS, V.M., PRATA, A., GAZZINELLI, G., COFFMAN, R.L. & CORREA-GUTIERREZ, Y. (2000). *Diagnostic Pathology of Parasitic Infections with Clinical Correlation* (2nd Edition), pp. 545-576. Oxford University Press, Oxford.

GALLI, S.J. & COSTA, J.J. (1995). Mast-cell-leukocyte cytokine cascades in allergic inflammation. *Allergy* **50**, 851-862.

GODFREY, R.C. (1975). Asthma and IgE levels in rural and urban communities of The Gambia. *Clinical Allergy* **5**, 201-207.

GROGAN, J.L., KREMSNER, P.G., DEELDER, A..M. & YAZDANBAKHSH, M. (1998). The effect of anti-IL-10 on proliferation and cytokine production in human schistosomiasis: fresh *versus* cryopreserved cells. *Parasite Immunology* **20**, 345-349.

GROGAN, J.L., KREMSNER, P.G., DEELDER, A.M.. & YAZDANBAKHSH, M. (1998). Antigen-specific proliferation and interferon-gamma and interleukin-5 production are down-regulated during Schistosoma haematobium infection. *Journal of Infectious Diseases* **177**, 1433-1437.

GRUNIG, G., CORRY, D.B., LEACH, M.W., SEYMOUR, B.W., KURUP, V.P. & RENNICK, D.M. (1997). Interleukin-10 is a natural suppressor of cytokine production and inflammation in a murine model of allergic bronchopulmonary aspergillosis. *Journal of Experimental Medicine* **185**, 1089-1099.

GUTIERREZ, Y. (2000). *Diagnostic Pathology of Parasitic Infections with Clinical Correlation* (2nd Edition), pp. 545-576. Oxford University Press, Oxford.

HAGAN, P., BLUMENTHAL, U.J., DUNN, D., SIMPSON, A.J. & WILKINS, H.A. (1991). Human IgE, IgG4 and resistance to reinfection with *Schistosoma haematobium*. *Nature* **17** (349), 243-245.

HANSEN, G., BERRY, G., DeKRUYFF, R.H. & UMETSU, D.T. (1999). Allergen-specific Th1 cells fail to counterbalance Th2 cell-induced airway hyperreactivity but cause severe airway inflammation. *Journal of Clinical Investigation* , 175-183.

HOFFMANN, K.F., CHEEVER, A.W. & WYNN, T.A. (2000). IL-10 and the dangers of immune polarization: excessive type 1 and type 2 cytokine responses induce distinct forms of lethal immunopathology in murine schistosomiasis. *Journal of Immunology* **164**, 6406-6416.

HUSSAIN, R., HOFSTETTER, M., GOLDSTONE, A., KNIGHT, W.B. & OTTESEN, E.A. (1983). IgE responses in human schistosomiasis. I. Quantitation of specific IgE by radioimmunoassay and correlation of results with skin test and basophil histamine release. *American Journal of Tropical Medicine and Hygiene* **32**, 1347-1355.

JARVIS, D., CHINN, S., LUCZYNSKA, C. & BURNEY, P. (1997). The association of family size with atopy and atopic disease. *Clinical and Experimental Allergy* **27**, 240-245.

JEANNIN, P., LECOANET, S., DELNESTE, Y., GAUCHAT, J.F. & BONNEFOY, J.Y. (1998). IgE versus IgG4 production can be differentially regulated by IL-10. *Journal of Immunology* **160**, 3555-3561.

JONULEIT, H., SCHMITT, E., SCHULER, G.., KNOP, J. & ENK, A.H. (2000). Induction of interleukin 10-producing, nonproliferating CD4(+) T cells with regulatory properties by repetitive stimulation with allogeneic immature human dendritic cells. *Journal of Experimental Medicine* **192**, 1213-1222.

JONULEIT, H., SCHMITT, E., STEINBRINK, K. & ENK, A.H. (2001). Dendritic cells as a tool to induce anergic and regulatory T cells. *Trends in Immunology* **22**, 394-400.

KALINSKI, P., HILKENS, C.M., WIERENGA, E.A. & KAPSENBERG, M.L. (1999). T-cell priming by type-1 and type-2 polarized dendritic cells: the concept of a third signal. *Immunology Today* **20**, 561-567.

KAPLAN, M.H., WHITFIELD, J.R., BOROS, D.L. & GRUSBY, M.J. (1998). Th2 cells are required for the *Schistosoma mansoni* egg-induced granulomatous response. *Journal of Immunology* **160**, 1850-1856.

KAPSENBERG, M.L., HILKENS, C..M., WIERENGA, E.A. & KALINSKI, P. (1999). The paradigm of type 1 and type 2 antigen-presenting cells. Implications for atopic allergy. *Clinical and Experimental Allergy* **29** Suppl 2, 33-36.

KAPSENBERG, M.L., HILKENS, C..M., VAN DER POUW KRAAN, T..C., WIERENGA, E.A. & KALINSKI, P. (2000). Atopic allergy: a failure of antigen-presenting cells to properly polarize helper T cells? *American Journal of Respiratory Critical Care and Medicine* **162**, S76-80.

KEATINGS, V.M., O'CONNOR, B.J., WRIGHT, L..G., HUSTON, D..P., CORRIGAN, C.J. & BARNES, P.J. (1997). Late response to allergen is associated with increased concentrations of tumor necrosis factor-alpha and IL-5 in induced sputum. *Journal of Allergy and Clinical Immunology* **99**, 693-698.

KIM, T.S., DEKRUYFF, R.H., RUPPER, R., MAECKER, H.T., LEVY, S. & UMETSU, D.T. (1997). An ovalbumin-IL-12 fusion protein is more effective than ovalbumin plus free recombinant IL-12 in inducing a T helper cell type 1-dominated immune response and inhibiting antigen-specific IgE production. *Journal of Immunology* **158**, 4137-4144.

KING, C.L., MAHANTY, S., KUMARASWAMI, V., ABRAMS, J.S., REGUNATHAN, J., JAYARAMAN, K., OTTESEN, E.A. & NUTMAN, T.B. (1993). Cytokine control of parasite-specific anergy in human lymphatic filariasis. Preferential induction of a regulatory T helper type 2 lymphocyte subset.*Journal of Clinical Investigations* **92**, 1667-7316.

KING, C.L., MEDHAT, A., MALHOTRA, I., NAFEH, M., HELMY, A., KHAUDARY, J., IBRAHIM, S., EL-SHERBINY, M., ZAKY, S., STUPI, R.J., BRUSTOSKI, K., SHEHATA, M. & SHATA, M.T. (1996). Cytokine control of parasite-specific anergy in human urinary schistosomiasis. IL-10 modulates lymphocyte reactivity. *Journal of Immunology* **156**, 4715-4721.

LARRICK, J.W., BUCKLEY, C.E. 3RD, MACHAMER, C.E., SCHLAGEL, G.D., YOST, J.A., BLESSING-MOORE, J. & LEVY, D. (1983). Does hyperimmunoglobulinemia-E protect tropical populations from allergic disease? *Journal of Allergy and Clinical Immunology* **71**, 184-188.

LUNDGREN, M., PERSSON, U., LARSSON, P., MAGNUSSON, C., SMITH, C.I., HAMMARSTROM, L. & SEVERINSON, E. (1989). Interleukin 4 induces synthesis of IgE and IgG4 in human B cells. *European Journal of Immunology* **19**, 1311-1315.

LYNCH, N.R., HAGEL, I., PEREZ, M., DI PRISCO, M.C., LOPEZ, R. & ALVAREZ, N. (1993). Effect of anthelmintic treatment on the allergic reactivity of children in a tropical slum. *Journal of Allergy and Clinical Immunology* **92**, 404-411.

LYNCH, N.R., LOPEZ, R.., ISTURIZ, G. & TENIAS-SALAZAR, E. (1983). Allergic reactivity and helminthic infection in Amerindians of the Amazon Basin. *International Archives of Allergy and Applied Immunology* **72**, 369-372.

MACAUBAS, C., SLY, P.D., BURTON, P., TILLER, K., YABUHARA, A., HOLT, B.J., SMALLACOMBE, T.B., KENDALL, G., JENMALM, M.C. & HOLT, P.G. (1999). Regulation of T-helper cell responses to inhalant allergen during early childhood. *Clinical and Experimental Allergy* **29**, 1223-1231.

MALHOTRA, I., MUNGAI, P.., WAMACHI, A., KIOKO, J.., OUMA, J..H., KAZURA, J.W. & KING, C.L. (1999). Helminth- and Bacillus Calmette-Guerin-induced immunity in children sensitized in utero to filariasis and schistosomiasis. *Journal of Immunology* **162**, 6843-6848.

MATRICARDI, P.M., ROSMINI, F., FERRIGNO, L., NISINI, R., RAPICETTA, M., CHIONNE, P., STROFFOLINI, T., PASQUINI, P. & D'AMELIO, R. (1997). Cross sectional retrospective study of prevalence of atopy among Italian military students with antibodies against hepatitis A virus. *British Medical Journal* **314**, 999-1003.

MEDHAT, A., SHEHATA, M., BUCCI, K., MOHAMED, S., DIEF, A.D., BADARY, S., GALAL, H., NAFEH, M. & KING, C.L. (1998). Increased interleukin-4 and interleukin-5 production in response to *Schistosoma haematobium* adult worm antigens correlates with lack of reinfection after treatment. *Journal of Infectious Diseases* **178**, 512-519.

MOHAMED-ALI, Q., ELWALI, N.E., ABDELHAMEED, A.A., MERGANI, A., RAHOUD, S., ELAGIB, K.E., SAEED, O.K., ABEL, L., MAGZOUB, M.M. & DESSEIN, A.J. (1999). Susceptibility to periportal (Symmers) fibrosis in human *Schistosoma mansoni* infections: Evidence that intensity and duration of infection, gender, and inherited factors are critical in disease progression. *Journal of Infectious Diseases* **180**, 1298-1306.

MOLA, P.W., FARAH, I.O.., KARIUKI, T.M., NYINDO, M., BLANTON, R.E. & KING, C.L. (1999). Cytokine control of the granulomatous response in *Schistosoma mansoni*-infected baboons: role of exposure and treatment. *Infection and Immunity* **67**,6565-6571.

MONTENEGRO, S.M., MIRANDA, P., MAHANTY, S., ABATH, F.G., TEIXEIRA, K.M., COUTINHO, E.M., BRINKMAN, J., GONCALVES, I., DOMINGUES, L..A., DOMINGUES, A.L., SHER A, & WYNN, T.A. (1999). Cytokine production in acute versus chronic human *Schistosomiasis mansoni*: the cross-regulatory role of interferon-gamma and interleukin-10 in the responses of peripheral blood mononuclear cells and splenocytes to parasite antigens. *Journal of Infectious Diseases* **179**, 1502-1514.

MWATHA, J.K., KIMANI, G., KAMAU, T., MBUGUA, G.G., OUMA, J.H., MUMO, J., FULFORD, A..J., JONES, F.M., BUTTERWORTH, A.E., ROBERTS, M.B. & DUNNE, D.W. (1998). High levels of TNF, soluble TNF receptors, soluble ICAM-1, and IFN-gamma, but low levels of IL-5, are associated with hepatosplenic disease in human *Schistosomiasis mansoni*. *Journal of Immunology* **160**, 1992-1999.

NAUS, C.W., KIMANI, G., OUMA, J.H., FULFORD, A.J., WEBSTER, M., VAN DAM, G.J., DEELDER, A.M., BUTTERWORTH. A.E. & DUNNE, D.W. (1999). Development of antibody isotype responses to Schistosoma mansoni in an immunologically naïve immigrant population: influence of infection duration, infection intensity, and host age. *Infection and Immunity* **67**, 3444-3451.

NDHLOVU P., CADMAN, H., VENNERVALD, B.J., CHRISTENSEN, N.O., CHIDIMU, M. & CHANDIWANA, S.K. (1996). Age-related antibody profiles in *Schistosoma haematobium* infections in a rural community in Zimbabwe. *Parasite Immunology* **18**, 181-191.

OHKAWARA, Y., LIM, K.G., XING, Z., GLIBETIC, M., NAKANO, K., DOLOVICH, J., CROITORU, K., WELLER, P.F. & JORDANA, M. (1996). CD40 expression by human peripheral blood eosinophils. *Journal of Clinical Investigations* **97**, 1761-1766.

ONKAMO, P., VAANANEN, S., KARVONEN, M. & TUOMILEHTO, J. (1999). Worldwide increase in incidence of Type I diabetes. The analysis of the data on published incidence trends. *Diabetologia* **42**, 1395-1403.

OUMA, J.H., FULFORD, A.J., KARIUKI, H.C., KIMANI, G., STURROCK, R..F., MUCHEMI, G., BUTTERWORTH, A.E. & DUNNE, D.W. (1998). The development of schistosomiasis mansoni in an immunologically naïve immigrant population in Masongaleni, Kenya. *Parasitology* **117**, 123-132.

PELLETIER, C., VARIN-BLANK, N., RIVERA, J., IANNASCOLI, B., MARCHAND, F., DAVID, B., WEYER, A. & BLANK, U. (1998). Fc epsilonRI-mediated induction of TNF-alpha gene expression in the RBL-2H3 mast cell line: regulation by a novel NF-kappaB-like nuclear binding complex. *Journal of Immunology* **161**, 4768-4776.

PLATTS-MILLS, T., VAUGHAN, J., SQUILLACE, S., WOODFOLK, J. & SPORIK, R. (2001). Sensitisation, asthma, and a modified Th2 response in children exposed to cat allergen: a population-based cross-sectional study. *Lancet* **357**, 752-756.

RANDOLPH, D.A., CARRUTHERS, C.J., SZABO, S.J., MURPHY, K.M. & CHAPLIN, D.D. (1999). Modulation of airway inflammation by passive transfer of allergen-specific Th1 and Th2 cells in a mouse model of asthma. *Journal of Immunology* **162**, 2375-2383.

RANDOLPH, D.A., STEPHENS, R., CARRUTHERS, C..J. & CHAPLIN, D.D. (1999). Cooperation between Th1 and Th2 cells in a murine model of eosinophilic airway inflammation. *Journal of Clinical Investigation* **104**, 1021-1029.

ROBERTS, M., BUTTERWORTH, A.E., KIMANI, G., KAMAU, T., FULFORD, A.J., DUNNE, D.W., OUMA, J.H. & STURROCK, R.F. (1993). Immunity after treatment of human schistosomiasis: association between cellular responses and resistance to reinfection. *Infection and Immunity* **61**, 4984-4993.

ROBINSON, D.S., HAMID, Q., YING, S., TSICOPOULOS, A., BARKANS, J., BENTLEY, A.M., CORRIGAN, C., DURHAM, S.R. & KAY, A.B. (1992). Predominant TH2-like bronchoalveolar T-lymphocyte population in atopic asthma. *New England Journal of Medicine* **326**, 298-304.

RONCAROLO, M.G., LEVINGS, M.K. & TRAVERSARI, C. (2001). Differentiation of T regulatory cells by immature dendritic cells. *Journal of Experimental Medicine* **193**, F5-9.

ROYER, B,, VARADARADJALOU, S., SAAS, P., GUILLOSSON, J.J., KANTELIP, J.P. & AROCK, M. (2001). Inhibition of IgE-induced activation of human mast cells by IL-10. *Clinical and Experimental Allergy* **31**, 694-704.

SABIN, E.A., ARAUJO, M.I., CARVALHO, E.M. & PEARCE, E.J. (1996), Impairment of tetanus toxoid-specific Th1-like immune responses in humans infected with *Schistosoma mansoni*. *Journal of Infectious Diseases* **173**, 269-272.

SAINI, S.S., MACGLASHAN, D.W. JR, STERBINSKY, S.A., TOGIAS, A., ADELMAN, D.C., LICHTENSTEIN, L.M. & BOCHNER, B.S. (1999). Down-regulation of human basophil IgE and FC epsilon RI alpha surface densities and mediator release by anti-IgE-infusions is reversible *in vitro* and *in vivo*. *Journal of Immunology* **162**, 5624-5630.

SCHOTTELIUS, A..J., MAYO, M.W., SARTOR, R.B. & BALDWIN, A.S. JR. (1999). Interleukin-10 signaling blocks inhibitor of kappaB kinase activity and nuclear factor kappaB DNA binding. *Journal of Biological Chemisry* **274**, 31868-31874.

SHAHEEN, S.O., AABY, P., HALL, A.J., BARKER, D.J., HEYES, C.B., SHIELL, A.W. & GOUDIABY, A. (1996). Measles and atopy in Guinea-Bissau. *Lancet* **347**, 1792-1796.

SHER, A. & COFFMAN, R.L. (1992). Regulation of immunity to parasites by T cells and T cell-derived cytokines. *Annual Reviews in Immunology* **10**, 385-409.

SHER, A., FIORENTINO, D.., CASPAR, P., PEARCE, E. & MOSMANN, T. (1991). Production of IL-10 by CD4+ T lymphocytes correlates with down-regulation of Th1 cytokine synthesis in helminth infection. *Journal of Immunology* **147**, 2713-2716.

SHER, A., GAZZINELLI, R.T., OSWALD, I.P., CLERICI, M., KULLBERG, M., PEARCE, E.J., BERZOFSKY, J.A., MOSMANN, T.R., JAMES,, S.L., MORSE, H.C., III & SHEARER, G.M. (1992). Role of T-cell derived cytokines in the downregulation of immune responses in parasitic and retroviral infection. *Immunological Reviews* **127**, 183-204.

SHIRAKAWA, T., ENOMOTO, T., SHIMAZU, S., HOPKIN, J.M. (1997). The inverse association between tuberculin responses and atopic disorder. *Science* **275**, 77-79.

STELMA, F.F., TALLA, I., POLMAN, K., NIANG, M., STURROCK, R.F., DEELDER, A.M. & GRYSEELS, B. (1993). Epidemiology of Schistosoma mansoni infection in a recently exposed community in northern Senegal. *American Journal of Trpical Medicine and Hygiene* **49**, 701-706.

STRACHAN, D.P.. (1989). Hay fever, hygiene, and household size. *British Medical Journal* **299**, 1259-1260.

TAKANASHI, S., HASEGAWA, Y., KANEHIRA, Y., YAMAMOTO, K., FUJIMOTO, K., SATOH, K. & OKAMURA, K. (1999). Interleukin-10 level in sputum is reduced in bronchial asthma, COPD and in smokers. *European Respiratory Journal* **14**, 309-314.

TAKANASKI, S., NONAKA, R., XING, Z., O'BYRNE, P., DOLOVICH, J. & JORDANA, M. (1994). Interleukin 10 inhibits lipopolysaccharide-induced survival and cytokine production by human peripheral blood eosinophils. *Journal of Experimental Medicine* **180**, 711-715.

TEDESCHI, A. & AIRAGHI, L. (2001). Common risk factors in type 1 diabetes and asthma. *The Lancet* **357**, 1622.

TEN HACKEN, N.H., OOSTERHOFF, Y., KAUFFMAN, H.F., GUEVARRA, L., SATOH, T., TOLLERUD, D.J. & POSTMA, D.S. (1998). Elevated serum interferon-gamma in atopic asthma correlates with increased airways responsiveness and circadian peak expiratory flow variation. *European Respiratory Journal* **11**, 312-316.

THE INTERNATIONAL STUDY OF ASTHMA AND ALLERGIES IN CHILDHOOD (ISAAC) STEERING COMMITTEE. (1998). Worldwide variation in prevalence of symptoms of asthma, allergic rhinoconjunctivitis, and atopic eczema: ISAAC. *Lancet* **351**, 1225-1232.

THOMAS, W.R.. & SMITH, W. (1999). Towards defining the full spectrum of important house dust mite allergens. *Clinical and Experimental Allergy* **29**, 1583-1587.

TSURUTA, L., LEE, H.J., MASUDA, E.S., YOKOTA, T., ARAI, N. & ARAI, K. (1995). Regulation of expression of the IL-2 and IL-5 genes and the role of proteins related to nuclear factor of activated T cells. *Journal of Allergy and Clinical Immunology* **96**, 1126-1135.

UMETSU, D.T. & DeKRUYFF, R.H. (1997). TH1 and TH2 CD4+ cells in human allergic diseases. *Journal of allergy and Clinical Immunology* **100**, 1-6.

UMETSU, D.T. & DeKRUYFF, R.H. (1999). Interleukin-10: The missing link in asthma regulation? *American Journal of Respiratory Cell and Molecular Biology* **21**, 562-563.

VAN DAM, G.J., STELMA, F.F., GRYSEELS, B., FALCAO FERREIRA, S.T., TALLA, I., NIANG, M., ROTMANS, J.P. & DEELDER, A.M. (1996). Antibody response patterns against Schistosoma mansoni in a recently exposed community in Senegal. *Journal of Infectious Diseases* **173**, 1232-1241.

VAN DEN BIGGELAAR, A..H.J., LOPUHAA, C., VAN REE, R., VAN DERZEE, J.S., JANS, J.., HOEK, A., MIGOMBET, B., LUCKNER, C., KREMSNER, P.G. & YAZDANBAKHSH, M. (2001). The prevalence of parasite infestation and house dust mite sensitization in Gabonese schoolchildren. *International Archives of Allergy and Immunology*, In press.

VAN DEN BIGGELAAR, A.H.J., VAN REE, R., RODRIGUES, L.C., LELL, B., DEELDER, A.M., KREMSNER, P.G. & YAZDANBAKHSH, M. (2000). Decreased atopy in children infected with *Schistosoma haematobium*: a role for parasite-induced interleukin-10. *Lancet* **356**, 1723-1727.

VIANA, I.R., SHER, A., CARVALHO, O.S., MASSARA, C.L., ELOI-SANTOS, S.M., PEARCE, E.J., COLLEY, D.G., GAZZINELLI, G. & CORREA-OLIVEIRA, R. (1994). Interferon-gamma production by peripheral blood mononuclear cells from residents of an area endemic for *Schistosoma mansoni*. *Transactions of the Royal Socieety of Tropical Medicine and Hygiene* **88**, 466-470.

VLAHOPOULOS, S., BOLDOGH, I., CASOLA, A. & BRASIER, A.R. (1999). Nuclear factor-kappaB-dependent induction of interleukin-8 gene expression by tumor necrosis factor alpha: evidence for an antioxidant sensitive activating pathway distinct from nuclear translocation. *Blood* **94**, 1878-1889.

WHELAN, M., HARNETT, M.M., HOUSTON, K..M.., PATEL, V., HARNETT, W. & RIGLEY, K..P. (2000). A filarial nematode-secreted product signals dendritic cells to acquire a phenotype that drives development of Th2 cells. *Journal of Immunology* **164**, 6453-6460.

WHO REPORT OF THE WHO INFORMAL CONSULTATION ON SCHISTOSOMIASIS CONTROL. (1999). World Health Organization, Geneva.

WILLIAMS, M.E., MONTENEGRO, S., DOMINGUES, A.L., WYNN, T.A., TEIXEIRA, K., MAHANTY, S., COUTINHO, A. & SHER, A. (1994). Leukocytes of patients with *Schistosoma mansoni* respond with a Th2 pattern of cytokine production to mitogen or egg antigens but with a Th0 pattern to worm antigens. *Journal of Infectious Diseases* **170**, 946-954.

WILSON, R.A., COULSON, P.S., BETTS, C., DOWLING, M.A. & SMYTHIES, L.E. (1996). Impaired immunity and altered pulmonary responses in mice with a disrupted interferon-gamma receptor gene exposed to the irradiated *Schistosoma mansoni* vaccine. *Immunology* **87**, 275-282.

WYNN, T.A., CHEEVER, A.W., JANKOVIC, D., POINDEXTER, R.W., CASPAR, P., LEWIS, F.A. & SHER, A. (1995). An IL-12-based vaccination method for preventing fibrosis induced by schistosome infection. *Nature* **376** (6541), 594-596.

WYNN, T.A., CHEEVER, A.W., WILLIAMS, M.E., HIENY, S., CASPAR, P., KUHN, R., MULLER, W. & SHER, A. (1998). IL-10 regulates liver pathology in acute murine *Schistosomiasis mansoni* but is not required for immune down-modulation of chronic disease. *Journal of Immunology* **160**, 4473-4480.

YAZDANBAKHSH, M., PAXTON, W.A., KRUIZE, Y.C., SARTONO, E., KURNIAWAN, A., VAN HET WOUT, A., SELKIRK, M.E., PARTONO, F. & MAIZELS, R.M. (1993). T cell responsiveness correlates differentially with antibody isotype levels in clinical and asymptomatic filariasis. *Journal of Infectious Diseases* **167**, 925-931.

YEMANEBERHAN, H., BEKELE, Z., VENN, A., LEWIS, S., PARRY, E. & BRITTON, J. (1997). Prevalence of wheeze and asthma and relation to atopy in urban and rural Ethiopia. *Lancet* **350**, 85-90.

REFERENCES FOR TABLE

1. BODNER C, GODDEN D, LITTLE J. *et al.* (1999). Antioxidant intake and adult-onset wheeze: a case control study. Aberdeen WHEASE Study Group. *The European Respiratory Journal* **13**, 22-30.
2. HIJAZI, N., ABALKHAIL, B. & SEATON, A. (2000). Diet and childhood asthma in a society in transition: a study in urban and rural Saudi Arabia. *Thorax* **55**, 775-779.
3. HODGE, L., SALOME, C.M., HUGHES, J.M., LIU-BRENNAN, D., RIMMER, J., ALLMAN, M., PANG, D., ARMOUR, C. & WOOLCOCK, A.J. (1998). Effect of dietary intake of omega-3 and omega-6 fatty acids on severity of asthma in children. *The European Respiratory Journal* **11**, 361-365.
4. HODGE, L., SALOME, C.M., PEAT, J.K., HABY, M.M., XUAN, W. & WOOLCOCK, A.J. (1996). Consumption of oily fish and childhood asthma risk. *The Medical Journal of Australia* **164**, 137-140.
5. WEILAND, S.K., VON MUTIUS, E., HUSING, A. & ASHER, M.I. (1999). Intake of trans fatty acids and prevalence of childhood asthma and allergies in Europe. ISAAC Steering Committee. *Lancet* **353** (9169), 2040-2041
6. DEMISSIE, K., ERNST, P., GRAY, DONALD K. & JOSEPH, L. (1996). Usual dietary salt intake and asthma in children: a case-control study. *Thorax* **51**, 59-63.
7. CAREY, O.J., LOCKE, C. & COOKSON, J.B. (1993). Effect of alterations of dietary sodium on the severity of asthma in men. *Thorax* **48**, 714-718.
8. VON MUTIUS, E., MARTINEZ, F.D., FRITZSCH, C., NICOLAI, T., ROELL, G. & THIEMANN, H.H. (1994). Prevalence of asthma and atopy in two areas of West and East Germany. *American Journal of Respiratory Critical Care and Medicine* **149**, 358-364

9. WONG, G.W., KO, F.W., LAU, T.S., LI, S.T., HUI, D., PANG, S..W., LEUNG, R., FOK, T.F. & Lai, C.K. (2001). Temporal relationship between air pollution and hospital admissions for asthmatic children in Hong Kong. *Clinical and Experimental Allergy* **31**, 565-569.
10. ANDERSON, H..R., PONCE DE LEON, A., BLAND, J.M., BOWER, J.S., EMBERLIN, J. & STRACHAN, D.P. (1998). Air pollution, pollens, and daily admissions for asthma in London 1987-92. *Thorax* **53**, 842-848.
11. GARTY, B.Z., KOSMAN, E., GANOR, E., BERGER, V., GARTY, L., WIETZEN, T., WAISMAN, Y., MIMOUNI, M. & WAISEL, Y. (1998). Emergency room visits of asthmatic children, relation to air pollution, weather, and airborne allergens. *Annals of Allergy, Asthma and Immunology* **81**, 563-570.
12. HAJNAL, B.L., BRAUN-FAHRLANDER, C., GRIZE, L., GASSNER, M., VARONIER, H.S., VUILLE, J.C., WUTHRICH, B. & SENNHAUSER. F.H. (1999). Effect of environmental tobacco smoke exposure on respiratory symptoms in children. SCARPOL Team. Swiss Study on Childhood Allergy and Respiratory Symptoms with Respect to Air Pollution, Climate and Pollen. *Schweizerische medizinische Wochenschrift* **129**, 723-730.
13. ODHIAMBO, J.A., NG'ANG'A, L.W., MUNGAI, M.W., GICHEHA, C.M., NYAMWAYA, J.K., KARIMI, F., MACKLEM, P.T. & BECKLAKE, M.R. (1998). Urban-rural differences in questionnaire-derived markers of asthma in Kenyan school children. *The European Respiratory Journal* **12**, 1105-12.
14. ZOCK, J.P., SUNYER, J., KOGEVINAS, M., KROMHOUT, H., BURNEY, P. & ANTO, J.M. (2001) Occupation, Chronic Bronchitis, and Lung Function in Young Adults. An international study. *American Journal of Respiratory Critical Care and Medicine***163**, 1572-1577.
15. WYLER, C., BRAUN-FAHRLANDER, C., KUNZLI, N., SCHINDLER, C., ACKERMANN-LIEBRICH, U., PERRUCHOUD, A.P., LEUENBERGER, P. & WUTHRICH B. (2000). Exposure to motor vehicle traffic and allergic sensitization. The Swiss Study on Air Pollution and Lung Diseases in Adults (SAPALDIA) Team. *Epidemiology* **11**, 450-456.
16. BRAUN-FAHRLANDER, C., VUILLE, J.C., SENNHAUSER, F.H., NEU, U., KUNZLE, T., GRIZE, L., GASSNER, M., MINDER, C., SCHINDLER, C., VARONIER, H.S. & WUTHRICH, B. (2000), Respiratory health and long-term exposure to air pollutants in Swiss schoolchildren. SCARPOL Team. Swiss Study on Childhood Allergy and Respiratory Symptoms with Respect to Air Pollution, Climate and Pollen. *American Journal of Respiratory Critical Care and Medicine* **155**, 1042-1049.
17. DIAZ-SANCHEZ, D., JYRALA, M., NG,, D., NEL, A. & SAXON, A. (2000). *In vivo* nasal challenge with diesel exhaust particles enhances expression of the CC chemokines rantes, MIP-1alpha, and MCP-3 in humans *Clinical Immunology* **97**, 140-145.
18. HUANG, S.L., SHIAO, G. & CHOU, P. (1999). Association between body mass index and allergy in teenage girls in Taiwan. *Clinical and Expimental Allergy* **29**, 323-329
19. STENIUS-AARNALIA, B., POUSSA, T., KVARNSTROM, J., GRONLUND, E.L., YLIKAHRI, M. & MUSTAJOKI, P. (2000). Immediate and long term effects of weight reduction in obese people with asthma: randomised controlled study. *British Medical Journal* **320**, 827-832
20. SHAHEEN, S.O., STERNE, J.A., MONTGOMERY, S.M. & AZIMA, H. (1999). Birth weight, body mass index and asthma in young adults. *Thorax* **54**, 396-402.
21. FIGUEROA-MUNOZ, J.I., CHINN, S. & RONA, R.J. (2001). Association between obesity and asthma in 4-11 year old children in the UK. *Thorax* **56**, 133-137.

22. LUDER, E., MELNIK, T.A. & DIMAIO, M. (1998). Association of being overweight with greater asthma symptoms in inner city black and Hispanic children. *Journal of Pediatrics* **132**, 699-703.
23. CHEN, Y., DALES, R., KREWSKI, D. & BREITHAUPT, K. (1999). Increased effects of smoking and obesity on asthma among female Canadians: the National Population Health Survey, 1994-1995. *American Journal of Epidemiology* **150**, 255-262
24. CAMARGO, C.A., WEISS, S.T., ZHANG, S., WILLETT, W.C. & SPEIZER, F.E. (1999). Prospective study of body mass index, weight change, and risk of adult-onset asthma in women. *Archives of Internal Medicine* **159**, 2582-2588.
25. CASTRO-RODRIGUEZ, J..A., HOLBERG, C..J., MORGAN, W..J., WRIGHT, A.L. & MARTINEZ, F.D. (2001). Increased Incidence of Asthmalike Symptoms in Girls Who Become Overweight or Obese during the School Years. *American Journal of Respiratory Critical Care and Medicine* **163**, 1344-1349.
26. VAN DEN BIGGELAAR, A.H.J., VAN REE. R., RODRIGUES, L.C., LELL, B., DEELDER, A.M., KREMSNER, P.G. & YAZDANBAKHSH, M. (2000). Decreased atopy in children infected with *Schistosoma haematobium*: a role for parasite-induced interleukin-10. *Lancet* **356** (9243), 1723-1727.
27. ARAUJO, M.I., LOPES, A.A., MEDEIROS, M., CRUZ, A.A., SOUSA-ATTA, L., SOLE, D. & CARVALHO, E.M. (2000). Inverse association between skin response to aeroallergens and *Schistosoma mansoni* infection. *International Archives of Allergy and Immunology* **123**, 145-148.
28. LYNCH, N.R., HAGEL, I., PEREZ, M., DI PRISCO, M.C., LOPEZ, R. & ALVAREZ, N. (1993) Effect of anthelmintic treatment on the allergic reactivity of children in a tropical slum. *The Journal of Allergy and Clinical Immunology* **92**, 404-411
29. SHIELD, J.M., SCRIMGEOUR, E.M. & VATERLAWS, A.L. (1980). Intestinal helminths in an adult hospital population in the Eastern Highlands of Papua New Guinea: relationship with anaemia, eosinophilia and asthma. *Papua New Guinea Medical Journal* **23**, 157-164.
30. SHAHEEN, S.O., AABY, P., HALL, A..J., BARKER, D.J., HEYES, C.B., SHIELL, A.W. & GOUDIABY, A. (1996). Measles and atopy in Guinea-Bissau. *Lancet* **347**(9018), 1792-1796
31. LEWIS, S.A. & BRITTON, J.R. (1998). Measles infection, measles vaccination and the effect of birth order in the aetiology of hay fever. *Clinical and Experimental Allergy* **28**, 1493-1500.
32. PAUNIO, M., HEINONEN, O.P., VIRTANEN, M., LEINIKKI, P., PATJA, A. & PELTOLA, H. (2000). Measles history and atopic diseases: a population-based cross-sectional study. *The Journal of the American Microbiological Association* **283**, 343-346.
33. WICKENS, K.L., CRANE, J., KEMP, T.J., LEWIS, S.J., D'SOUZA, W.J., SAWYER, G.M., STONE, M.L., TOHILL, S.J., KENNEDY, J.C., SLATER, T.M. & PEARCE, N.E. (1999). Family size, infections, and asthma prevalence in New Zealand children. *Epidemiology* **10**, 699-705.
34. DOLD, S., HEINRICH, J., WICHMANN, H.E. & WJST, M. (1998). Ascaris-specific IgE and allergic sensitization in a cohort of school children in the former East Germany. *The Journal of Allergy and Clinical Immunology* **102**, 414-420
35. SELASSIE, F.G., STEVENS, R.H., CULLINAN, P., PRITCHARD, D., JONES, M., HARRIS, J., AYRES, J.G. & NEWMAN TAYLOR, A.J. (2000). Total and specific IgE (house dust mite and intestinal helminths) in asthmatics and controls from Gondar, Ethiopia. *Clinical and Experimental Allergy* **30**, 356-358

36. BUIJS, J., BORSBOOM, G., RENTING, M., HILGERSOM, W..J., VAN WIERINGEN, J.C., JANSEN, G. & NEIJENS, J. (1997). Relationship between allergic manifestations and Toxocara seropositivity: a cross-sectional study among elementary school children. *The European Respiratory Journal* **10**, 1467-1475
37. HAKIM, S..L., THADASAVANTH, M., SHAMILAH, R.H. & YOGESWARI, S. (1997). Prevalence of *Toxocara canis* antibody among children with bronchial asthma in Klang Hospital, Malaysia. *Transactions of the Royal Society of Tropical Medicine and Hygiene* **91**, 528.
38. OTEIFA, N..M., MOUSTAFA, M.A. & ELGOZAMY, B.M. (1998). Toxocariasis as a possible cause of allergic diseases in children. *Journal of the Egyptian Society of Parasitology* **28**, 365-372.
39. SHIRAKAWA, T., ENOMOTO, T., SHIMAZU, S. & HOPKIN, J.M. (1997). The inverse association between tuberculin responses and atopic disorder. *Science* **275**, 77-79.
40. VON HERTZEN, L., KLAUKKA, T., MATTILA, H. & HAAHTELA, T. (1999). Mycobacterium tuberculosis infection and the subsequent development of asthma and allergic conditions. *The Journal of Allergy and Clinical Immunology* **104**, 1211-1214.
41. VON MUTIUS, E., PEARCE, N., BEASLEY, R., CHENG, S., VON EHRENSTEIN, O., BJORKSTEN, B. & WEILAND, S. (2000). International patterns of tuberculosis and the prevalence of symptoms of asthma, rhinitis, and eczema. *Thorax* **55**, 449-453.
42. AABY, P., SHAHEEN, S.O., HEYES, C.B., GOUDIABY, A., HALL, A.J., SHIELL, A.W., JENSEN, H. & MARCHANT, A. (2000). Early BCG vaccination and reduction in atopy in Guinea-Bissau. *Clinical and Experimental Allergy* **30**, 644-650.
43. ALM, J.S., LILJA, G., PERSHAGEN, G. & SCHEYNIUS, A. (1997). Early BCG vaccination and development of atopy. *Lancet* **350** (9075), 400-403.
44. OMENAAS, E., JENTOFT, H.F., VOLLMER, W..M., BUIST, A.S. & GULSVIK, A. (2000). Absence of relationship between tubeculin reactivity and atopy in BCG vaccinated young adults. *Thorax* **55**, 454-458.
45. MATRICARDI, P.M., ROSMINI, F., FERRIGNO, L., NISINI, R., RAPICETTA, M., CHIONNE, P., STROFFOLINI, T., PASQUINI, P. & D'AMELIO, R. (1997). Cross sectional retrospective study of prevalence of atopy among Italian military students with antibodies against hepatitis A virus. *British Medical Journal* **314**, 999-1003.
46. MATRICARDI, P.M., ROSMINI, F., RIONDINO, S., FORTINI, M., FERRIGNO, L., RAPICETTA, M. & BONINI, S. (2000). Exposure to foodborne and orofecal microbes versus airborne viruses in relation to atopy and allergic asthma: epidemiological study. *British Medical Journal* **320**, 412-417.
47. BODNER, C., ANDERSON, W.J., REID, T.S. & GODDEN, D.J. (2000). Childhood exposure to infection and risk of adult onset wheeze and atopy. *Thorax* **55**, 383-387.
48. ILLI, S., VON MUTIUS, E., LAU, S., BERGMANN, R., NIGGEMANN, B., SOMMERFELD, C. & WAHN, U. (2001). Early childhood infectious diseases and the development of asthma up to school age: a birth cohort study. *British Medical Journal* **322**, 390-395.
49. MATRICARDI, P.M., FRANZINELLI, F., FRANCO, A., CAPRIO, G., MURRU, F., CIOFFI, D., FERRIGNO, L., PALERMO, A., CICCARELLI, N. & ROSMINI, F. (1998). Sibship size, birth order, and atopy in 11,371 Italian young men. *The Journal of Allergy and Clinical Immunology* **101**, 439-444.

50. FORASTIERE, F., AGABITI, N., CORBO, G.M., DELL'ORCO, V., PORTA, D., PISTELLI, R., LEVENSTEIN, S. & PERUCCI, C.A. (1997). Socioeconomic status, number of siblings, and respiratory infections in early life as determinants of atopy in children. *Epidemiology* **8**, 566-570.
51. STRACHAN, D.P. (1989). Hay fever, hygiene, and household size. *British Medical Journal* **299**, 1259-1260.
52. JARVIS, D., CHINN, S., LUCZYNSKA, C. & BURNEY, P. (1997). The association of family size with atopy and atopic disease. *Clinical and Experimental Allergy* **27**, 240-245.
53. BODNER, C., GODDEN, D. & SEATON, A. (1998). Family size, childhood infections and atopic diseases. The Aberdeen WHEASE Group. *Thorax* **53**, 28-32.
54. BALL, T.M., CASTRO-RODRIGUEZ, J.A., GRIFFITH, K.A., HOLBERG, C.J., MARTINEZ, F.D. & WRIGHT, A.L. (2000). Siblings, day-care attendance, and the risk of asthma and wheezing during childhood. *The New England Journal of Medicine* **343**. 538-543.
55. KRAMER, U., HEINRICH, J., WJST, M. & WICHMANN, H.E. (1999). Age of entry to day nursery and allergy in later childhood. *Lancet* **353**, 450-454.
56. ALM, J.S., SWARTZ, J., LILJA, G., SCHEYNIUS, A. & PERSHAGEN, G. (1999). Atopy in children of families with an anthroposophic lifestyle. *Lancet* **353**, 1485-1488.
57. YEMANEBERHAN, H., BEKELE, Z., VENN, A., LEWIS, S., PARRY, E. & BRITTON, J. (1997). Prevalence of wheeze and asthma and relation to atopy in urban and rural Ethiopia. *Lancet* **350**, 85-90.
58. SUNYER, J., TORREGROSA, J., ANTO, J.M., MENENDEZ, C., ACOSTA, C., SCHELLENBERG, D., ALONSO, P.L. & KAHIGWA, E. (2000). The association between atopy and asthma in a semirural area of Tanzania (East Africa). *Allergy* **55**, 762-766.
59. FANIRAN, A.O., PEAT, J.K. & WOOLCOCK, A.J. (1999). Prevalence of atopy, asthma symptoms and diagnosis, and the management of asthma: comparison of an affluent and a non-affluent country. *Thorax* **54**, 606-610.
60. LEUNG, R., HO, P., LAM, C.W. & LAI, C.K. (1997). Sensitization to inhaled allergens as a risk factor for asthma and allergic diseases in Chinese population. *The Journal of Allergy and Clinical Immunology* **99**, 594-599.
61. KILPELAINEN, M., TERHO, E.O., HELENIUS, H. & KOSKENVUO, M. (2000). Farm environment in childhood prevents the development of allergies. *Clinical and Experimental Allergy* **30**, 201-208.
62. VON EHRENSTEIN, O.S., VON MUTIUS, E., ILLI, S., BAUMANN, L., BOHM, O. & VON KRIES, R. (2000). Reduced risk of hay fever and asthma among children of farmers. *Clinical and Experimental Allergy* **30**, 187-193.
63. RIEDLER, J., EDER, W., OBERFELD, G. & SCHREUER, M. (2000). Austrian children living on a farm have less hay fever, asthma and allergic sensitization. *Clinical and Experimental Allergy* **30**, 194-200.
64. BRAUN-FAHRLANDER, C., GASSNER, M., GRIZE, L., NEU, U., SENNHAUSER, F.H., VARONIER, H.S., VUILLE, J.C. & WUTHRICH, B. (1999). Prevalence of hay fever and allergic sensitization in farmer's children and their peers living in the same rural community. SCARPOL team. Swiss Study on Childhood Allergy and Respiratory Symptoms with Respect to Air Pollution. *Clinical and Experimental Allergy* **29**, 28-34.
65. WICKENS, K., PEARCE, N., CRANE, J. & BEASLEY, R. (1999). Antibiotic use in early childhood and the development of asthma. *Clinical and Experimental Allergy* **29**, 766-771.

66. GEREDA, J.E., LEUNG, D.Y., THATAYATIKOM, A., STREIB, J.E., PRICE, M.R., KLINNERT, M.D. & LIU, A.H. (2000). Relation between house-dust endotoxin exposure, type 1 T-cell development, and allergen sensitisation in infants at high risk of asthma. *Lancet* **355**, 1680-1683.
67. BOTTCHER, M.F., NORDIN, E.K., SANDIN. A., MIDTVEDT. T. & BJORKSTEN, B. (2000). Microflora-associated characteristics in faeces from allergic and nonallergic infants. *Clinical and Experimental Allergy* **30**, 1590-1596.
68. KALLIOMAKI, M., SALMINEN, S., ARVILOMMI, H., KERO, P., KOSKINEN, P. & ISOLAURI, E. (2001). Probiotics in primary prevention of atopic disease: a randomised placebo-controlled trial. *Lancet* **357**, 1076-1079.
69. RENZI, P.M., TURGEON, J.P., MARCOTTE, J.E., DRBLIK, S.P., BERUBE, D., GAGNON, M.F. & SPIER, S. (1999). Reduced interferon-gamma production in infants with bronchiolitis and asthma. *American Journal of Respiratory Critical Care and Medicine* **159**, 1417-1422.
70. SIGURS, N., BJARNASON, R., SIGURBERGSSON, F. & KJELLMAN, B. (2000). Respiratory syncytial virus bronchiolitis in infancy is an important risk factor for asthma and allergy at age 7. *American Journal of Respiratory Critical Care and Medicine* **161**, 1501-1507.

Chapter 16
GEOHELMINTHS, HIV/AIDS AND TB

Gadi Borkow and Zvi Bentwich
R. Ben-Ari Institute of Clinical Immunology and AIDS Center, Kaplan Medical Center, Hebrew University Hadassah Medical School, Rehovot 87100, Israel
e-mail: Gadi_Borkow@clalit.org.il

1. INTRODUCTION

Tuberculosis (TB) and AIDS are the two worst world epidemics of infectious diseases. According to a recent WHO report, the global prevalence of *Mycobacterium tuberculosis* (MTB) infection in 1997 was 32% (1.86 billion people), the number of new cases totaled ~8 million, 16 million had active disease and close to 2 million people died from it (Dye *et al.* 1999). Regarding HIV infection, according to the estimate of the United Nations Program on AIDS (UNAIDS), more than 36 million people are presently infected with HIV-1, over 23 millions have already died from AIDS and more than 100 million people will be carrying the virus in less than 10 years! (Piot *et al.* 2001). There is a striking concordance in the distribution of HIV and TB, the highest incidence of both infections occurring in Sub Saharan Africa and Southeast Asia (Dye *et al.* 1999; Piot *et al.* 2001).

Next to TB, geohelminthic infections are probably the most common infections in all of the developing countries. Altogether, a *quarter* of the world's population is infested with helminths, the estimated number of infected people being over one and a half billion (Bundy & De Silva, 1998; De Silva, Chan & Bundy, (1997). The remarkable similarity in the geographic distribution of geohelminthic infections, and that of HIV and TB, particularly in Africa (Joint United Nations Programme on HIV/AIDS and World Health Organization, 2001) (Figure 16.1), raises the question of possible causal relationships between these infections.

In the following review we shall try to summarize the current knowledge on the effects of chronic helminthic infections on the immune system of the host, and the possible implication of these changes on susceptibility of the host to HIV and TB infection, on the natural course of

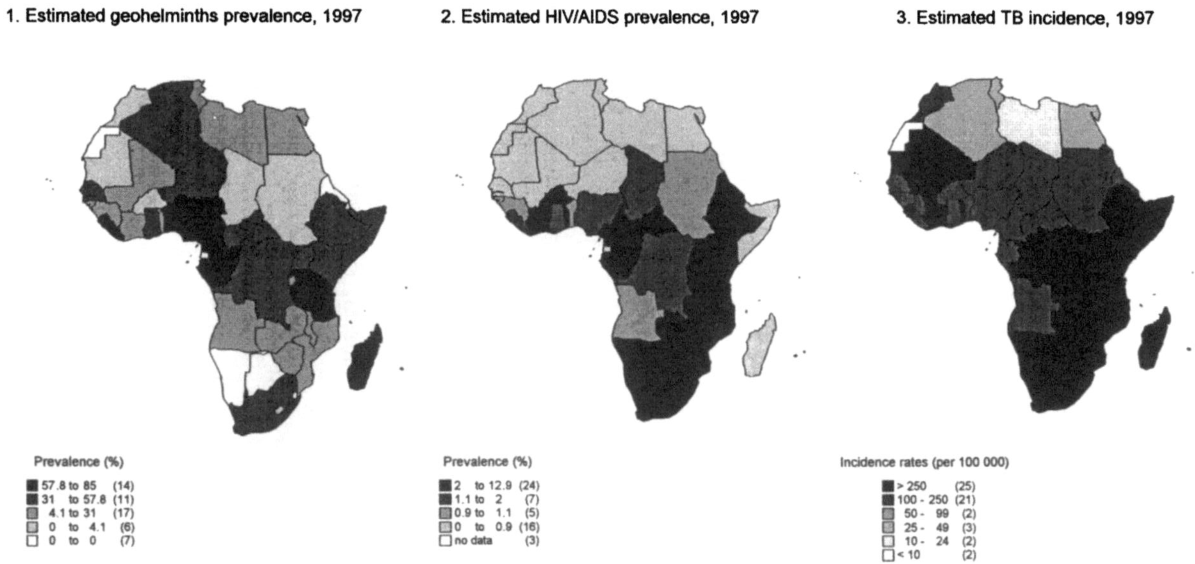

Fig 16.1. Similar geographic distribution of geohelminths, HIV and MTB in Africa (Adapted from Bundy, Sher & Michael, 2000).

HIV and TB, and on the ability of the host to develop protective immunity to HIV and TB following vaccination.

2. HELMINTHIC INFECTIONS RESULT IN CHRONIC IMMUNE ACTIVATION AND A DOMINANT TH2 CYTOKINE PROFILE

All geohelminthic infections are associated with strong chronic immune changes that have some common features to them. The most recognized is the dominant TH2 immune profile, with blood eosinophilia, high serum IgE, and a TH2 cytokine profile with increased secretion of interleukin- (IL)-4, IL-5 and IL-10 (Bentwich *et al.* 1996; Falcao *et al.* 1998; Maizels *et al.* 1993; Malaquias *et al.* 1997; Yazdanbakhsh, 1999). This cytokine profile may vary, either during the same infection, such as a switch from T-helper type 1 (TH1) to T-helper type 2 (TH2) during *Schistosoma* infection, or in different helminthic infections, such as in filariasis (Maizels *et al.* 1993; Yazdanbakhsh, 1999). There are also conflicting reports as to TH1 cytokine secretion, IL-2 and interferon (INF)-γ, during these infections, though it seems that their secretion is not always decreased below normal levels (Correa-Oliveira *et al.* 1998; Maizels *et al.* 1993; Pearce *et al.* 1998; Yazdanbakhsh, 1999). Less known and less appreciated is the wide immune activation and dysbalance that have been observed in association with helminthic infections (Bentwich *et al.* 1996; Elson *et al.* 1994; Leroy *et al.* 1997). These changes may have a major impact on the host ability to respond immunologically and they consist of the following (Bentwich *et al.* 1997; Kalinkovich *et al* 1998; Weisman *et al.* 1999):

- changes in peripheral lymphocyte populations with marked increase in the proportion of activated (HLA-DR+) and memory (CD45RO+) T cells, and in the proportion of peripheral blood mononuclear cells (PBMC) undergoing spontaneous apoptosis, and at the same time a significant decrease in the proportion of naive (CD45RA+) and CD8+CD28+ T cells (Bentwich *et al.* 1996).
- impaired immune response with decreased delayed type skin hypersensitivity and impaired cell proliferation to recall antigen (Borkow *et al.* 2000; Sabin *et al.* 1996);

- impaired signal transduction and anergy (Borkow *et al.* 2000; Maizels *et al.* 1991; Sabin *et al.* 1996; Villa & Kuhn, 1996) manifested by several features:

    - defective or no early transmembrane signaling (phosphorylation and/or dephosphorylation of tyrosine kinases);
    - deficient degradation of phosphorylated IkBα;
    - lack or attenuated phosphorylation of MAPK kinases, such as ERK1/2 and p38;
    - increased expression of the cytotoxic T lymphocyte-associated antigen 4 (CTLA-4), which is a negative modulator of immune effector mechanisms and cell proliferation (Thompson & Allison, 1997), and some restoration of such proliferative responses, after CTLA-4 blocking;
    - increased proportion of cells expressing the chemokine receptors CCR5 and CXCR4 with lower levels of chemokine secretion (RANTES and MIP-1α), by CD8+ cells (Kalinkovich *et al.* 1999; 2001).

That these changes are indeed the result of helminthic infections and not just the result of poor nutrition, hygiene or other environmental factors, is strongly supported by our own observations on the reversibility of all these immune derangements - both the cytokine profile and the chronic activation, following eradication of the helminth infections (Bentwich *et al.* 1997; Borkow *et al.* 2001; Kalinkovich *et al.* 1998). Most importantly, we were able to compare two groups of Ethiopian immigrants that arrived in Israel at the same time, which were both highly infected with helminths, had the same degree of immune activation on arrival, and settled in the same region in Israel. Because eradication of helminths took place in only one of the two groups, it was indeed revealing, that the immune derangements reverted to normal only in the helminth-free group while persisting in the helminth infected group, despite the similar new environmental conditions that applied to both groups, i.e. improved nutrition, hygiene, etc (Bentwich, Kalinkovich & Weisman, 1995; Bentwich *et al.* 1996, 1998).

The ability of the host to mount an immune response and the nature of that response, are greatly determined by the preexisting state of the immune system. Thus, the TH2 skewed immune profile associated with the helminthic infections, influences the host's immune response towards a TH2 type of response, as observed by several investigators (Actor *et al.* 1993; Bentwich, Kalinkovich & Weisman, 1995; Correa-Oliveira *et al.* 1998; Infante-Duarte & Kamradt, 1999; Maizels *et al.* 1993; Maizels & Holland, 1998; Sher *et al.* 1992; Wolday *et al.* 1999; Yazdanbakhsh, 1999):

- In the presence of a dominant TH2 profile, the immune response to other antigens, is skewed towards a TH2 type of response.
- The ability to mount a cellular response is impaired in *Schistosoma*-infected (TH2 dominant) animals and the generation of HIV specific cytotoxic T lymphocytes (CTL) is impaired in *Schistosoma*-infected mice.
- Suppressed immune response and anergy accompanies chronic helminthic infection.
- The specific immune response to geohelminths diminishes with progression of the infection and with helminth load. Taken together, all these findings clearly indicate that the immune system of thehelminth-infested host is profoundly changed and therefore is expected to behave quite differently from that of the uninfected host.

3. TH1 CYTOKINE DEPENDENT CELLULAR IMMUNITY AND PROTECTION FROM HIV AND MTB INFECTIONS

The role of TH1 cells and TH2 cells in controlling the immune response is well established. While cytokines produced from TH1 cells induce a cellular immune response, cytokines produced from TH2 cells induce a humoral immune response (Abbas, Murphy & Sher, 1996). These two cell types cross–regulate each other and thus, cytokines produced by one TH subset can suppress the production and/or activity of the other subset (Mosmann & Coffman, 1989).

The most effective mechanism for the control of MTB infection is by a TH1 immune response. After infection with MTB, more than 90% of individuals do not develop overt TB (Ottenhoff, Kumararatne & Casanova, 1998). A strong IFN-γ response to MTB is present in individuals who contain the infection, while this response is blunted in individuals with active TB disease (Ottenhoff *et al.* 1998; Torres *et al.* 1998). Individuals lacking IFN-γ or IL-12 receptors are highly susceptible to poorly pathogenic mycobacterium species, providing genetic evidence for the importance of a

TH1 response for controlling mycobacterium infection, and suggesting that there are no redundant protective immune mechanisms that can compensate for these deficiencies (Ottenhoff, Kumararatne & Casanova, 1998). In contrast to TH1 lymphocytes, TH2 lymphocytes, which produce IL-4 and IL-10, do not contribute to antimycobacterial immunity (Singleton *et al.* 1997). When PBMC from TB patients co-infected with HIV are exposed to MTB *in vitro*, they produce less IFN-γ but similar amounts of IL-4 and IL-10, as compared with lymphocytes from HIV-negative patients with TB (Oscherwitz *et al.* 1997), indicating that reduced TH1 response in HIV-infected patients contributes to their susceptibility to TB.

The role of the TH1/TH2 types in the pathogenesis of HIV has also been studied extensively (reviewed in Clerici & Shearer, 2001; Romagnani & Maggi, 1994). Though there is no general agreement as to the role of these responses in every phase of the infection, there are some important findings that clearly bear on the response type in different stages:

- Activated CTL are responsible for the initial clearance of the primary viremia and probably for maintaining low viremia during the asymptomatic phase of the infection (Oscherwitz *et al.* 1997; Singleton *et al.* 1997; Torres *et al.* 1998).
- Progression of the infection is accompanied by a TH1 to TH2 switch, with a reduction in the number of TH1 clones and an increase in the number of T-helper type 0 (TH0)/TH2 clones (Clerici & Shearer, 1993; Maggi *et al.* 1994; Romagnani, Maggi & Del Prete, 1994).
- TH1 functions are correlated with better survival and slower progression (Clerici *et al.* 1992; Romagnani & Maggi, 1994).
- TH0 cells (non-differentiated cells) or TH2 cloned cells show increased susceptibility for HIV infection and replication (Maggi *et al.* 1994).
- Progression may be correlated to reduction of cellular immunity, together with higher permissiveness of TH0/TH2 cells to HIV infection (Maggi *et al.* 1994). Hence, protection from HIV infection may also be associated with an effective TH1 cellular defense. The best evidence is found in individuals that have been exposed to HIV and yet remained HIV seronegative while having specific HIV cellular immunity (Clerici *et al.* 1992, 1994; Fowke *et al.* 1996; Langlade-Demoyen *et al.* 1994; Looney, 1994; Pinto *et al* 1995; Plummer *et al.* 1999), and HIV seronegative infants born to HIV infected mothers and having HIV specific CTL activity (De Maria, Cirillo & Moretta, 1994). The

importance of cellular immunity in conferring protection from infection has also been shown in several studies of protective vaccination to SIV in primates (Boyer *et al.* 1996; McMichael & Rowland-Jones, 2001; Putkonen *et al.* 1997).

It is thus believed that generation of HIV or MTB specific cellular immunity by a vaccine, via a TH1 immune response, is a major prerequisite for any protective HIV or MTB vaccine (Harboe, 1998; Heilman & Baltimore, 1998).

4. INTERACTION BETWEEN GEOHELMINTHIC INFECTIONS AND HIV AND TUBERCULOSIS

We have previously suggested that the chronic immune activation and the TH2 immune profile caused by helminthic infections are major factors in the pathogenesis of AIDS in Africa, which may account for the different behavior of the epidemic in Africa - its rapid spread and probably its faster progression (Bentwich, Kalinkovich & Weisman, 1995). We have then argued that these immune changes caused by the helminthic infection may also affect the spread and parallel alarming growth of tuberculosis in the developing countries (Bentwich *et al.* 1999). Though the issue of faster progression of HIV infection in Africa is controversial, and there is a paucity of controlled studies on the natural course of HIV in Africa (Piot *et al.* 2001), there are studies from other developing countries in Asia and the Caribbean, which clearly demonstrate faster progression of HIV infection in these countries (Bentwich *et al.* 1995; 1997; 1998; Weisman *et al.* 1999).

Overall our hypothesis is supported by the several following observations:

- Similar immune activation and dysregulation of peripheral T cell populations has recently been observed in other parts of Africa and in India, where helminthic infections are endemic (Ghosh *et al.* 1996; Messele *et al.* 1999).
- The similar distribution and mutual enhancement of HIV and tuberculosis, as well as the reactivation of latent tuberculosis, occurs mostly in the poor populations where helminthic infections are extremely common (Beyers *et al.* 1996; Coovadia, Jeena & Wilkinson, 1998).

- The chronic immune activation due to helminthic infections is associated with increased expression of HIV co-receptors, both CCR5 and CXCR4, as well as with increased susceptibility for HIV infection *in vitro* (Gopinath *et al.* 2000; Kalinkovich *et al.* 1999. 2001; Shapira-Nahor *et al.* 1998).
- Plasma HIV viral load is higher in people living in Sub-Saharan Africa, where helminth infections are extremely prevalent (Dyer *et al.* 1998).
- Faster progression to AIDS has been documented in Africa and Asia in areas endemic for helminths (Anzala *et al.* 1995; Dyer *et al.* 1998; Jean *et al.* 1997; Srikanth *et al.* 2000) and becomes similar to western rate, once helminthic infections are eradicated (Weisman *et al.* 1999).
- Helminthic load (number of eggs excreted in the stool) is correlated with increased HIV plasma viral load (Borkow *et al.* 2001 and Wolday *et al.* manuscript submitted).
- Eradication of helminthic infection may result in significant reduction of HIV plasma viral load, though this may probably depend on the type and load of helminth infection (Wolday *et al.* 2001).
- Cellular immunity to tuberculin is severely impaired in areas highly endemic for tuberculosis and highly infested with helminths and may be regained after helminth eradication (Coovadia, Jeena & Wilkinson, 1998; Elias *et al.* 2001a).
- Leishmaniasis is associated with increased HIV plasma viral load, which decreases following treatment of Leishmaniasis (Preiser *et al.* 1996).
- BCG vaccination is poorly protective against tuberculosis in most developing countries where helminthic infections are endemic and widespread (Elias *et al.* 2001,a,b).
- The response to anti-tuberculosis treatment is accompanied by a decrease in eosinophilia, possibly reflecting the dependence of successful treatment on Th2 to Th1 switch and helminth eradication (Adams *et al.* 1999; Elias *et al.* 2001,a,b; Ohrui *et al.* 2000; Scanga & Le Gros, 2000).

5. PROTECTIVE HIV AND MTB IMMUNITY AND THE HELMINTH INFECTIONS

It is clear that the efficacy of HIV and TB candidate vaccines will have to be tested in human field trials that can only take place in Africa and Asia, in areas with a high incidence of HIV and MTB infections. However, potentially good vaccines may fail in such clinical trial if examined in the immune scenario presently pertaining in the developing world. It is quite clear that the host immune background in developing countries is biased towards a TH2 profile and that most individuals are in a chronic immune activation state, both of which are accounted for to a large extent by the chronic geohelminthic infections. Our findings that signal transduction in such individuals is impaired and that their immune cells can respond poorly to stimuli, such as to PPD, suggest that their capacity to elicit an immune response following vaccination will be heavily encumbered. The failure of BCG vaccination in Africa and Asia to confer protective immunity to TB is consistent with this idea. It therefore becomes essential to take this major issue into consideration for any protective vaccine development.

Several key questions remain to be answered. For example, can the pre-existing pronounced TH2 background be shifted to a TH0 or TH1 bias prior to vaccination, and thus allow for the generation of cellular immunity by HIV or TB vaccines? Would eradication of helminths by itself be enough to bring back the immune profile to a TH0, or allow easier manipulation towards a TH1 profile? What are the kinetics of the changes of the immune profile following eradication of helminth infection, and the conditions necessary for them to persist? It is important to determine whether modulation of the immune response, such as by adjuvants, is possible in the presence of helminths and following their eradication.

DNA vaccination has received much attention recently as a possible broad-based, inexpensive approach to vaccine development (Gurunathan, Klinman & Seder, 2000). Importantly, immunization with DNA, in addition to coding for a specified antigen, has a potent TH1-type adjuvant effect that is effective with virtually any type of antigen (Davis, 2000; Krieg & Davis, 2001; Roman *et al.* 1997). DNA immunization generates MHC class 1-restricted CTLs, as well as TH1 cells secreting predominantly IFN-γ. The type of immune response induced is strongly affected by non-coding immunostimulatory DNA sequences (ISS) encoded within the plasmid, which are centered around unmethylated CpG basepairs. Thus, in contrast to conventional protein vaccines, an ISS-enriched DNA vaccine can generate a dominant TH1 response (Raz *et al.* 1996; Roman *et al.* 1997). The activity of CpG DNA can be mimicked with synthetic oligodeoxynucleotides (ODN), which, when added to a vaccine, such as protein, greatly boost the TH1 dependent immune responses (Davis, 2000; Krieg & Davis, 2001).

The TH1 promoting adjuvant characteristic of DNA may thus have potential applications in HIV or Tuberculosis vaccination, by inducing a cellular rather than a humoral response to HIV or MTB co-administered protein and/or DNA encoded immunogens. Thus, several groups are currently developing DNA vaccines against HIV and MTB (Fomsgaard, 1999).

The capacity of pDNA to elicit a specific TH1 immune response in the face of a pre-existing TH2-type immune response, such as during parasitic infection, has been recently addressed in our laboratory (Ayash-Rashkovsky *et al.* 2001). We used *Schistosoma*-infected mice and immune responses to β-galactosidase (β-gal) as a model to test the induction of a specific Th1 immune response by pDNA encoding β-gal on a dominant pre-existing Th2 immune profile. We found that intradermal immunization with pDNA encoding β-gal of *Schistosoma*-infected mice (thereby exhibiting a dominant Th2 immune bias) induced a strong TH1 anti-β-gal response, as opposed to immunization with β-gal alone. Importantly, the established protective TH2 immune response to schistosomes was not compromised. Furthermore, by using the same *Schistosoma*-infected mice model, we found that immunization of the mice, with a whole, killed, gp120-depleted, HIV-1 antigen in incomplete Freund's adjuvant combined with CpG ODN [1826], elicited a strong TH1 response to the core HIV-1 antigen (p24) with one thousand-fold higher titers of IgG2a antibodies (indicative of a TH1 response), enhanced cell proliferation to HIV-1 antigen, and increased secretion of RANTES (a β-chemokine) and IFN-γ. In contrast to these HIV-1 specific immune responses, the general TH2 immune background and the protective TH2 immune response to *Schistosoma* was not altered in the CpG ODN immunized animals.

The utility of CpG ODN as adjuvants for vaccines designed to prevent TH-2 dependent immunopathology has also been shown by others (Chiaramonte *et al.* 2000). Taken together, these results lend strong support to the possibility of using pDNA or CpG ODN as a TH1-inducing adjuvant in combination with candidate HIV vaccines when immunizing populations with a strong pre-existing TH2 immune profile, even without altering their immune background and eradication of the helminth infections.

6. CONCLUSIONS

We have presented the results of the interaction between helminthic infection, HIV and tuberculosis, with a major emphasis on the immune changes accompanying helminthic infections and their implications for these interactions. The major common denominator to these interactions is the immune activation and the skewed cytokine profile which the helminthic infections cause and which markedly affects the ability of the host to cope with either HIV or tuberculosis. The most important conclusion that can be drawn at this stage, is that the suppression of helminthic infections, may have a major impact on the spread and progression of HIV infection and tuberculosis, as well as on the success of protective vaccines against both. This conclusion applies particularly to the developing countries, where all these infections are so common and where access to antiretroviral therapy is not available. It is however critical to show if suppression of helminthic infections leads to decreased incidence of HIV and tuberculosis, will it be accompanied by decreased HIV viral load and progression of the infection, and just as importantly, whether generation of protective immunity to HIV and tuberculosis will be improved after deworming. Since the public health case for deworming has already been demonstrated by its effectiveness in enhancing the development of children (see Chapters 3 and 4), large scale eradication of helminthic infections throughout the developing world in the context of the AIDS and tuberculosis epidemics, should be seriously considered and implemented, even if the consequences are only probable or partially positive.

REFERENCES

ABBAS, A. K., MURPHY, K.M. & SHER, A. (1996). Functional diversity of helper T lymphocytes. *Nature* **383**, 787-793.

ACTOR, J.K, SHIRAI, M., KULLBERG, M.C., BULLER, R.M., SHER, A. & BERZOFSKY, J.A. (1993). Helminth infection results in decreased virus-specific CD8+ cytotoxic T-cell and Th1 cytokine responses as well as delayed virus clearance. *Proceedings of the National Academy of Sciences (USA)* **90**, 948-952.

ADAMS, J.F., SCHOLVINCK, E.H., GIE, R.P., POTTER, P.C., BEYERS, N. & BEYERS, A.D. (1999). Decline in total serum IgE after treatment for tuberculosis. *Lancet* **353**, 2030-2033.

ANZALA, O.A., NAGELKERKE, N.J., BWAYO, J.J., HOLTON, D., MOSES, S., NGUGI, E.N., NDINYA-ACHOLA, J.O. & PLUMMER, F.A. (1995). Rapid progression to disease in African sex workers with human immunodeficiency virus type 1 infection. *Journal of Infectious Diseases* **171**, 686-689.

AYASH-RASHKOVSKY, M., WEISMAN, Z., ZLOTNIKOV, S., RAZ, E., BENTWICH, Z. & BORKOW, G. (2001). Induction of antigen-specific Th1-biased immune responses by plasmid DNA in schistosoma-infected mice with a preexistent dominant Th2 immune profile. *Biochemical and Biophysical Research Communications* **282**, 1169-1176.

BENTWICH, Z., KALINKOVICH, A. & WEISMAN, Z. (1995). Immune activation is a dominant factor in the pathogenesis of African AIDS. *Immunology Today* **16**, 187-191.

BENTWICH, Z. KALINKOVICH, A., WEISMAN, Z., BORKOW, G., BEYERS, N. & BEYERS, A.D. (1999). Can eradication of helminthic infections change the face of AIDS and tuberculosis? *Immunology Today* **20**, 485-487.

BENTWICH, Z. KALINKOVICH, A., WEISMAN, Z. & GROSSMAN, Z. (1998). Immune activation in the context of HIV infection. *Clinical and Experimental Immunology* **111**, 1-2.

BENTWICH, Z., WEISMAN, Z., GROSSMAN, Z., GALAI, N. & KALINKOVICH, A. (1997). Pathogenesis of AIDS in Africa - Lessons from the Ethiopian immigrants in Israel. *Immunologist* **5**, 211-226.

BENTWICH, Z. WEISMAN, Z., MOROZ, C., BAR-YEHUDA, S. & KALINKOVICH, A. (1996). Immune dysregulation in Ethiopian immigrants in Israel: relevance to helminth infections? *Clinical and Experimental Immunology* **103**, 239-243.

BEYERS, N. GIE, R.P., ZIETSMAN, H.L., KUNNEKE, M., HAUMAN, J., TATLEY, M. & DONALD, P.R. (1996). The use of a geographical information system to evaluate the distribution of tuberculosis in a high-incidence community. *South African Medical Journal* **86**, 40-1, 44.

BORKOW, G. LENG, Q., WEIZMAN, Z., STEIN, M., GALAI, N., KALINKOVICH, A. & BENTWICH, Z. (2000). Chronic immune activation associated with intestinal helminth infections results in impaired signal transduction and anergy. *Journal of Clinical Investigation* **106**, 1053-1060.

BORKOW, G. WEIZMAN, Z., LENG, Q., STEIN, M., KALINKOVICH, A., WOLDAY, D. & BENTWICH, Z. (2001). Helminths, Human Immunodeficiency Virus and Tuberculosis. *Scandinavian Journal of Infectious Diseases* In press.

BOYER, J.D., WANG, B., UGEN, K.E., AGADJANYAN, M., JAVADIAN, A., FROST, P., DANG, K., CARRANO, R.A., CICCARELLI, R., CONEY, L., WILLIAMS, W.V. & WEINER, D.B. (1996). *In vivo* protective anti-HIV immune responses in non-human primates through DNA immunization. *Journal of Medical Primatology* **25**, 242-250.

BUNDY, D.A. & DE SILVA, N.R. (1998). Can we deworm this wormy world? *British Medical Bulletin* **54**, 421-432.

BUNDY, D.A., Sher, A. & MICHAEL, E. (2000). Good worms or bad worms : Do worm infections affect the epidemiological patterns of other diseases ? *Parasitology Today* **16**, 273-274.

CHIARAMONTE, M.G., HESSE, M., CHEEVER, A.W. & WYNN, T.A. (2000). CpG oligonucleotides can prophylactically immunize against Th2-mediated schistosome egg-induced pathology by an IL-12-independent mechanism. *Journal of Immunology* **164**, 973-985.

CLERICI, M. & SHEARER, G.M. (2001). The TH1-TH2 hypothesis of HIV infection: new insights. *Immunology Today,* **12**, 575-581.

CLERICI, M., GIORGI, J.V., CHOU, C.C., GUDEMAN, V.K., ZACK, J.A., GUPTA, P., HO, H.N., NISHANIAN, P.G., BERZOFSKY, J.A. & SHEARER, G.M. (1992). Cell-mediated immune response to human immunodeficiency virus (HIV) type 1 in seronegative homosexual men with recent sexual exposure to HIV-1. *Journal of Infectious Diseases* **165**, 1012-1019.

CLERICI, M., LEVIN, J.M., KESSLER, H.A., HARRIS, A., BERZOFSKY, J.A., LANDAY, A.L. & SHEARER, G.M. (1994). HIV-specific T-helper activity in seronegative health care workers exposed to contaminated blood. *Journal of the American Microbiological Association* **271**, 42-46.

CLERICI, M. & SHEARER, G.M. (1993). A TH1-->TH2 switch is a critical step in the etiology of HIV infection. *Immunology Today* **14**, 107-111.

COOVADIA, H. M., JEENA, P. & WILKINSON, D. (1998). Childhood human immunodeficiency virus and tuberculosis co-infections: reconciling conflicting data. *International Journal of Tuberculosis and Lung Disease* **2**, 844-851.

CORREA-OLIVEIRA, R., MALAQUIAS, L.C., FALCAO, P.L., VIANA, I.R., BAHIA-OLIVEIRA, L.M., SILVEIRA, A.M., FRAGA, L.A., PRATA, A., COFFMAN, R.L., LAMBERTUCCI, J.R., CUNHA-MELO, J.R., MARTINS-FILHO, O.A., WILSON, R.A. & GAZZINELLI, G. (1998). Cytokines as determinants of resistance and pathology in human Schistosoma mansoni infection. *Brazilian Journal of Medical Biological Research* **31**, 171-177.

DAVIS, H.L., (2000). Use of CpG DNA for enhancing specific immune responses. *Current Topics in Microbiology and Immunology* **247**, 171-183.

DE MARIA, A., CIRILLO, C. & MORETTA, L. (1994). Occurrence of human immunodeficiency virus type 1 (HIV-1)-specific cytolytic T cell activity in apparently uninfected children born to HIV- 1-infected mothers. *Journal of Infectious Diseases* **170**, 1296-1299.

DE SILVA, N.R., CHAN, M.S. & BUNDY, D.A. (1997). Morbidity and mortality due to ascariasis: re-estimation and sensitivity analysis of global numbers at risk. *Tropical Medicine and International Health* **2**, 519-528.

DYE, C., SCHEELE, S., DOLIN, P., PATHANIA, V. & RAVIGLIONE, M.C. (1999). Consensus statement. Global burden of tuberculosis: estimated incidence, prevalence, and mortality by country. WHO Global Surveillance and Monitoring Project. *Journal of the American Microbiological Association* **282**, 677-686.

DYER, J.R., KAZEMBE, P., VERNAZZA, P.L., GILLIAM, B.L., MAIDA, M., ZIMBA, D., HOFFMAN, I.F., ROYCE, R.A., SCHOCK, J.L., FISCUS, S.A., COHEN, M.S. & ERON, J.J., JR. (1998). High levels of human immunodeficiency virus type 1 in blood and semen of seropositive men in sub-Saharan Africa. *Journal of Infectious Diseases* **177**, 1742-1746.

ELIAS, D., ASHMARE, G., PETROS, B. & BRITTON, S. (2001). Deworming and resistance against TB. *Ethiopian Medicine Journal* **37**, 3-7.

ELIAS, D., WOLDAY, D., AKUFFO, H., PETROS, B., BRONNER, U. & BRITTON, S. (2001). Effect of deworming on human T cell responses to mycobacterial antigens in helminth-exposed individuals before and after bacille Calmette- Guerin (BCG) vaccination. *Clinical and Experimental Immunology* **123**, 219-225.

ELSON, L.H., SHAW, S., VAN LIER, R.A. & NUTMAN, T.B. (1994). T cell subpopulation phenotypes in filarial infections: CD27 negativity defines a population greatly enriched for Th2 cells. *International Immunology* **6**, 1003-1009.

FALCAO, P.L., MALAQUIAS, L.C., MARTINS-FILHO, O.A., SILVEIRA, A.M., PASSOS, V.M., PRATA, A., GAZZINELLI, G., COFFMAN, R.L. & CORREA-OLIVEIRA, R. (1998). Human Schistosomiasis mansoni: IL-10 modulates the in vitro granuloma formation. *Parasite Immunology* **20,** 447-454.

FOMSGAARD, A. (1999). HIV-1 DNA vaccines. *Immunology Letters* **65,** 127-131 .

FOWKE, K.R., NAGELKERKE, N.J., KIMANI, J., SIMONSEN, J.N., ANZALA, A.O., BWAYO, J.J., MACDONALD, K.S., NGUGI, E.N. & PLUMMER, F.A. (1996). Resistance to HIV-1 infection among persistently seronegative prostitutes in Nairobi, Kenya. *Lancet* **348,** 1347-1351.

GHOSH, M.K., GHOSH, A.K., ADDY, M., NANDY, A. & GHOSE, A.C. (1996). Subpopulations of T lymphocytes in the peripheral blood and lymph nodes of Indian kala-azar patients. *Medical Microbiology and Immunology (Berlin)* **185,** 183-187.

GOPINATH, R., OSTROWSKI, M., JUSTEMENT, S.J., FAUCI, A.S. & NUTMAN, T.B. (2000). Filarial infections increase susceptibility to human immunodeficiency virus infection in peripheral blood mononuclear cells in vitro. *Journal of Infectious Diseases* **182,** 1804-1808.

GURUNATHAN, S., KLINMAN, D. M. & SEDER, R. A. (2000). DNA vaccines: immunology, application, and optimization. *Annual Reviews in Immunology* **18,** 927-974.

HARBOE, M. (1998). Protective immunity to intracellular parasites--a focus for vaccine research and a challenge to immunization programmes. *Developments in Biological Standardization (Basel)* **92,** 145-147.

HEILMAN, C.A. & BALTIMORE, D. (1998). HIV vaccines - where are we going? *Nature Medicine* **4,** 532-534.

INFANTE-DUARTE, C. & KAMRADT, T. (1999). Th1/Th2 balance in infection. *Seminars in Immunopathology* **21,** 317-338.

JEAN, S.S., REED, G.W., VERDIER, R.I., PAPE, J.W., JOHNSON, W.D. & WRIGHT, P.F. (1997). Clinical manifestations of human immunodeficiency virus infection in Haitian children. *Pediatric Infectious Disease Journal* **16,** 600-606.

JOINT UNITED NATIONS PROGRAMME ON HIV/AIDS AND WORLD HEALTH ORGANIZATION. (2001). Report on the global HIV/AIDS epidemic.

KALINKOVICH, A., BORKOW, G., WEISMAN, Z., TSIMANIS, A., STEIN, M. & BENTWICH, Z. (2001). Increased ccr5 and cxcr4 expression in Ethiopians living in Israel: environmental and constitutive factors. *Clinical Immunology* **100,** 107-117.

KALINKOVICH, A., WEISMAN, Z., GREENBERG, Z., NAHMIAS, J., EITAN, S., STEIN, M. & BENTWICH, Z. (1998). Decreased CD4 and increased CD8 counts with T cell activation is associated with chronic helminth infection. *Clinical and Experimental Immunology* **114,** 414-421.

KALINKOVICH, A., WEISMAN, Z., LENG, Q., BORKOW, G., STEIN, M., GREENBERG, Z., ZLOTNIKOV, S., EITAN, S. & BENTWICH, Z. (1999). Increased CCR5 expression with decreased beta chemokine secretion in Ethiopians: relevance to AIDS in Africa. *Journal of Human Virology* **2,** 283-289.

KRIEG, A.M. & DAVIS, H.L. (2001). Enhancing vaccines with immune stimulatory CpG DNA. *Current Opinions in Molecular Therapy* **3,** 15-24.

LANGLADE-DEMOYEN, P., NGO-GIANG-HUONG, N., FERCHAL, F. & OKSENHENDLER, E. (1994). Human immunodeficiency virus (HIV) nef-specific cytotoxic T lymphocytes in noninfected heterosexual contact of HIV-infected patients. *Journal of Clinical Investigation* **93,** 1293-1297.

LEROY, E., BAIZE, S., WAHL, G., EGWANG, T.G. & GEORGES, A.J. (1997). Experimental infection of a nonhuman primate with Loa loa induces transient strong immune activation followed by peripheral unresponsiveness of helper T cells. *Infection and Immunity* **65**, 1876-1882.

LOONEY, D.J. (1994). Immune responses to human immunodeficiency virus type 1 in exposed but uninfected individuals: protection or chance? *Journal of Clinical Investigation* **93**, 920.

MAGGI, E., MAZZETTI, M., RAVINA, A., ANNUNZIATO, F., DE CARLI, M., PICCINNI, M.P., MANETTI, R., CARBONARI, M., PESCE, A.M. & DEL PRETE, G. (1994). Ability of HIV to promote a TH1 to TH0 shift and to replicate preferentially in TH2 and TH0 cells. *Science* **265**, 244-248.

MAIZELS, R.M., BUNDY, D.A., SELKIRK, M.E., SMITH, D.F. & ANDERSON, R.M. (1993). Immunological modulation and evasion by helminth parasites in human populations. *Nature* **365**, 797-805.

MAIZELS, R.M. & HOLLAND, M.J. (1998). Parasite immunology: pathways for expelling intestinal helminths. *Current Biology* **8**, R711-R714.

MAIZELS, R.M., KURNIAWAN, A., SELKIRK, M.E. & YAZDANBAKHSH, M. (1991). Immune responses to filarial parasites. *Immunology Letters* **30**, 249-254.

MALAQUIAS, L.C., FALCAO, P.L., SILVEIRA, A.M., GAZZINELLI, G., PRATA, A., COFFMAN, R.L., PIZZIOLO, V., SOUZA, C.P., COLLEY, D.G. & CORREA-OLIVEIRA, R. (1997). Cytokine regulation of human immune response to Schistosoma mansoni: analysis of the role of IL-4, IL-5 and IL-10 on peripheral blood mononuclear cell responses. *Scandinavian Journal of Immunology* **46**, 393-398.

McMICHAEL, A.J. & ROWLAND-JONES, S.L. (2001). Cellular immune responses to HIV. *Nature* **410**, 980-987.

MESSELE, T., ABDULKADIR, M., FONTANET, A.L., PETROS, B., HAMANN, D., KOOT, M., ROOS, M.T., SCHELLEKENS, P.T., MIEDEMA, F. & RINKE DE WIT, T.F. (1999). Reduced naive and increased activated CD4 and CD8 cells in healthy adult Ethiopians compared with their Dutch counterparts. *Clinical and Experimental Immunology* **115**, 443-450.

MOSMANN, T.R. & COFFMAN, R.L. (1989). TH1 & TH2 cells: different patterns of lymphokine secretion lead to different functional properties. *Annual Reviews in Immunology* **7**, 145-173.

OHRUI, T., ZAYASU, K., SATO, E., MATSUI, T., SEKIZAWA, K. & SASAKI, H. (2000). Pulmonary tuberculosis and serum IgE. *Clinical and Experimental Immunology* **122**, 13-15.

OSCHERWITZ, T., TULSKY, J.P., ROGER, S., SCIORTINO, S., ALPERS, A., ROYCE, S. & LO, B. (1997). Detention of persistently nonadherent patients with tuberculosis. *Journal of the American Microbiological Association* **278**, 843-846.

OTTENHOFF, T.H., KUMARARATNE, D. & CASANOVA, J.L. (1998). Novel human immunodeficiencies reveal the essential role of type-I cytokines in immunity to intracellular bacteria. *Immunology Today* **19**, 491-494.

PEARCE, E.J., LA FLAMME, A., SABIN, E. & BRUNET, L.R. (1998). The initiation and function of Th2 responses during infection with Schistosoma mansoni. *Advances in Experimental Medical Biology* **452**, 67-73.

PINTO, L.A., SULLIVAN, J., BERZOFSKY, J.A., CLERICI, M., KESSLER, H.A., LANDAY, A.L. & SHEARER, G.M. (1995). ENV-specific cytotoxic T lymphocyte responses in HIV seronegative health care workers occupationally exposed to HIV-contaminated body fluids. *Journal of Clinical Investigation* **96,** 867-876.

PIOT, P., BARTOS, M., GHYS, P.D., WALKER, N. & SCHWARTLANDER, B. (2001). The global impact of HIV/AIDS. *Nature* **410,** 968-973.

PLUMMER, F.A., BALL, T.B., KIMANI, J. & FOWKE, K.R. (1999). Resistance to HIV-1 infection among highly exposed sex workers in Nairobi: what mediates protection and why does it develop? *Immunological Letters* **66,** 27-34.

PREISER, W., CACOPARDO, B., NIGRO, L., BRANER, J., NUNNARI, A., DOERR, H.W. & WEBER, B. (1996). Immunological findings in HIV-*Leishmania* coinfection. *Intervirology* **39,** 285-288.

PUTKONEN, P., MAKITALO, B., BOTTIGER, D., BIBERFELD, G. & THORSTENSSON, R. (1997). Protection of human immunodeficiency virus type 2-exposed seronegative macaques from mucosal simian immunodeficiency virus transmission. *Journal of Virology* **71.** 4981-4984.

RAZ, E., TIGHE, H., SATO, Y., CORR, M., DUDLER, J.A., ROMAN M., SWAIN, S.L., SPIEGELBERG, H.L. & CARSON, D.A. (1996). Preferential induction of a Thl immune response and inhibition of specific IgE antibody formation by plasmid DNA immunization. *Proceedings of the National Academy of Science (USA)* **93,** 5141-5145.

ROMAGNANI, S. & MAGGI, E. (1994). Thl versus Th2 responses in AIDS. *Current Opinions in Immunology* **6,** 616-622.

ROMAGNANI, S., MAGGI, E. & DEL PRETE, G. (1994). HIV can induce a TH1 to TH0 shift, and preferentially replicates in CD4+ T-cell clones producing TH2-type cytokines. *Research in Immunology* **145,** 611-617.

ROMAN, M., MARTIN-OROZCO, E., GOODMAN, J.S., NGUYEN, M.D., SATO, Y., RONAGHY, A., KORNBLUTH, R.S., RICHMAN, D.D., CARSON, D.A. & RAZ, E. (1997). Immunostimulatory DNA sequences function as T helper-1-promoting adjuvants. *Nature Medicine* **3,** 849-854.

SABIN, E.A., ARAUJO, M.I., CARVALHO, E.M. & PEARCE, E.J. (1996). Impairment of tetanus toxoid-specific Thl-like immune responses in humans infected with *Schistosoma mansoni. Journal of Infectious Diseases* **173,** 269-272.

SCANGA, C.B. & LE GROS, G. (2000). Development of an asthma vaccine: research into BCG. *Drugs* **59,** 1217-1221.

SHAPIRA-NAHOR, O., KALINKOVICH, A., WEISMAN, Z., GREENBERG, Z., NAHMIAS, J., SHAPIRO, M., PANET, A. & BENTWICH, Z. (1998). Increased susceptibility to HIV-1 infection of peripheral blood mononuclear cells from chronically immune-activated individuals. *AIDS* **12,** 1731-1733.

SHER, A., GAZZINELLI, R.T., OSWALD, I.P., CLERICI, M., KULLBERG, M., PEARCE, E.J., BERZOFSKY, J.A., MOSMANN, T.R., JAMES, S.L. & MORSE, H.C., III (1992). Role of T-cell derived cytokines in the downregulation of immune responses in parasitic and retroviral infection. *Immunology Reviews* **127,** 183-204.

SINGLETON, L., TURNER, M., HASKAL, R., ETKIND, S., TRICARICO, M. & NARDELL, E. (1997). Long-term hospitalization for tuberculosis control. Experience with a medical-psychosocial inpatient unit. *Journal of the American Microbiological Association* **278,** 838-842.

SRIKANTH, P., CASTILLO, R.C., SRIDHARAN, G., JOHN, T.J., ZACHARIAH, A., MATHAI, D. & SCHWARTZ, D.H. (2000). Increase in plasma IL-10 levels and rapid loss of CD4+ T cells among HIV-infected individuals in south India. *International Journal of STD and AIDS* **11,** 49-51.

THOMPSON, C.B. & ALLISON, J.P., (1997). The emerging role of CTLA-4 as an immune attenuator. *Immunity* **7**, 445-450.

TORRES, M., HERRERA, T., VILLAREAL, H., RICH, E.A. & SADA, E. (1998). Cytokine profiles for peripheral blood lymphocytes from patients with active pulmonary tuberculosis and healthy household contacts in response to the 30-kilodalton antigen of Mycobacterium tuberculosis. *Infection and Immunity* **66**, 176-180.

VILLA, O.F. & KUHN, R.E. (1996). Mice infected with the larvae of *Taenia crassiceps* exhibit a Th2-like immune response with concomitant anergy and downregulation of Th1- associated phenomena. *Parasitology* **112**, 561-570.

WEISMAN, Z., KALINKOVICH, A., BORKOW, G., STEIN, M., GREENBERG, Z. & BENTWICH, Z. (1999). Infection by different HIV-1 subtypes (B and C) results in a similar immune activation profile despite distinct immune backgrounds. *Journal of Acquired Immune Deficiency Syndrome* **21**, 157-163.

WOLDAY, D., BERHE, N., AKUFFO, H. & BRITTON, S. (1999). Leishmania-HIV interaction: immunopathogenic mechanisms. *Parasitology Today* **15**, 182-187.

WOLDAY, D. MAAYAN, S., MIRIAM, G. and BENTWICH, Z. (2000). Eradication of helminthic infection decreases HIV plasma viral load in dually-infected people. Abstract 157. 7th Conference on Retrovirus and Opportunistic Infections. San Francisco, USA.

YAZDANBAKHSH, M. (1999). Common features of T cell reactivity in persistent helminth infections: lymphatic filariasis and schistosomiasis. *Immunology Letters* **65**, 109-115.

INDEX

D

MIX
Papier aus verantwortungsvollen Quellen
Paper from responsible sources
FSC® C105338

If you have any concerns about our products,
you can contact us on
ProductSafety@springernature.com

In case Publisher is established outside the EU,
the EU authorized representative is:
Springer Nature Customer Service Center GmbH
Europaplatz 3, 69115 Heidelberg, Germany

Printed by Libri Plureos GmbH
in Hamburg, Germany